AF389481

Modeling, Simulation and Data Processing for Additive Manufacturing

Modeling, Simulation and Data Processing for Additive Manufacturing

Editor

Mika Salmi

MDPI • Basel • Beijing • Wuhan • Barcelona • Belgrade • Manchester • Tokyo • Cluj • Tianjin

Editor
Mika Salmi
Department of Mechanical
Engineering
Aalto University
Espoo
Finland

Editorial Office
MDPI
St. Alban-Anlage 66
4052 Basel, Switzerland

This is a reprint of articles from the Special Issue published online in the open access journal *Materials* (ISSN 1996-1944) (available at: www.mdpi.com/journal/materials/special_issues/additive_manufacturing_AM).

For citation purposes, cite each article independently as indicated on the article page online and as indicated below:

LastName, A.A.; LastName, B.B.; LastName, C.C. Article Title. *Journal Name* **Year**, *Volume Number*, Page Range.

ISBN 978-3-0365-2949-3 (Hbk)
ISBN 978-3-0365-2948-6 (PDF)

Contents

About the Editor

Mika Salmi

Mika Salmi works as Research Director at Aalto University Digital Design Laboratory and Staff scientist at the Department of Mechanical Engineering. He did his Ph.D. in 2013, focusing on medical applications of additive manufacturing in surgery and dental care. Salmi has published more than 60 scientific and technical papers from industrial and medical additive manufacturing applications. In his research, he takes advantage of to whole AM workflow and recent development in materials, simulation, design for AM, and post-processing, ending the actual application. Currently, he is involved in projects related to 3D printing of spare parts and medical applications, 4D printing, and researching industrial additive manufacturing applications.

Preface to "Modeling, Simulation and Data Processing for Additive Manufacturing"

Additive manufacturing (AM) or, more commonly, 3D printing is one of the fundamental elements of Industry 4.0. and the fourth industrial revolution. It has shown its potential example in the medical, automotive, aerospace, and spare part sectors. Personal manufacturing, complex and optimized parts, short series manufacturing and local on-demand manufacturing are some of the current benefits. Businesses based on AM have experienced double-digit growth in recent years. Accordingly, we have witnessed considerable efforts in developing processes and materials in terms of speed, costs, and availability. These open up new applications and business case possibilities all the time, which were not previously in existence.

Most research has focused on material and AM process development or effort to utilize existing materials and processes for industrial applications. However, improving the understanding and simulation of materials and AM process and understanding the effect of different steps in the AM workflow can increase the performance even more. The best way of benefitting from AM is to understand all the steps related to that—from the design and simulation to additive manufacturing and post-processing ending the actual application.

The objective of this Special Issue was to provide a forum for researchers and practitioners to exchange their latest achievements and identify critical issues and challenges for future investigations on "Modeling, Simulation and Data Processing for Additive Manufacturing". The Special Issue consists of 10 original full-length articles on the topic.

Mika Salmi
Editor

Editorial

Modeling, Simulation and Data Processing for Additive Manufacturing

Mika Salmi

Department of Mechanical Engineering, Aalto University, 02150 Espoo, Finland; mika.salmi@aalto.fi

Citation: Salmi, M. Modeling, Simulation and Data Processing for Additive Manufacturing. *Materials* **2021**, *14*, 7755. https://doi.org/10.3390/ma14247755

Received: 17 August 2021
Accepted: 13 December 2021
Published: 15 December 2021

Publisher's Note: MDPI stays neutral with regard to jurisdictional claims in published maps and institutional affiliations.

Additive manufacturing or, more commonly, 3D printing is one of the fundamental elements of Industry 4.0. and the fourth industrial revolution. It has shown its potential example in medical and dentistry, automotive, aerospace, and spare parts [1–5]. Personal manufacturing, complex and optimized parts, short series manufacturing and local on-demand manufacturing are some of the current benefits. Development of process and materials in terms of speed, costs and availability open new business cases all the time. Most of the research has focused on material and AM process development or effort to utilize existing materials and processes for industrial applications. However, improving the understanding and simulation of materials and AM process and understanding the effect of different steps in the AM workflow can increase the performance even more. The best way of benefit of AM is to understand all the steps related to that—from the design and simulation to additive manufacturing and post-processing ending the actual application.

This Special Issue consists of 10 original full-length articles on modeling, simulation and data processing for AM. Li et al. [6] studied finite element modeling of a novel lattice bimetallic composite comprising 316L stainless steel and a functional dissolvable aluminum alloy. Samples were fabricated and characterized, and experimental, finite element analysis (FEA) and digital image correlation (DIC) results were compared. The dissolvable aluminum showed higher Young's modulus, yield stress, and ultimate stress than the lattice and composite, but less elongation. Moreover, they demonstrated FEA and DIC efficient methods to simulate, analyze, and verify the experimental results. Sikan et al. [7] developed a finite element model for electron beam additive manufacturing of Ti-6Al-4V to understand the metallurgical and mechanical aspects of the process. Thin wall plates of 3 mm thickness were fabricated to validate the simulation results and ensure the reliability of the developed model. The thermal predictions of the model, when validated experimentally, gave a low average error of 3.7%. The model proved to be highly successful for predicting the cooling rates, grain morphology, and microstructure. The maximum deviations observed in the mechanical predictions of the model were as low as 100 MPa in residual stresses and 0.05 mm in distortion.

Singh et al. [8] used the finite element method to understand the influences of laser power and scanning speed on the heat flow and melt-pool dimensions in the powder bed fusion process for titanium and Inconel 718. A transient 3D finite-element model was developed to perform a quantitative comparative study to examine the temperature distribution and disparities in melt-pool behaviors under similar processing conditions. The temperature and melt-pool increase as laser power moves in the same layer and when new layers are added. The same is observed when the laser power increases. The opposite is observed for increasing scanning speed while keeping other parameters constant. Luo et al. [9] established a multi-layer and multi-track finite element model of 24CrNiMo alloy steel for powder bed fusion. The distribution and evolution of temperature and stress fields and the influence of process parameters on them were systematically studied. The results showed that the peak temperature increases from 2153 °C to 3105 °C, and the residual stress increases from 335 MPa to 364 MPa by increasing laser power from 200 W to 300 W; the peak temperature decreases from 2905 °C to 2405 °C, and the residual stress increases from 327 MPa to 363 MPa at scanning speeds from 150 mm/s to 250 mm/s; the peak temperature increases from 2621 °C to 2914 °C and the residual stress decreases from 354 MPa to 300 MPa at preheating temperatures of 25 °C and 400 °C, respectively.

Rupal et al. [10] systematically designed a benchmark for geometric tolerance characterization that can present tolerances in three principal planar directions. The benchmark was simulated using the finite element method, made with the LPBF process from stainless steel (316L), and the geometric tolerances were characterized. The effect of base plate removal on the geometric tolerances was quantified. Simulation and experimental results were compared to understand tolerance variations using different process parameters such as base plate removal, orientation, and size. Budzik et al. [11] presented a quality control methodology for additive manufacturing products made of polymer materials. The methodology varies depending on the intended use. Depending on the use of the models, the quality control process is divided into three stages: data control, manufacturing control, and post-processing control. When selecting materials, the 3D printing process and measurement methods, the purpose of the model and economic aspects should be taken into account. All products do not require high accuracy and durability.

Kusoglu et al. [12] studied nanoparticle additivation effects on LPBF of metals and polymers. They formed a theoretical concept for an inter-laboratory study design considering the process chain, including research data management. Macioł et al. [13] compared automatically detected precipitates in LPBF of Inconel 625 with a combination of the complementary electron techniques such as the chemical composition performed by EDS. Image processing methods and statistical tools were applied to maximize information gain from data with a low signal-to-noise ratio, keeping human interactions on a minimal level. The proposed algorithm allowed for the automatic detection of precipitates.

Kuschmitz et al. [14] used machine-learning techniques to compute acoustic material parameters from the material's micro-scale geometry for the additively manufactured porous sound absorbers. Laboratory measurements data of the flow resistivity and absorption coefficient were used to train two different machine learning models, an artificial neural network and a k-nearest neighbor approach. Both models could predict acoustic parameters from the specimen's micro-scale with reasonable accuracy. Salmi [15] reviewed additive manufacturing processes and materials in medical applications and process workflow for different uses. Based on the findings, directed energy deposition is rarely utilized in implants and sheet lamination for medical models or phantoms. Powder bed fusion, material extrusion and VAT photopolymerization are utilized in all categories. Material jetting is not used for implants and biomanufacturing, and binder jetting is not utilized for tools, instruments and parts for medical devices. The most common materials are thermoplastics, photopolymers and metals such as titanium alloys.

Funding: This research received no external funding.

Institutional Review Board Statement: Not applicable.

Informed Consent Statement: Not applicable.

Data Availability Statement: Not applicable.

Acknowledgments: All the published articles were refereed through at least a double-blind peer-review process. As Guest Editor, I would like to thank all the authors for their excellent contributions and especially the reviewers for their valuable and critical comments that significantly improved the papers' quality. Finally, I would like to thank the staff of Materials journals, in particular, Ariel Zhou, for her efforts and kind assistance.

Conflicts of Interest: The authors declare no conflict of interest.

References

1. Salmi, M.; Tuomi, J.; Sirkkanen, R.; Ingman, T.; Makitie, A. Rapid tooling method for soft customized removable oral appliances. *Open Dent. J.* **2012**, *6*, 85–89. [CrossRef] [PubMed]
2. Mäkitie, A.; Paloheimo, K.S.; Björkstrand, R.; Salmi, M.; Kontio, R.; Salo, J.; Yan, Y.; Paloheimo, M.; Tuomi, J. Medical applications of rapid prototyping—Three-dimensional bodies for planning and implementation of treatment and for tissue replacement. *Duodecim* **2010**, *126*, 143–151. [PubMed]
3. Leal, R.; Barreiros, F.; Alves, L.; Romeiro, F.; Vasco, J.; Santos, M.; Marto, C. Additive manufacturing tooling for the automotive industry. *Int. J. Adv. Manuf. Technol.* **2017**, *92*, 1671–1676. [CrossRef]
4. Kestilä, A.; Nordling, K.; Miikkulainen, V.; Kaipio, M.; Tikka, T.; Salmi, M.; Auer, A.; Leskelä, M.; Ritala, M. Towards space-grade 3D-printed, ALD-coated small satellite propulsion components for fluidics. *Addit. Manuf.* **2018**, *22*, 31–37. [CrossRef]
5. Chekurov, S.; Salmi, M. Additive manufacturing in offsite repair of consumer electronics. *Phys. Procedia* **2017**, *89*, 23–30. [CrossRef]
6. Li, X.; Ghasri-Khouzani, M.; Bogno, A.-A.; Liu, J.; Henein, H.; Chen, Z.; Qureshi, A.J. Investigation of Compressive and Tensile Behavior of Stainless Steel/Dissolvable Aluminum Bimetallic Composites by Finite Element Modeling and Digital Image Correlation. *Materials* **2021**, *14*, 3654. [CrossRef] [PubMed]
7. Sikan, F.; Wanjara, P.; Gholipour, J.; Kumar, A.; Brochu, M. Thermo-Mechanical Modeling of Wire-Fed Electron Beam Additive Manufacturing. *Materials* **2021**, *14*, 911. [CrossRef] [PubMed]
8. Singh, S.N.; Chowdhury, S.; Nirsanametla, Y.; Deepati, A.K.; Prakash, C.; Singh, S.; Wu, L.Y.; Zheng, H.Y.; Pruncu, C. A Comparative Analysis of Laser Additive Manufacturing of High Layer Thickness Pure Ti and Inconel 718 Alloy Materials Using Finite Element Method. *Materials* **2021**, *14*, 876. [CrossRef] [PubMed]
9. Luo, X.; Zhao, M.; Li, J.; Duan, C. Numerical Study on Thermodynamic Behavior during Selective Laser Melting of 24CrNiMo Alloy Steel. *Materials* **2020**, *13*, 45. [CrossRef] [PubMed]
10. Rupal, B.S.; Singh, T.; Wolfe, T.; Secanell, M.; Qureshi, A.J. Tri-Planar Geometric Dimensioning and Tolerancing Characteristics of SS 316L Laser Powder Bed Fusion Process Test Artifacts and Effect of Base Plate Removal. *Materials* **2021**, *14*, 3575. [CrossRef] [PubMed]
11. Budzik, G.; Woźniak, J.; Paszkiewicz, A.; Przeszłowski, Ł.; Dziubek, T.; Dębski, M. Methodology for the Quality Control Process of Additive Manufacturing Products Made of Polymer Materials. *Materials* **2021**, *14*, 2202. [CrossRef] [PubMed]
12. Kusoglu, I.M.; Huber, F.; Doñate-Buendía, C.; Rosa Ziefuss, A.; Gökce, B.; T Sehrt, J.; Kwade, A.; Schmidt, M.; Barcikowski, S. Nanoparticle Additivation Effects on Laser Powder Bed Fusion of Metals and Polymers—A Theoretical Concept for an Inter-Laboratory Study Design All Along the Process Chain, Including Research Data Management. *Materials* **2021**, *14*, 4892. [CrossRef] [PubMed]
13. Maciol, P.; Falkus, J.; Indyka, P.; Dubiel, B. Towards Automatic Detection of Precipitates in Inconel 625 Superalloy Additively Manufactured by the L-PBF Method. *Materials* **2021**, *14*, 4507. [CrossRef] [PubMed]
14. Kuschmitz, S.; Ring, T.P.; Watschke, H.; Langer, S.C.; Vietor, T. Design and Additive Manufacturing of Porous Sound Absorbers—A Machine-Learning Approach. *Materials* **2021**, *14*, 1747. [CrossRef] [PubMed]
15. Salmi, M. Additive Manufacturing Processes in Medical Applications. *Materials* **2021**, *14*, 191. [CrossRef] [PubMed]

materials

Article

Investigation of Compressive and Tensile Behavior of Stainless Steel/Dissolvable Aluminum Bimetallic Composites by Finite Element Modeling and Digital Image Correlation

Xiuhui Li [1], Morteza Ghasri-Khouzani [1], Abdoul-Aziz Bogno [2], Jing Liu [2], Hani Henein [2], Zengtao Chen [1,*] and Ahmed Jawad Qureshi [1,*]

[1] Department of Mechanical Engineering, University of Alberta, Edmonton, AB T6G 1H9, Canada; xiuhui1@ualberta.ca (X.L.); ghasrikh@ualberta.ca (M.G.-K.)
[2] Department of Chemical and Materials Engineering, University of Alberta, Edmonton, AB T6G 1H9, Canada; bogno@ualberta.ca (A.-A.B.); jing22@ualberta.ca (J.L.); hhenein@ualberta.ca (H.H.)
* Correspondence: zengtao@ualberta.ca (Z.C.); ajquresh@ualberta.ca (A.J.Q.)

Abstract: This study reports fabrication, mechanical characterization, and finite element modeling of a novel lattice structure based bimetallic composite comprising 316L stainless steel and a functional dissolvable aluminum alloy. A net-shaped 316L stainless steel lattice structure composed of diamond unit cells was fabricated by selective laser melting (SLM). The cavities in the lattice structure were then filled through vacuum-assisted melt infiltration to form the bimetallic composite. The bulk aluminum sample was also cast using the same casting parameters for comparison. The compressive and tensile behavior of 316L stainless steel lattice, bulk dissolvable aluminum, and 316L stainless steel/dissolvable aluminum bimetallic composite is studied. Comparison between experimental, finite element analysis (FEA), and digital image correlation (DIC) results are also investigated in this study. There is no notable difference in the tensile behavior of the lattice and bimetallic composite because of the weak bonding in the interface between the two constituents of the bimetallic composite, limiting load transfer from the 316L stainless steel lattice to the dissolvable aluminum matrix. However, the aluminum matrix is vital in the compressive behavior of the bimetallic composite. The dissolvable aluminum showed higher Young's modulus, yield stress, and ultimate stress than the lattice and composite in both tension and compression tests, but much less elongation. Moreover, FEA and DIC have been demonstrated to be effective and efficient methods to simulate, analyze, and verify the experimental results through juxtaposing curves on the plots and comparing strains of critical points by checking contour plots.

Keywords: selective laser melting (SLM); lattice structure; bimetallic composite; mechanical properties; finite element analysis (FEA); digital image correlation (DIC); hybrid manufacturing

Citation: Li, X.; Ghasri-Khouzani, M.; Bogno, A.-A.; Liu, J.; Henein, H.; Chen, Z.; Qureshi, A.J. Investigation of Compressive and Tensile Behavior of Stainless Steel/Dissolvable Aluminum Bimetallic Composites by Finite Element Modeling and Digital Image Correlation. *Materials* **2021**, *14*, 3654. https://doi.org/10.3390/ma14133654

Academic Editor: Mika Salmi

Received: 1 May 2021
Accepted: 8 June 2021
Published: 30 June 2021

1. Introduction

Recently, lattice structures have attracted the attention of many researchers due to properties such as light weight, high strength, energy absorption, reduced material consumption, and biocompatibility. Lattice structures are formed mathematically or geometrically by spatial arrangement and combination of a grouping of unit cells. Most researchers focus on the mechanical properties, such as compression and tension behavior [1–8], fracture behavior [9,10], fatigue behavior [11,12], and shear response [13], and biocompatibility [14–16] of these cells. Research has also been dedicated to design methods of lattice structures, including creating functionally graded porous structures [4,17–19], panel or sandwich-shaped lattice structures [20–22], and the mathematically designing algorithm [23–28].

Most work done on lattice structures has been about unit cells. Researchers were more likely to conduct experiments on normal unit cells formed by the spatial arrangements of struts. However, some of them shed light on complicated unit cells, whose composition components conform to specific mathematical algorithms, such as gyroid [1,3,5,18,19], Schwarz diamond [17,29] called TPMS (triply periodic minimal surfaces), and plate lattices [25]. Compression and tension tests were applied in studying F2CC,Z (face-centered cubic with Z-struts), hollow spherical unit cells by Kohnen et al. [30], and concluded that the mechanical properties for F2CC,Z are better than hollow spherical. Contuzzi et al. [31] studied F2CC,Z structure, and compressive testing using two samples of different volume fractions and concluded that increasing strut thickness is more significant than introducing reinforcement in the lattice structure. Rehme et al. [32] investigated not only F2CC,Z, but also FCC (face-centered cubic) and F2BCC,Z (body-centered and face-centered cubic combined with Z-struts) structures. The difference between these three face-centered cubic unit cells can be seen in Figure 1a,b,e. BCC (body-centered cubic), BCC,Z (body-centered cubic with Z-struts), gyroid and rhombic were also analyzed [2,3,12,29,33–35] through compressive, tensile, and fracture testing. They concluded that F2CC,Z has a higher load capacity, and gyroid can be very useful in applications requiring high stiffness. Peto et al. [36] and Park et al. [4] also gave an eye on other kinds of unit cells, which are relatively uncommon and not widely applied, and finally found that CD (cubic diamond) exhibited higher strength compared to others. An image of some unit cells mentioned above is shown in Figure 1. All of these are self-supported for 3D printing except FCC and CD.

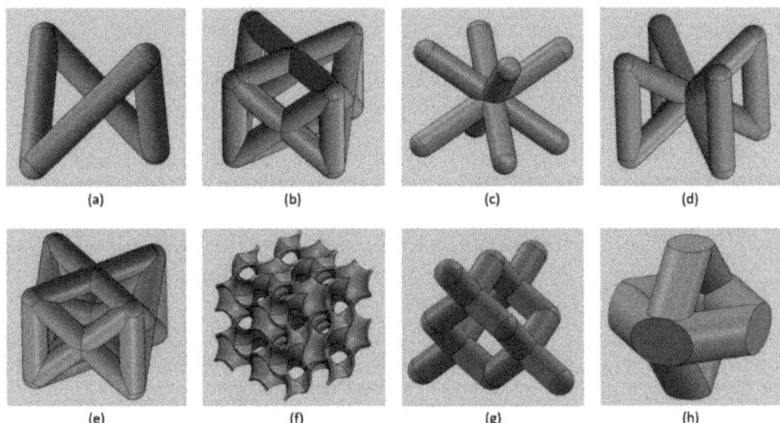

Figure 1. Unit cells in lattice structures: (**a**) FCC; (**b**) F2CC,Z; (**c**) BCC; (**d**) BCC,Z; (**e**) F2BCC,Z; (**f**) gyroid; (**g**) CD; (**h**) Ansys Space-Claim™ diamond.

Among all unit cells, diamond unit cells are considered the best choice for structures with strength requirements. With predictions of the Gibson-Ashby model, research done by Maconachie et al. [29] evidenced that diamond lattice structures exhibit larger relative strength and relative modulus in the same volume fraction of lattice. However, traditional diamond unit cells, namely CD unit cells, are not self-supported, which might cause some problems in fabrication through additive manufacturing; hence, another type of diamond unit cell inspired by ANSYS Space-Claim™ is plotted in Figure 1h. This diamond unit cell was shown in the lattice auto-generating feature in the Space-Claim, yet there is limited research literature on its properties. Consequently, this diamond unit cell was chosen for the lattice structure in our study.

The manufacturing method of lattice structures has also received widespread attention with metal additive manufacturing (MAM) being a feasible option given the complexity of the geometry. MAM can directly print a geometry layer by layer on a substrate from the bottom to up by metal material feedstock. The sample can be printed from a computer-aided design (CAD), although there are some limitations of samples to be printed in terms of size and geometry for different machines. Selective laser melting (SLM) is one of the categories of MAM. In SLM, thin layers of atomized fine metal powder are evenly distributed using a coating mechanism onto a substrate plate. Then, each layer of the part geometry is fused by selectively melting the powder, which is achieved with a high-power laser beam. Some researchers investigate the defects of the structures fabricated by SLM or AM (additive manufacturing). It was noted that struts waviness, strut oversizing or strut thickness variation is prevalent on lattice structures fabricated by SLM [9,37–41] with horizontal struts features showing more severe geometric imperfections than vertical struts and diagonal struts [9,39,41,42]. Moreover, vertical struts were found to be thinner than as-designed ones [9,37,39], and the magnitude of strut oversizing can change the failure mode from one to another [9]. SLM parameters also affect the mechanical properties of lattice structures [3,43]. Horizontal struts are the first to fracture, indicating they are experiencing greater stress than neighboring struts [41,44].

Although there are some flaws in the structure fabricated by SLM, evidence shows that SLM lattice structures manufactured from stainless steel powder have excellent mechanical performance [32]. Microstructural and mechanical characterizations of duplex stainless steel UNS S31803 processed by SLM was conducted by Hengsbach et al. [45], who validated the successful fabrication of duplex stainless steel processed via SLM. In addition to duplex stainless steel, 316L stainless steel also has been favored by researchers. The mechanical properties and deformation behavior of 316L stainless steel lattice structures fabricated by SLM were studied [30,46], as well as fracture toughness [3]. Bimetallic lattice composites have slowly been gaining interest with researchers. This latticed composite contains two parts, namely the lattice and the matrix, in which another material is filled into the lattice gaps. There is also much research on the microstructure and mechanical properties of bimetallic lattice structures manufactured by SLM, such as CuSn/18Ni300 bimetallic porous structures [47], and A356/316L interpenetrating phase composites [48,49], in which [49] investigated the mechanical properties of PrintCast composites through finite element analysis (FEA), coupled with digital image correlation (DIC) to capture the deformation and failure processes.

FEA is commonly used for simulating the experimental process and validating testing results. Researchers usually conducted FEA for performance evaluation [50–52], structure design [53], investigating configurational effects [54], and studying the failure mechanism [55,56]. Digital image correlation (DIC) is a 3D, full-field, non-contact optical technique to measure contour, deformation, vibration, and strain on almost any material. DIC setting is essential for investigating strain rate by analyzing captured images, and it is also apparent to show elongation changing along with the experiments processing. Limited research was done for analyzing deformation and strain evolution applying DIC on stainless steel such as 316L [30,49,57]. Mostly, the focus has been on studying titanium alloy Ti6Al4V [58–61]. Other investigations into displacement, velocities, and stress measurements using DIC were also done on polymers [62], glass fibers [63], and other materials [64,65].

In this study, FEA and DIC are used to investigate the mechanical properties of 316L stainless steel/dissolvable aluminum bimetallic composites, which are vital for simulating and recording experimental processes. 316L stainless steel lattice structures formed by the unit cell shown in Figure 1h were built using the SLM method, and a molten aluminum alloy infiltrated the 316L stainless steel lattice gaps to create the bimetallic composite. Mechanical properties were analyzed thoroughly by both tension and compression tests, and the experimental results were compared with those from FEA to validate its effectiveness. Simultaneously, the DIC system was also applied to capture strain distribution and verify the FEA results. The following section provides the details of materials and methods used. Section Three describes the FEA simulation model and experimental validation for individual lattices and filler structure. This is followed by Section Four, which details the FEA simulation and experimental validation of bimetallic composite strcutures. Finally, Section Five provides conclusions.

2. Materials and Methods

2.1. Manufacturing

To study the mechanical properties of materials and structures, both compression and tension tests were performed. Hence, bulk samples, lattice samples and bimetallic composite samples were required for both tension and compression tests. Stainless steel 316L, lattice samples were printed through an EOS M290 machine (EOS, Krailling, Germany), while a proprietary aluminum alloy supplied by the industrial partner was used for the filled-in matrix part of composite by casting. Bulk aluminum samples were also fabricated by casting.

Compression samples of lattice were in the shape of a cube with a length of 12.5 mm. The tension samples of lattice were a dog-bone shape, whose dimensions conformed to ASTM E8M standard [66], with a gauge length of 50 mm and a gauge width of 12.5 mm. The lattice structure unit cell's strut diameter is 2 mm, which is the same for both the compression and tension samples. Failure of the tension samples should occur in the gauge zone rather than the interface between the diamond lattice part and the solid gripping part, which is the location of stress concentration. Therefore, fillets were designed on the junction interface of grips to reduce the concentrated stress and avoid failure in this area. The 0.75 mm fillets of the tension sample and the compression sample are displayed in Figure 2. The chemical composition of gas atomized 316L stainless steel powder for the SLM process is listed in Table 1. Tension lattice dog-bone samples were fabricated in a horizontal orientation to the building plate (a hot-rolled mild steel panel with a dimension of 252 mm × 252 mm × 25 mm). EOS Company recommended processing parameters were applied for the 316L stainless steel, and the detailed parameters are listed in [67].

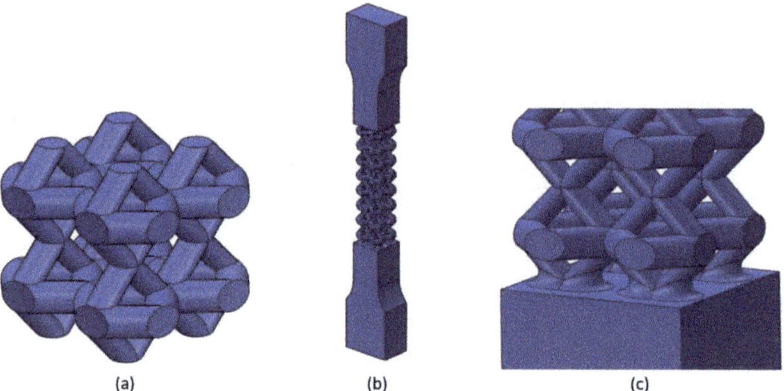

(a)   (b)   (c)

Figure 2. Computer-aided design models (CAD) of the Space-Claim diamond lattice structure parts: (a) compression model; (b) tension dog-bone model; (c) fillets in the interface of dog-bone model.

Table 1. Chemical composition of 316L stainless steel powder used as the feedstock material for the AM process (wt.%).

Chemical Composition	C	Cr	Mn	Mo	N	Ni	O	S	Si	Fe
Value (wt.%)	0.03	17.9	2.0	2.4	0.1	13.9	0.04	0.01	0.75	Balance

Bimetallic composite samples were manufactured based on the lattice ones. For both compression and tension composite samples, dissolvable aluminum alloy was filled into the lattice structure gaps and formed a matrix part of the composite by the casting process. The chemical composition of dissolvable aluminum is shown in Table 2, and the details for the casting process are presented in Section 2.2 of [67]. Bulk aluminum samples were also fabricated under the same casting condition.

Table 2. The chemical composition of the aluminum alloy used for casting (wt.%).

Chemical Composition	Fe	Ag	Ga	Cu	Mg	Al
Value (wt.%)	0.6	2.1	2.0	2.6	4.1	Balance

Microstructure analysis for the specimens can be found in Section 2.3 of [67]. An image of all the experimental samples is presented in Figure 3.

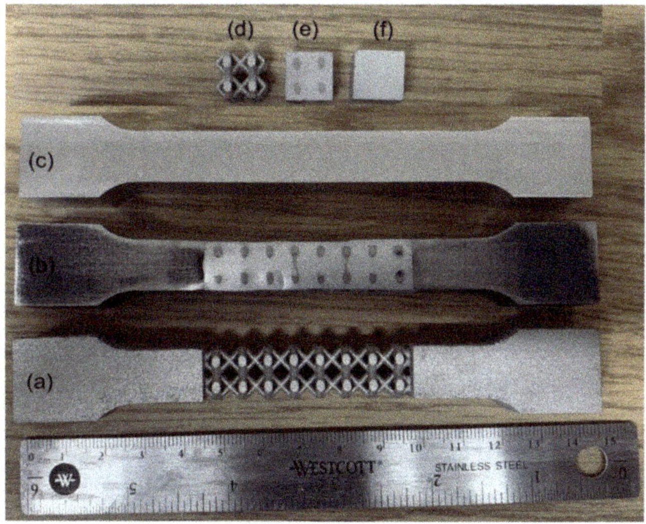

Figure 3. An image of the experimental samples: (**a**) stainless steel lattice dog-bone; (**b**) stainless steel/aluminum composite dog-bone; (**c**) bulk aluminum dog-bone; (**d**) stainless steel lattice cube; (**e**) stainless steel/aluminum composite cube; (**f**) bulk aluminum cube.

2.2. DIC System Setting

In our experiments, VIC-Snap commercial software (V8, manufactured by Correlated Solutions, Inc., Irmo, SC, USA) was used to capture images, and VIC-3D commercial software (V8, manufactured by Correlated Solutions, Inc., Irmo, SC, USA) was applied to process the images.

Two Allied Vision Technology (AVT) Pike F421b cameras (resolution of 2048 (H) × 2048 (V), sensor size: type 1.2, (Allied Vision Technologies GmbH, Stadtroda, Germany), equipped with two Nikon 28–85 mm F-mount lenses by two C to F-mount adapters (for lenses, Nikon, Tokyo, Japan), which allow for the adjusting of aperture, focus, and zoom, were mounted on a tripod and used in the experiments. Both two lenses provide an average magnification of 10 pixel/mm. One of the cameras was precisely positioned with its lens perpendicular to the focused surface of the lattice sample during the experiments. The other camera's lens was positioned at 25° to the primary camera. The testing images were captured at the rate of one frame per second, with each frame capturing a compression displacement at around 8 μm and a tension displacement around 33 μm according to the loading speed of 0.5 and 2 mm/min, respectively. The specimens were sprayed with black and white paint (Rust-Oleum, Evanston, IL, USA) to form a scattered speckle pattern on the focused surface with an average diameter of speckles of about 1.3 mm (approximately 5 pixels). Before capturing testing images, a calibration target card with 8 × 8 dots was imaged simultaneously by rotating to different angles in both cameras to calibrate the system in one step thoroughly.

2.3. Mechanical Testing

Uniaxial compression and tension tests at room temperature were conducted on all the experimental specimens. The displacement-controlling mode was applied on all the tests using a servo-hydraulic mechanical testing system (MTS 810, MTS, Eden Prairie, MN, USA). The cross-head speed was 0.5 mm/min for compression tests and 2 mm/min for tension tests, leading to an initial strain rate of 6.673×10^{-4} s^{-1} for both compression and tension experiments. For more details of the mechanical testing, please refer to Section 2.4 of [67].

3. FEA Simulation and Experimental Validation of Individual Lattice, and Bulk Structures

3.1. FEA Procedure

The FE analysis was conducted using the commercial FE code ABAQUS™/Explicit (2019 version, Dassault Systemes, Vélizy-Villacoublay, France) [68], with simulation models generated using SolidWorks (V2020, Dassault Systemes, Vélizy-Villacoublay, France) [69]. Comparing to ABAQUS™/Standard, ABAQUS™/Explicit solver can better solve the convergent problems for models with complex configurations, especially for lattice structures. Furthermore, it can also readily analyze problems with complicated contact interaction between the independent bodies [49] for the bimetallic lattice structures clarified in Section 4.

The simulation model needs to be imported into ABAQUS™ before conducting the FE analysis. Then, the material parameters such as Young's modulus, Poisson's ratio for elasticity, and "true stress" vs. "plastic strain" values for plasticity in the ABAQUS™ property-material module are set up. The plasticity "true stress" vs. "plastic strain" pairs of values for 316L stainless steel were obtained from [70], while data for aluminum alloy were obtained from the bulk aluminum experiments. After setting up the materials, assigning the specific material to the model configuration accordingly, for example, 316L stainless steel, was given to the lattices while aluminum was given to the bulk aluminum models.

For compression model boundary conditions, the bottom end (one surface for bulk models, four small surfaces for lattice models) was fixed for all the six degrees of freedom (U1 = U2 = U3 = UR1 = UR2 = UR3 = 0). Simultaneously, a reference point was generated on the top and coupled with the top end (one surface for bulk models, four small surfaces for lattice models), with five degrees of freedom fixed (U1 = U3 = UR1 = UR2 = UR3 = 0) and one remained (U2) for the loading. A velocity of 0.5 mm/min was then applied to the top reference point in the U2 direction. Note that the applying velocity should not be consistent from the beginning of the analysis until the end. Based on the actual experiment, the loading speed shall change gradually from 0 mm/min initially, to the maximum in the middle, then drop back to 0 mm/min in the end, at which time the average rate would be 0.5 mm/min. In this case, the amplitude of velocity gradually changed throughout the whole loading process. As for tension models, similarly, the bottom end of the dog-bone gripping area was fixed for all degrees of freedom (U1 = U2 = U3 = UR1 = UR2 = UR3 = 0), while a velocity of 2 mm/min was applied to the reference point on the top in the U2 direction (U1 = U3 = UR1 = UR2 = UR3 = 0).

The last step before running the FE analysis was meshing. The free linear tetrahedral 3D stress element (C3D4 element type) was selected for both compression and tension lattice models and tension bulk dog bones, while the structured linear hexahedral 3D stress element (C3D8 element type) without reduced integration was used for compression bulk samples. Note that C3D4 was also used on the gripping block areas of tension lattice models to assure consistency with the lattice part. The mesh size for compression lattice samples is 0.5 mm, and 1 mm for all other models. For the compression bulk 316L stainless steel model, the compression bulk aluminum alloy model, and the 316L stainless steel lattice model, the number of elements are 2197, 2197 and 47,336, respectively, with node numbers of 2744, 2744, and 10,895. For the tension bulk 316L stainless steel model, tension bulk aluminum model, and tension 316L stainless steel lattice model, the numbers of elements are 158,001, 158,001, and 188,681, respectively, with nodes numbers of 30,622, 30,622, and 40,588.

Figures 4 and 5 show deformation contour plots for bulk 316L stainless steel, bulk aluminum, and 316L stainless steel lattice under both compressive and tensile conditions. Stresses shown in the plots were all von Mises stress averaging at 75%. The value 75% here means that if the relative difference between contributions that a specific node gets from its neighboring elements is less than 75%, these contributing values are averaged [68]. The local effects on Figure 4a,b might come from the contact boundary condition applied. The rigid plate is used to apply the compressive load to the sample. When the deformation reaches the highest level in compression, friction between the rigid surface and the sample surface will lead to "sticking condition" which leads to much higher result as seen in the model results. This however only accounts for a very limited range of the whole load carrying area. As a result, the actual stress used to represent the bulk behavior of the compression sample is much less than the 1110 MPa as shown. The same situation applies to the Figure 4b. It is also evident that 316L stainless steel is much stronger and can afford more stress than aluminum under both compressive and tensile conditions. Moreover, compressive strength is almost the same as tensile strength for the lattice sample since there is no significant difference between their ultimate stress in the deformed contour plots.

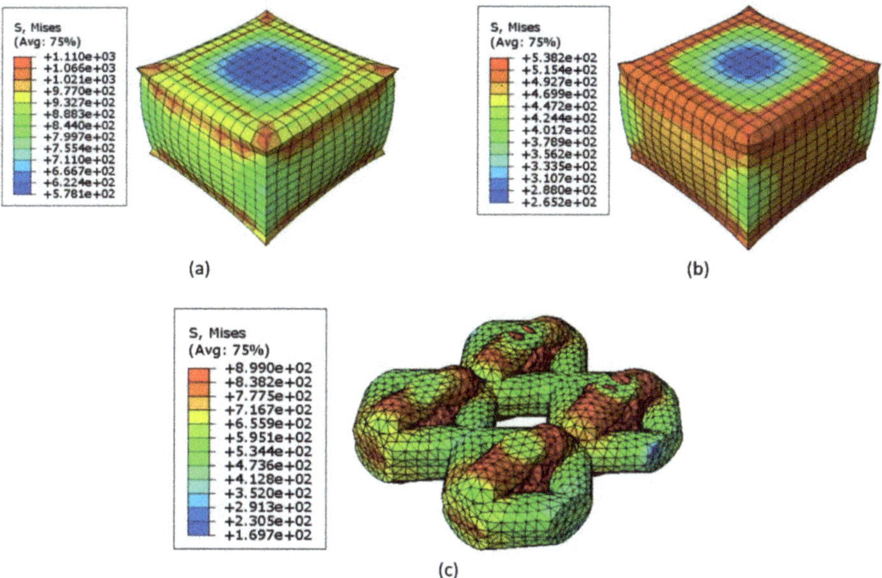

Figure 4. Deformation contour plots of FEA for compression samples: (**a**) bulk 316L stainless steel cube; (**b**) bulk dissolvable aluminum cube; (**c**) 316L stainless steel lattice.

After getting the contour plot, the reaction force and displacement of the top reference point of each model were exported from ABAQUS™ to an excel sheet. The engineering stress (σ_E) and engineering strain (ε_E) were obtained using the equations below:

$$\sigma_E = \frac{The\ reaction\ force\ (\mathrm{N})}{The\ failure\ cross\ section\ area\ (\mathrm{mm}^2)},\ (\mathrm{MPa}) \tag{1}$$

$$\varepsilon_E = \frac{The\ displacement\ (\mathrm{mm})}{The\ sample\ (\mathrm{gauge})\ length} \tag{2}$$

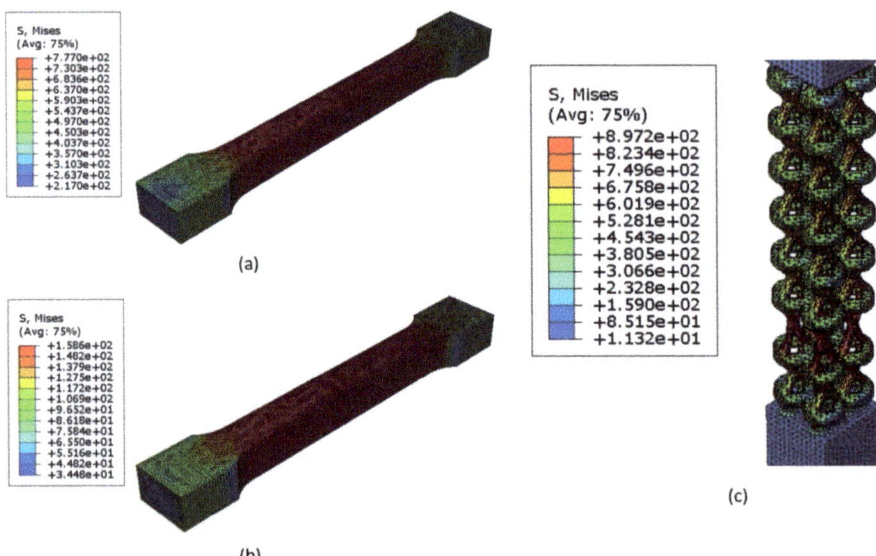

Figure 5. Deformation contour plots of FEA for tension samples: (**a**) bulk 316L stainless steel dog-bone; (**b**) bulk dissolvable aluminum dog-bone; (**c**) 316L stainless steel lattice dog-bone.

The compression model is a cube of 12.5 mm in each direction, and the gauge length for all tension models is 50 mm. The cross-section area for both compression and tension bulk models is 156.25 mm^2 (12.5 mm $\times$ 12.5 mm). However, as the cross-section area varies throughout the whole length of lattice samples, the average cross-section area size of 60.99 mm^2 is adopted with a maximum of 109.42 mm^2 and a minimum of 12.56 mm^2. Figure 6 shows the positions of maximum and minimum areas of the lattice using the compression one as the example.

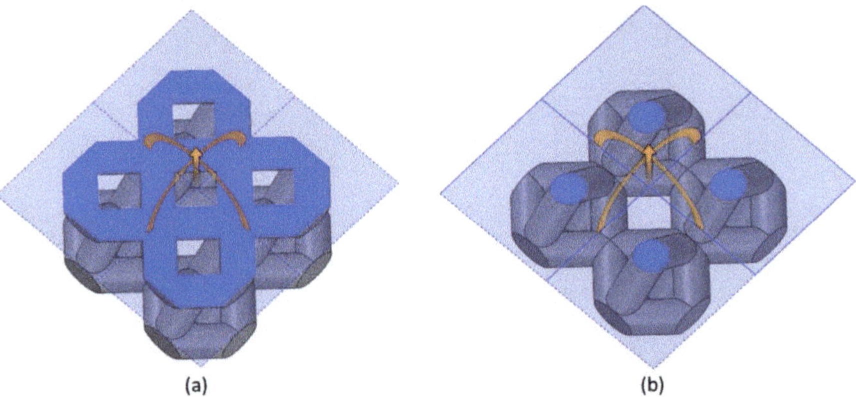

Figure 6. Maximum and minimum areas of the compression lattice model: (**a**) maximum area and (**b**) minimum area.

Using the formulas below, we can convert the engineering stress (σ_E) and engineering strain (ε_E) to true stress (σ_T) and true strain (ε_T):

$$\varepsilon_T = \ln(1 + \varepsilon_E) \tag{3}$$

$$\sigma_T = \sigma_E(1 + \varepsilon_E) \tag{4}$$

The "true stress" vs. "true strain" plots for FE compression and tension tests are shown in Figures 7 and 8. The experimental work will be discussed in Section 3.2, and the comparison will be made between the FEA and experimental results to verify the consistency.

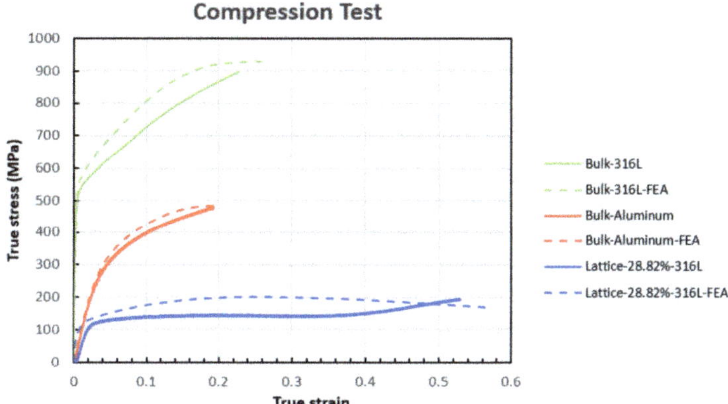

Figure 7. Comparison between experimental and FEA results of bulk 316L stainless steel, bulk dissolvable aluminum, and 316L stainless steel lattice for the compression test.

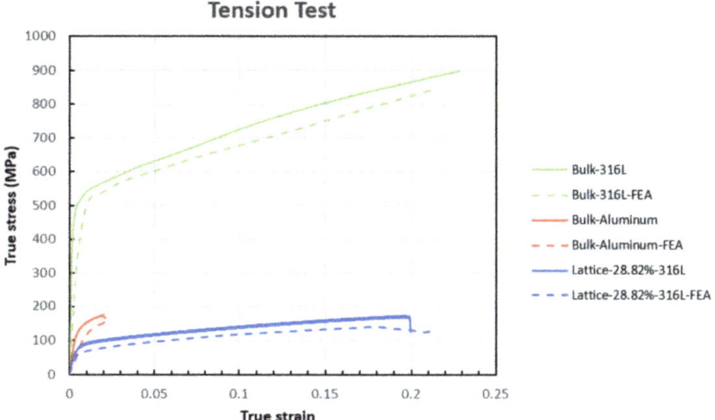

Figure 8. Comparison between experimental and FEA results of bulk 316L stainless steel, bulk dissolvable aluminum, and 316L stainless steel lattice for the tension test.

3.2. Experimental Validation of FEA Results

The experimental 316L stainless steel data was obtained from [70]. Overlapping the FEA compression plot in Section 4.1 to this experimental plot, we then obtained the final comparison plot between the FEA result and experimental result for all bulk and lattice specimens shown in Figure 7. We can see that for the three materials, the FEA results and experimental results are in conformance with each other, with average calculated numerical deviations of 9.8 and 5.0% for yield stress and ultimate compressive stress, respectively. Although the general shape of the Lattice-28.82%-316L with Lattice-28.82%-316L-FEA curves are in agreement, in certain areas, the curves show a difference. This difference becomes more apparent as the plastic deformation increases. These differences occur due to unavoidable manufacturing and material defects, such as microporosity, surface roughness, deviation from the nominal dimensions, and the offset of the strut axes from the ideal axes. These variations will affect the mechanical strength of samples. Macrostructure based finite element model as presented in this work has not integrated these defects. Therefore, consequently, the FEA results are overestimated compared to the corresponding experimental results. As the specific sample is a lattice structure with a high volumetric void ratio, these errors seem higher. However, as is apparent in Figure 7, as the void volume ratio decreases, these errors also significantly decrease. These errors also decrease as the plastic deformation progresses towards the end where the sample densification occurs. Moreover, it is also obvious that the yield and ultimate compressive stress of 316L stainless steel lattices are less than those of both the bulk aluminum and the bulk 316L stainless steel, which means the strength of the lattice with a volume fraction of 28.82% is significantly less than the solid samples due to low volume fractions. The ultimate compressive stress, which represents the compressive strength of the lattice, can be significantly enhanced by increasing the lattice strut diameter [31]. Furthermore, the cracks in the microstructure of the lattice can also explain the much lower yield stress and compressive strength.

Moreover, Figure 7 shows that the compression test for bulk aluminum stopped much earlier than the 316L stainless steel lattice counterpart. This is due to the test being stopped at the load limit (100 kN) of the mechanical testing machine before the specimen failure, while the 316L stainless steel sample collapsed before the test stopped. Three significant deformation stages, which are the elastic stage, plateau stage and densification stage, are shown in the 316L stainless steel compressive curve compared with the bulk aluminum. Initially, lattice struts were in an elastic deformation stage under the compressive load. Then, the struts approached the yield point, and the plastic stage began, which is indicated as the plateau stage. In the plateau stage, the strut nodes were dramatically squeezed, and plastic hinges formed. Finally, the densification started since the struts were continuously compressed to the point where some were broken, while others were closely squeezed against each other.

Identically, the experimental 316L stainless steel data were also collected from [70]. In order to be consistent with the compression result and further compare with the FEA result, all the experimental engineering values were transformed to the true values by using Equations (3) and (4). Similarly, mapping the FEA tension plot in Section 4.1 to this experimental plot, we then obtained the final tension plot between the FEA result and experimental result for all bulk and lattice specimens shown in Figure 8. This plot also validates that the FEA results agree with the experimental, with average calculated numerical deviations of 2.1 and 8.9% for yield stress and ultimate tensile stress. Likewise, the yield stress and tensile strength of the 316L stainless steel lattice are much lower than the other two bulk models. Increasing the strut diameter to achieve a bigger volume fraction will also improve the tension property.

Unlike the compression testing, which has three deformation stages, the 316L stainless steel lattice just experienced the initial elastic stage and the elongational plastic stage, followed by fracture failure with a sudden drop in stress eventually. Moreover, the tensile behavior of the bulk aluminum exhibits an apparent difference from the other two, with a higher Young's modulus than the lattice but much less elongation than the other two. This is because aluminum is more brittle and has lower resistance to the tensile loading than 316L stainless steel, making it much easier to fracture with shorter elongation. In contrast, the diamond lattice configuration achieved a much-extended elongation and can be widely used in the energy absorption structure.

3.3. Experimental Validation with DIC Results

As for the comparison between the experimental and DIC results, we discuss the compression bulk aluminum and tension 316L stainless steel dog-bone lattice samples for brevity. A detailed view of bulk aluminum compression experimental curve is shown in Figure 9. Three unique points, namely the yielding point, the point in the plastic region, and the point in the hardening region, were marked out with their true strain and true stress values. The corresponding DIC images to these points are shown in Figure 10.

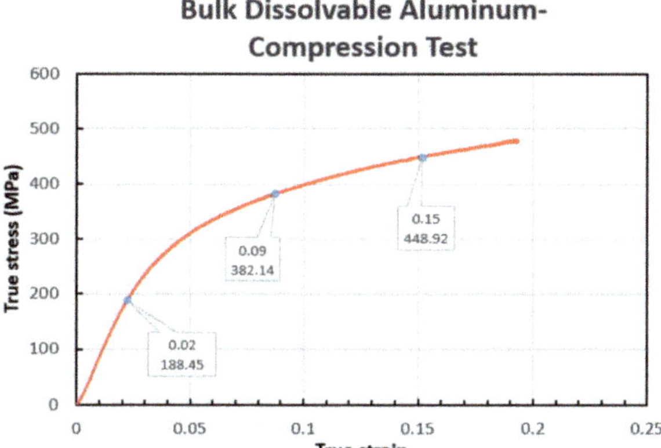

Figure 9. The experimental result of compression bulk dissolvable aluminum cube with three unique points marked out with true stress and true strain.

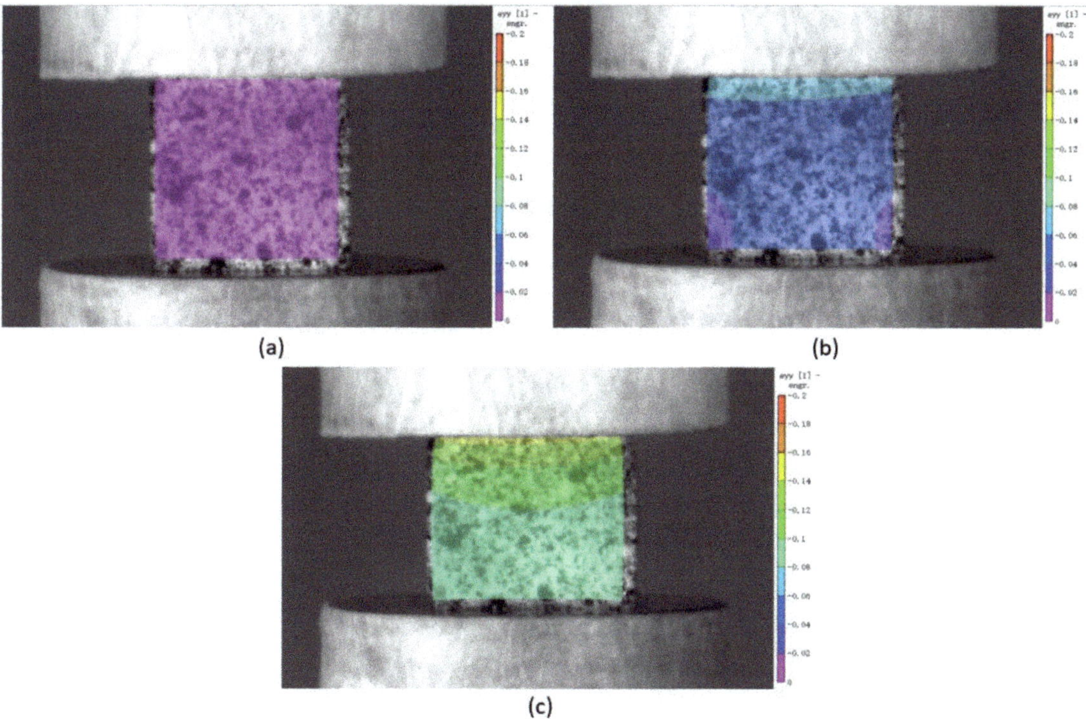

(a)

(b)

(c)

Figure 10. DIC frames of the three points marked out in the bulk dissolvable aluminum compression curve: (**a**) 34 s; (**b**) 131 s; (**c**) 228 s.

The scale bar is listed on the right side of each picture, with the strain range of −0.2 to 0 (negative values represent the compression test). From the frames, we can see that the color symbolizing engineering strain changes with loading progression, and the experimental results match the value range as the frames plotted. Figure 10a shows a uniform strain distribution as there is no severe displacement but with the increase in displacement, clear and uneven distribution can be observed in the subsequent Figure 10b,c. Similarly, four particular points, namely the yielding point, the turning point, the point in the plastic region, and the point before the curve drop, are marked out on the tension test experimental curve of the 316L stainless steel dog-bone lattice in Figure 11, with corresponding DIC images shown in Figure 12 in an increasing strain sequence, with strain ranging from 0–0.2. Figure 12a–d show the DIC images corresponding to the four points on the stress strain curve, obtained through the tensile testing machine using an extensometer. DIC shows slight uneven distribution of the strain within the sample gauge length. The highest strain obtained from DIC matches the result from extensometer well. It can also be observed from Figure 12c,d that the strain at the end of the lattice, where it attaches to the solid part of the sample is uneven and much less. This is in accordance with the expectation as the strain decreases with the increasing part density.

Figure 11. The experimental result of tension 316L stainless steel dog-bone lattice with four unique points marked out with true stress and true strain.

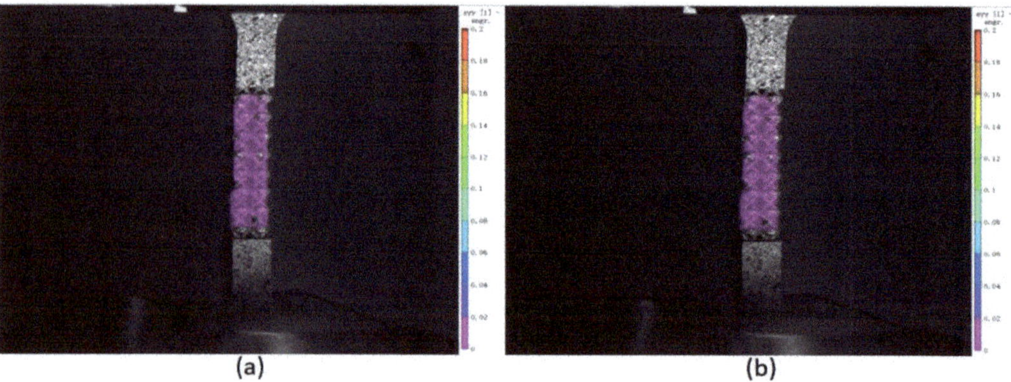

Figure 12. *Cont.*

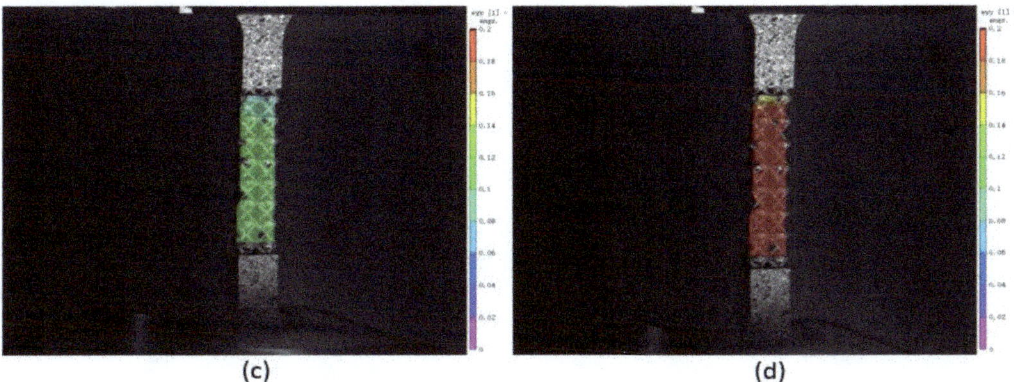

Figure 12. DIC frames of the four points marked out in the tension 316L stainless steel dog-bone lattice curve: (a) 6 s; (b) 18 s; (c) 150 s; (d) 295 s.

4. FEA Simulation and Experimental Validation of Bimetallic SS316L-Aluminum Alloy Bimetallic Composite

4.1. FEA Procedure

For FEA modeling of the bimetallic composite, two separate models were constructed in SolidWorks™ and imported and combined in ABAQUS™. ABAQUS™/Explicit (2019 version) solver was used in this work as it is appropriate to solve problems involving two models contacting each other. Separate models of both compression composite and tension composite created in SolidWorks are shown in Figure 13.

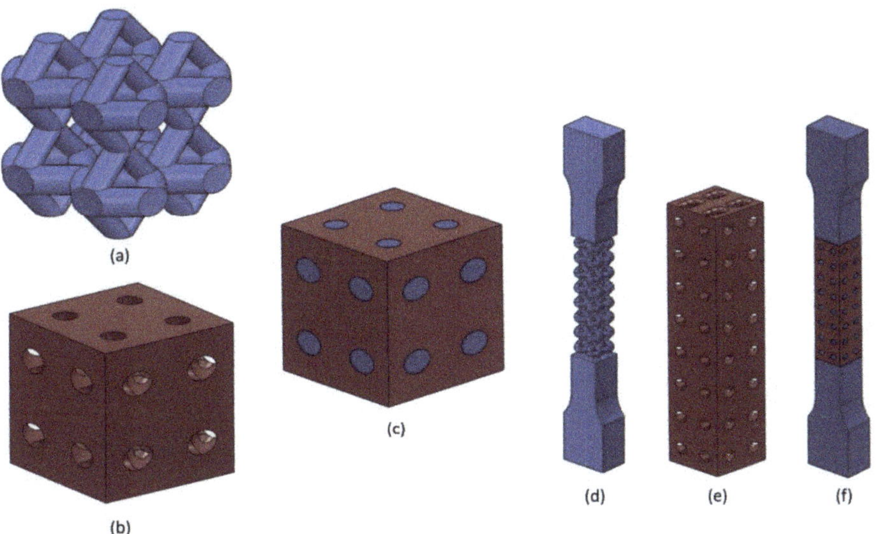

Figure 13. CAD models of the composite parts: (a) lattice part for the compression composite; (b) matrix part for the compression composite; (c) the compression composite; (d) lattice part for the tension composite; (e) matrix part for the tension composite; (f) the tension composite.

Similar to procedure in Section 3.1, the materials were assigned to the corresponding part of the composite after importing the models into ABAQUS™. Materials for both compression composite and tension composite are the same, namely 316L stainless steel for the lattice part, and aluminum for the filled-in matrix part. Next, separate models were assembled into one composite pattern, and the geometry centers of both the lattice part and the matrix part were ensured to coincide. Setting up interaction between two objects of a composite is critical in ABAQUS™ FEA. Based on the microstructural analysis of the interface as reported in [67], it is observed that there is no cohesive bonding between the two parts, and therefore, a "hard contact" interaction of the 316L/aluminum interface was generated in ABAQUS™. Two surface sets were established, with one set of the outer surfaces of the lattice, and the other of the inner surfaces of the matrix, to be selected for creating the surface interaction. No penetration in the normal direction is assumed, and isotropic friction with a coefficient of 0.3 in the tangential direction is applied without elastic slip and any other shear stress for both the compression and tension composite patterns. Finally, a reference point is created on the top surface and coupled with the top cover for applying the load.

The boundary conditions for both compression and tension composites are the same as the models for bulk and lattice experiments. The bottom end was fixed for all the six degrees of freedom (U1 = U2 = U3 = UR1 = UR2 = UR3 = 0), and the top reference point was held for five degrees of freedom except for U2 (U1 = U3 = UR1 = UR2 = UR3 = 0). A gradually changed velocity of an average of 0.5 mm/min was applied on the reference point for the compression sample, while 2 mm/min for the tension, maintaining consistency with the experiments. Figures of boundary conditions for compression and tension composites are omitted here since there is no significant difference with those shown in Section 3.1.

The free linear tetrahedral 3D stress element (C3D4 element type) was applied to both the lattice and matrix part of compression and tension composites. It is worth noting that the gripping block areas of the tension composite dog-bone also used C3D4, which is identical to the tension lattice dog-bone meshing. The mesh size for the compression composite was 0.5 and 1 mm for the tension composite. Moreover, there are overall 152,845 and 327,547 elements, and 32,891, and 70,978 nodes for the whole compression and tension composites, respectively.

Figure 14 gives the deformation contour plots of two composites. Stresses shown in the plots were all von Mises stress averaging at 75% of elongation. We can see that the composite is severely deformed under the compressive loading, and the matrix part is in light-green color, which means it afforded the load and played an essential role in resisting the load. In contrast, the tension composite matrix is almost in the blue color. Compared with the scale bar, we know that the insignificant load transferred to the matrix. This is due to a lack of interface fusion due to continuous cracks in the 316L/aluminum interface preventing the load transfer from the lattice to the matrix.

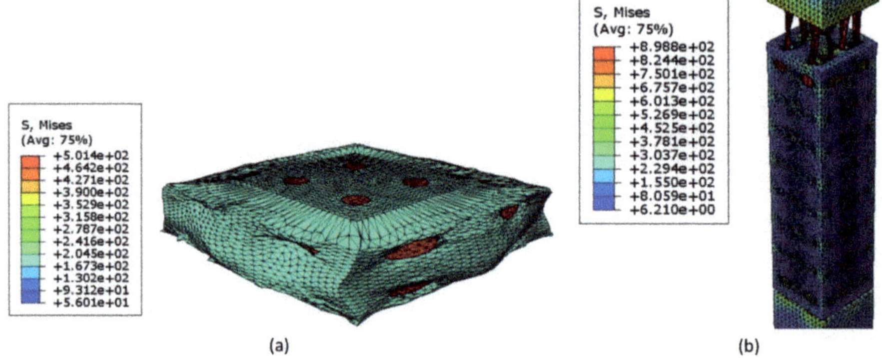

Figure 14. Deformation contour plots of FEA for composite samples: (**a**) compression composite cube and (**b**) tension composite dog-bone.

"Engineering stress" and "engineering strain" were then collected from the reaction force and displacement exported from ABAQUS™ using Equations (1) and (2), and corresponding "true stress" and "true strain" were calculated by Equations (3) and (4). The sample length was 12.5 mm for the compression composite, while 50 mm (gauge length) for the tension composite. The cross-section area was 156.25 mm^2 (12.5 mm × 12.5 mm) for the compression; however, this is not the case for the tension.

The "true stress" vs. "true strain" plots for compression and tension composite FEA results are shown as dashed black lines in Figures 15 and 16, respectively in Section 4.2 for comparison. Similarly, the experimental work will also be discussed, and the comparison will be made between the FEA and experimental results to verify the consistency.

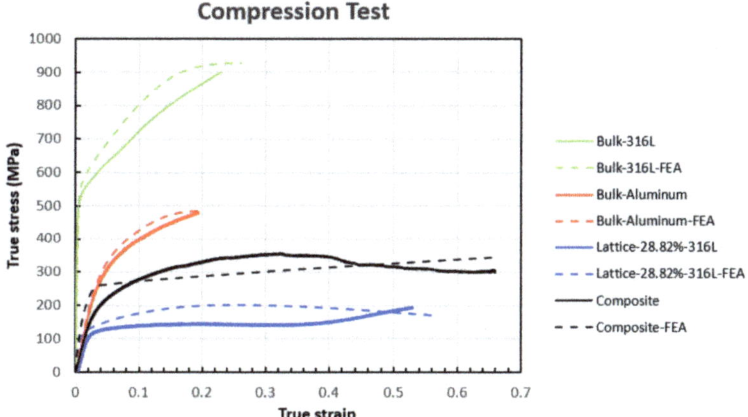

Figure 15. Comparison between experimental and FEA results of bulk 316L stainless steel, bulk dissolvable aluminum 316L stainless steel lattice, and 316L stainless steel/dissolvable aluminum composite for the compression test.

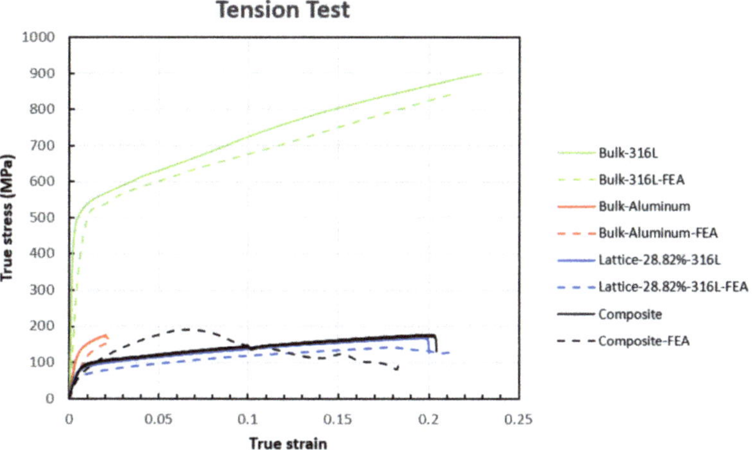

Figure 16. Comparison between experimental and FEA results of bulk 316L stainless steel, bulk dissolvable aluminum 316L stainless steel lattice, and 316L stainless steel/dissolvable aluminum composite for the tension test.

4.2. Experimental Validation of FEA Results

"True stress" vs. "True strain" curves of experimental results of the composite at room temperature as well as FEA results are plotted with other results of bulk and lattice samples in Figures 15 and 16 for compression and tension tests, respectively.

In terms of composite tests, there is a lack of bonding between the aluminum matrix and the SS316 lattice. This lack of material bonding plays a role in the experimental results of the compression as well as tension samples. This interface in the FEA is modeled as a hard contact with a corresponding friction coefficient. This coefficient is a constant value in the model. In the experimental tests, based upon the nature of test, i.e., compression, or tension, the interface between the two materials evolves as a function of strain and loading condition. Based on these differences it can be observed that the FEA results underestimate the compression and overestimate the tension. However, despite these, the calculated numerical deviation of 2.0% for the ultimate compressive stress confirms that the FEA simulation shows a good accuracy. Moreover, it is also apparent from the plot that the yielding and ultimate compressive strength has been significantly enhanced from the lattice shown in blue to the composite shown in black due to the filled-in matrix part. Nonetheless, the mechanical properties of the composite are less than the bulk aluminum properties shown in red. This can be addressed by increasing the volume fraction of the lattice. Using the rule of mixtures, this would result in composite properties between the lower bound of bulk aluminum and the upper bound of bulk 316L stainless steel.

Composite compression and tension experimental curves were taken out of the plots shown in Figures 17 and 18. For the compression test, as clarified in Section 3.3, three unique points, namely the yield point, a point in the plastic region, and a point in the hardening region, were marked out with their true strain and true stress values, and the corresponding frames captured by the DIC system are shown in Figure 19. In contrast, for the tensile test, four points, namely the yield point, a point in the plastic region, a point before the first curve dip, and the last point that the DIC effectively tracked, were marked out, and the DIC results were shown in Figure 20. The corresponding time calculated for the compression test was 35, 179, and 383 s, while 9, 21, 54, and 101 s for the tension test.

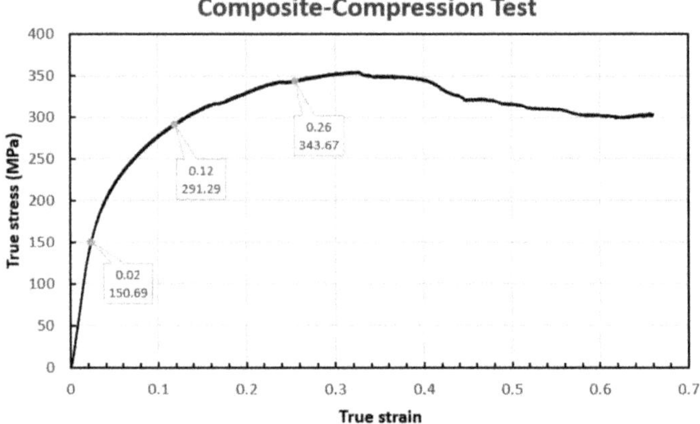

Figure 17. The experimental result of compression composite cube with three unique points marked out with true stress and true strain.

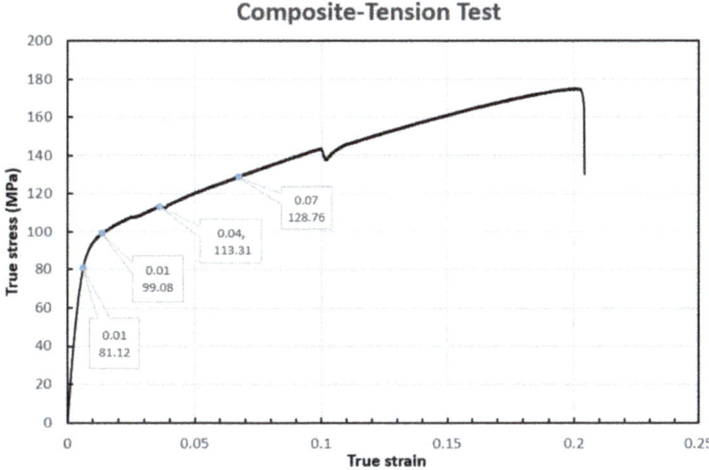

Figure 18. The experimental result of tension composite dog-bone with four unique points marked out with true stress and true strain.

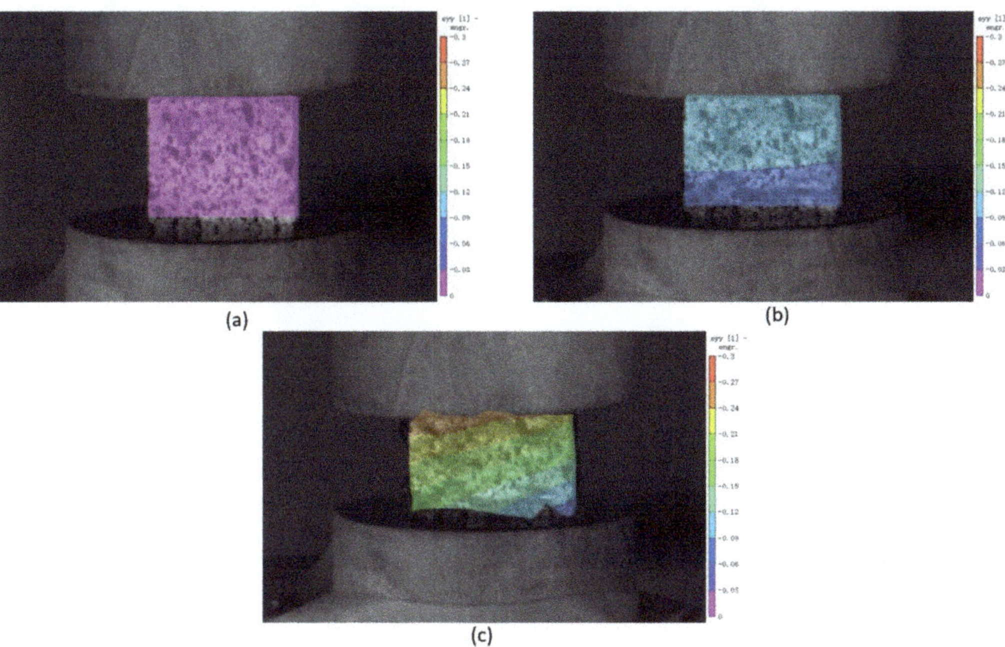

(a)

(b)

(c)

Figure 19. DIC frames of the three points marked out in the composite compression curve: (**a**) 35 s; (**b**) 179 s; (**c**) 383 s.

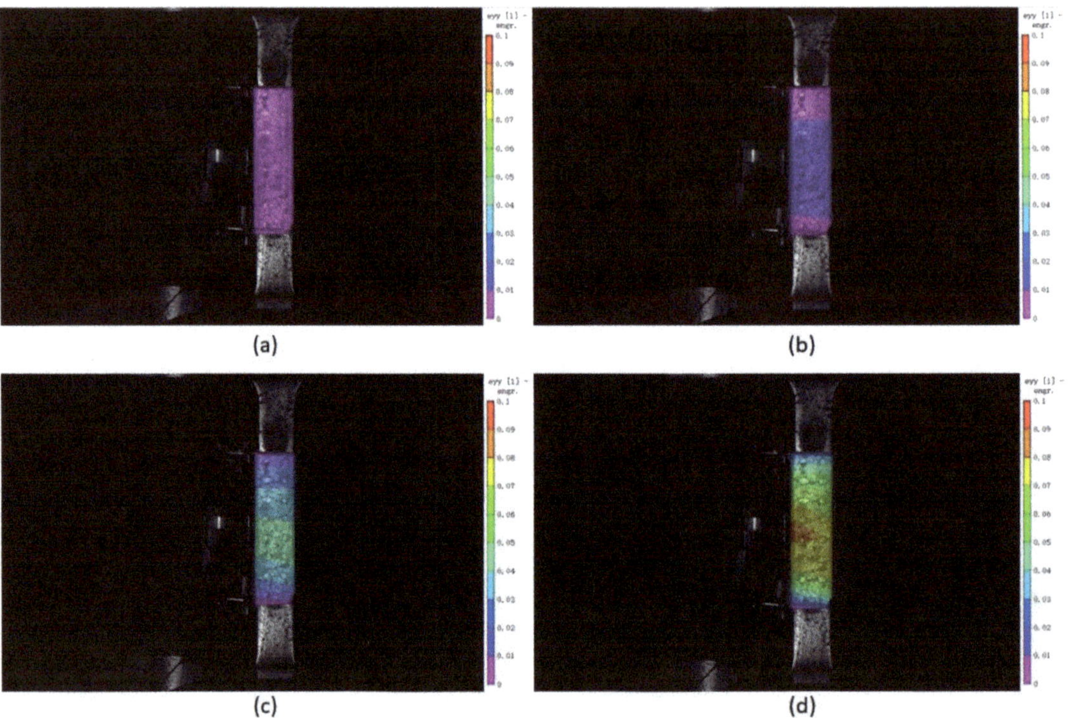

Figure 20. DIC frames of the four points marked out in the composite tension curve: (**a**) 9 s; (**b**) 21 s; (**c**) 54 s; (**d**) 101 s.

The tension results are different from the compression curves, where two distinct regions can be found in the experimental results, the elastic region, and plastic region, after which a sudden drop is shown, indicating the rupture of the sample. It is significant to note that the tensile curves for the 316L stainless steel lattice and bimetallic lattice are similar. This indicates that the aluminum matrix does not play an essential role due to lack of bonding. Similar to the compression results, the bulk 316L stainless steel and bulk aluminum possess higher yield stress and ultimate tensile stress, and both tensile curves of the 316L lattice and composite do not even surpass the curve of bulk aluminum. However, the dissolvable aluminum presents a much lower elongation comparing to the other three samples. The trivial difference between the experimental and FEA data for all four pairs validates the simulation results, including the numerical calculated deviation of 2.0% for the ultimate stress of the tension composite. The ABAQUS™ simulation curve for the bimetallic composite generally matches the results from Cheng et al. [49].

4.3. Experimental Validation with DIC Results

The DIC data for compression and tension tests of composite samples reveal that the strain pattern is uneven along the length of the sample. This is in departure from the DIC test results for the bulk aluminum, as well as the SS316 lattice structure, which showed a more even strain distribution as compared to the composite samples.

A strain range of −0.3 to 0 (Figure 19) was exhibited in the compression and 0 to 0.1 (Figure 20) in the tension. The strain behavior of the compression composite represented by the color coding was very similar to the bulk dissolvable aluminum. However, slight differences were observed for the tension composite. The strain growth was observed to grow gradually from the center to both sides, initially from 0 shown as purple color in the first frame to about 0.07 with orange color appearing in the middle part of the last frame.

Experimental strain results of the curve plots (Figure 17 for compression and Figure 19 for tension) match the value range plotted in the frames for both the compression and tension composite samples.

5. Conclusions

The work presented in this study provides a novel and original method to model and simulate bimetallic lattice structures. Bimetallic lattice structures are an emerging field of materials that harness the properties of their constituent materials and provide a meta material capable of engineered functional response. The capability to engineer these meta materials makes them an ideal candidate for applications in biomedical, aerospace, defence, space, and oil and gas industries. The bimetallic composite combination studied and modeled in this specific research work also possesses functional properties due to the dissolvable aluminum alloy matrix, which allows a part of the composite to dissolve while retaining its cellular, lattice-based structure. By investigating the compressive and tensile behaviour of 316L stainless steel lattice, bulk dissolvable aluminum alloy, and 316L stainless steel/dissolvable aluminum bimetallic composite, the following conclusions can be obtained:

1. The developed FEA model is an acceptable simulation for the experimental work. After validating the effectiveness of ABAQUS™ FEA simulation on the current experiments, the simulation can be used to explore different volume fractions of base lattice and filler to obtain desired properties without the need for extensive experiments. For bulk and lattice samples, the average calculated numerical deviations between experimental and FEA results in this study for yield stress and ultimate stress are 9.8 and 5.0% for compressive tests and 2.1 and 8.9% for tensile tests, respectively. For composite samples, the average calculated numerical deviations for ultimate stress are 2.0% for both compressive and tensile experiments. Further improvements to the model can be made by integrating the manufacturing dimensional variations as well as manufacturing induced material imperfections.
2. 316L stainless steel has better compressive properties and higher resistance to the tensile loading than dissolvable aluminum alloy, which is more brittle with less elongation.
3. In the tension test, due to lack of bonding, the load does not transfer from the 316L stainless steel lattice to aluminum alloy. However, the aluminum alloy part plays an indispensable role in the compression test and enhances the composite's compression strength compared to the lattice itself.
4. The elastic modulus, yield stress, and ultimate stress of both the 316L stainless steel lattice and bimetallic composite were lower than the bulk aluminum, proving that the performance of the lattice and composite with a volume fraction of 28.82% is still not that satisfactory. Increasing the strut diameter of lattice to achieve a higher volume fraction is expected to enhance the mechanical properties, including both compressive and tensile strengths.

Author Contributions: Data curation, formal analysis, investigation, software, validation, visualization, and writing—original draft preparation, X.L.; data curation, formal analysis, investigation, resources, validation, visualization, and writing—review and editing, M.G.-K.; investigation, methodology, resources, supervision, validation, and writing-review and editing, A.-A.B.; supervision, validation, and writing-review and editing, J.L.; conceptualization, funding acquisition, methodology, resources, supervision, validation, and writing-review and editing, H.H.; conceptualization, methodology, software, supervision, validation, and writing—review and editing, Z.C.; conceptualization, funding acquisition, methodology, project administration, resources, supervision, validation, and writing—review and editing, A.J.Q. All authors have read and agreed to the published version of the manuscript.

Funding: This research was funded by NSERC, grant number EGP 543505-19 Qureshi, and Alberta Innovates CASBE, grant number G2020000060 Q, and the APC was funded by the same grants.

Institutional Review Board Statement: Not applicable.

Informed Consent Statement: Not applicable.

Data Availability Statement: Data available on request due to restrictions eg privacy or ethical.

Acknowledgments: The authors thanks Alberta Innovates, the Natural Science and Engineering Research Council of Canada (NSERC), and Boss Completions Ltd. for financial support, as well as Kelsey Kalinski of Boss Completions. Ltd. for valuable suggestions.

Conflicts of Interest: The authors declare no conflict of interest. They also declare that they have no known competing financial interests or personal relationships, which seem to affect the work reported in this article.

References

1. Yan, C.; Hao, L.; Hussein, A.; Young, P.; Raymont, D. Advanced lightweight 316L stainless steel cellular lattice structures fabricated via selective laser melting. *Mater. Des.* **2014**, *55*, 533–541. [CrossRef]
2. McKown, S.; Shen, Y.; Brookes, W.K.; Sutcliffe, C.J.; Cantwell, W.J.; Langdon, G.S.; Nurick, G.N.; Theobald, M.D. The quasi-static and blast loading response of lattice structures. *Int. J. Impact Eng.* **2008**, *35*, 795–810. [CrossRef]
3. Alsalla, H.; Hao, L.; Smith, C. Fracture toughness and tensile strength of 316L stainless steel cellular lattice structures manufactured using the selective laser melting technique. *Mater. Sci. Eng. A* **2016**, *669*, 1–6. [CrossRef]
4. Park, J.H.; Park, K. Compressive behavior of soft lattice structures and their application to functional compliance control. *Addit. Manuf.* **2020**, *33*, 101148. [CrossRef]
5. Yánez, A.; Herrera, A.; Martel, O.; Monopoli, D.; Afonso, H. Compressive behaviour of gyroid lattice structures for human cancellous bone implant applications. *Mater. Sci. Eng. C* **2016**, *68*, 445–448. [CrossRef]
6. Kellogg, R.A.; Russell, A.M.; Lograsso, T.A.; Flatau, A.B.; Clark, A.E.; Wun-Fogle, M. Tensile properties of magnetostrictive iron-gallium alloys. *Acta Mater.* **2004**, *52*, 5043–5050. [CrossRef]
7. Rossiter, J.D.; Johnson, A.A.; Bingham, G.A. Assessing the Design and Compressive Performance of Material Extruded Lattice Structures. 3D Print. *Addit. Manuf.* **2020**, *7*, 19–27. [CrossRef]
8. Campanelli, S.L.; Contuzzi, N.; Ludovico, A.D.; Caiazzo, F.; Cardaropoli, F.; Sergi, V. Manufacturing and characterization of Ti6Al4V lattice components manufactured by selective laser melting. *Materials* **2014**, *7*, 4803–4822. [CrossRef] [PubMed]
9. Liu, L.; Kamm, P.; García-Moreno, F.; Banhart, J.; Pasini, D. Elastic and failure response of imperfect three-dimensional metallic lattices: The role of geometric defects induced by Selective Laser Melting. *J. Mech. Phys. Solids* **2017**, *107*, 160–184. [CrossRef]
10. Geng, L.; Wu, W.; Sun, L.; Fang, D. Damage characterizations and simulation of selective laser melting fabricated 3D re-entrant lattices based on in-situ CT testing and geometric reconstruction. *Int. J. Mech. Sci.* **2019**, *157*, 231–242. [CrossRef]
11. Zargarian, A.; Esfahanian, M.; Kadkhodapour, J.; Ziaei-Rad, S.; Zamani, D. On the fatigue behavior of additive manufactured lattice structures. *Theor. Appl. Fract. Mech.* **2019**, *100*, 225–232. [CrossRef]
12. Peng, C.; Tran, P.; Nguyen-Xuan, H.; Ferreira, A.J.M. Mechanical performance and fatigue life prediction of lattice structures: Parametric computational approach. *Compos. Struct.* **2020**, *235*, 111821. [CrossRef]
13. Moongkhamklang, P.; Deshpande, V.S.; Wadley, H.N.G. The compressive and shear response of titanium matrix composite lattice structures. *Acta Mater.* **2010**, *58*, 2822–2835. [CrossRef]
14. De Wild, M.; Ghayor, C.; Zimmermann, S.; Rüegg, J.; Nicholls, F.; Schuler, F.; Chen, T.H.; Weber, F.E. Osteoconductive Lattice Microarchitecture for Optimized Bone Regeneration. 3D Print. *Addit. Manuf.* **2019**, *6*, 40–49. [CrossRef]
15. Egan, P.; Wang, X.; Greutert, H.; Shea, K.; Wuertz-Kozak, K.; Ferguson, S. Mechanical and Biological Characterization of 3D Printed Lattices. 3D Print. *Addit. Manuf.* **2019**, *6*, 73–81. [CrossRef]
16. Melancon, D.; Bagheri, Z.S.; Johnston, R.B.; Liu, L.; Tanzer, M.; Pasini, D. Mechanical characterization of structurally porous bio-materials built via additive manufacturing: Experiments, predictive models, and design maps for load-bearing bone replacement implants. *Acta Biomater.* **2017**, *63*, 350–368. [CrossRef] [PubMed]
17. Al-Ketan, O.; Lee, D.W.; Rowshan, R.; Abu Al-Rub, R.K. Functionally graded and multi-morphology sheet TPMS lattices: Design, manufacturing, and mechanical properties. *J. Mech. Behav. Biomed. Mater.* **2020**, *102*, 103520. [CrossRef] [PubMed]
18. Liu, F.; Mao, Z.; Zhang, P.; Zhang, D.Z.; Jiang, J.; Ma, Z. Functionally graded porous scaffolds in multiple patterns: New design method, physical and mechanical properties. *Mater. Des.* **2018**, *160*, 849–860. [CrossRef]
19. Li, D.; Liao, W.; Dai, N.; Dong, G.; Tang, Y.; Xie, Y.M. Optimal design and modeling of gyroid-based functionally graded cellular structures for additive manufacturing. *CAD Comput. Aided Des.* **2018**, *104*, 87–99. [CrossRef]
20. Azzouz, L.; Chen, Y.; Zarrelli, M.; Pearce, J.M.; Mitchell, L.; Ren, G.; Grasso, M. Mechanical properties of 3-D printed truss-like lattice biopolymer non-stochastic structures for sandwich panels with natural fibre composite skins. *Compos. Struct.* **2019**, *213*, 220–230. [CrossRef]
21. Fan, H.; Yang, W.; Bin, W.; Yan, Y.; Qiang, F.; Zhuang, Z. Design and Manufacturing of a Composite Lattice Structure Reinforced by Continuous Carbon Fibers. *Tsinghua Sci. Technol.* **2006**, *5*, 1–5. [CrossRef]
22. Ye, G.; Bi, H.; Chen, L.; Hu, Y. Compression and Energy Absorption Performances of 3D Printed Polylactic Acid Lattice Core Sandwich Structures. 3D Print. *Addit. Manuf.* **2019**, *6*, 333–343. [CrossRef]

23. Arabnejad, S.; Pasini, D. Mechanical properties of lattice materials via asymptotic homogenization and comparison with alternative homogenization methods. *Int. J. Mech. Sci.* **2013**, *77*, 249–262. [CrossRef]

24. Wang, Y.; Xu, H.; Pasini, D. Multiscale isogeometric topology optimization for lattice materials. *Comput. Methods Appl. Mech. Eng.* **2017**, *316*, 568–585. [CrossRef]

25. Andersen, M.N.; Wang, F.; Sigmund, O. On the competition for ultimately stiff and strong architected materials. *Mater. Des.* **2021**, *198*, 109356. [CrossRef]

26. Alzahrani, M.; Choi, S.K.; Rosen, D.W. Design of truss-like cellular structures using relative density mapping method. *Mater. Des.* **2015**, *85*, 349–360. [CrossRef]

27. Ai, L.; Gao, X.L. Metamaterials with negative Poisson's ratio and non-positive thermal expansion. *Compos. Struct.* **2017**, *162*, 70–84. [CrossRef]

28. Deng, F.; Nguyen, Q.K.; Zhang, P. Multifunctional liquid metal lattice materials through hybrid design and manufacturing. *Addit. Manuf.* **2020**, *33*, 101117. [CrossRef]

29. Maconachie, T.; Leary, M.; Lozanovski, B.; Zhang, X.; Qian, M.; Faruque, O.; Brandt, M. SLM lattice structures: Properties, performance, applications and challenges. *Mater. Des.* **2019**, *183*, 108137. [CrossRef]

30. Köhnen, P.; Haase, C.; Bültmann, J.; Ziegler, S.; Schleifenbaum, J.H.; Bleck, W. Mechanical properties and deformation behavior of additively manufactured lattice structures of stainless steel. *Mater. Des.* **2018**, *145*, 205–217. [CrossRef]

31. Contuzzi, N.; Campanelli, S.L.; Casavola, C.; Lamberti, L. Manufacturing and characterization of 18Ni marage 300 lattice components by selective laser melting. *Materials* **2013**, *6*, 3451–3468. [CrossRef]

32. Rehme, O.; Emmelmann, C. *Virtual and Rapid Manufacturing*; Taylor & Francis/Balkema: Leiden, The Netherlands, 2008.

33. Hanzl, P.; Zetková, I.; Daňa, M. Uniaxial tensile load of lattice structures produced by metal additive manufacturing. *Manuf. Technol.* **2019**, *19*, 228–231. [CrossRef]

34. Li, C.; Lei, H.; Liu, Y.; Zhang, X.; Xiong, J.; Zhou, H.; Fang, D. Crushing behavior of multi-layer metal lattice panel fabricated by selective laser melting. *Int. J. Mech. Sci.* **2018**, *145*, 389–399. [CrossRef]

35. Sola, A.; Defanti, S.; Mantovani, S.; Merulla, A.; Denti, L. Technological Feasibility of Lattice Materials by Laser-Based Powder Bed Fusion of A357.0. 3D Print. *Addit. Manuf.* **2020**, *7*, 1–7. [CrossRef]

36. Peto, M.; Ramirez-Cedillo, E.; Uddin, M.J.; Rodriguez, C.A.; Siller, H.R. Mechanical behavior of lattice structures fabricated by direct light processing with compression testing and size optimization of unit cells. *ASME Int. Mech. Eng. Congr. Expo. Proc.* **2019**, *3*, 1–10. [CrossRef]

37. Dallago, M.; Zanini, F.; Carmignato, S.; Pasini, D.; Benedetti, M. Effect of the geometrical defectiveness on the mechanical properties of SLM biomedical Ti6Al4V lattices. *Procedia Struct. Integr.* **2018**, *13*, 161–167. [CrossRef]

38. El Elmi, A.; Melancon, D.; Asgari, M.; Liu, L.; Pasini, D. Experimental and numerical investigation of selective laser melting-induced defects in Ti-6Al-4V octet truss lattice material: The role of material microstructure and morphological variations. *J. Mater. Res.* **2020**, *35*, 1900–1912. [CrossRef]

39. Dallago, M.; Raghavendra, S.; Luchin, V.; Zappini, G.; Pasini, D.; Benedetti, M. Geometric assessment of lattice materials built via Selective Laser Melting. *Mater. Today Proc.* **2019**, *7*, 353–361. [CrossRef]

40. Sharma, P.; Pandey, P.M. Morphological and mechanical characterization of topologically ordered open cell porous iron foam fabricated using 3D printing and pressureless microwave sintering. *Mater. Des.* **2018**, *160*, 442–454. [CrossRef]

41. Dressler, A.D.; Jost, E.W.; Miers, J.C.; Moore, D.G.; Seepersad, C.C.; Boyce, B.L. Heterogeneities dominate mechanical performance of additively manufactured metal lattice struts. *Addit. Manuf.* **2019**, *28*, 692–703. [CrossRef]

42. Bagheri, Z.S.; Melancon, D.; Liu, L.; Johnston, R.B.; Pasini, D. Compensation strategy to reduce geometry and mechanics mismatches in porous biomaterials built with Selective Laser Melting. *J. Mech. Behav. Biomed. Mater.* **2017**, *70*, 17–27. [CrossRef]

43. Tsopanos, S.; Mines, R.A.W.; McKown, S.; Shen, Y.; Cantwell, W.J.; Brooks, W.; Sutcliffe, C.J. The influence of processing parameters on the mechanical properties of selectively laser melted stainless steel microlattice structures. *J. Manuf. Sci. Eng. Trans. ASME* **2010**, *132*, 0410111–04101112. [CrossRef]

44. Seepersad, C.C.; Allison, J.A.; Dressler, A.D.; Boyce, B.L.; Kovar, D. *An Experimental Approach for Enhancing the Predictability of Mechanical Properties of Additively Manufactured Architected Materials with Manufacturing-Induced Variability*; Elsevier Ltd: Amsterdam, The Netherlands, 2020.

45. Hengsbach, F.; Koppa, P.; Duschik, K.; Holzweissig, M.J.; Burns, M.; Nellesen, J.; Tillmann, W.; Tröster, T.; Hoyer, K.P.; Schaper, M. Duplex stainless steel fabricated by selective laser melting—Microstructural and mechanical properties. *Mater. Des.* **2017**, *133*, 136–142. [CrossRef]

46. Zhong, T.; He, K.; Li, H.; Yang, L. Mechanical properties of lightweight 316L stainless steel lattice structures fabricated by selective laser melting. *Mater. Des.* **2019**, *181*, 108076. [CrossRef]

47. Zhang, M.; Yang, Y.; Wang, D.; Song, C.; Chen, J. Microstructure and mechanical properties of CuSn/18Ni300 bimetallic porous structures manufactured by selective laser melting. *Mater. Des.* **2019**, *165*, 107583. [CrossRef]

48. Pawlowski, A.E.; Cordero, Z.C.; French, M.R.; Muth, T.R.; Keith Carver, J.; Dinwiddie, R.B.; Elliott, A.M.; Shyam, A.; Splitter, D.A. Damage-tolerant metallic composites via melt infiltration of additively manufactured preforms. *Mater. Des.* **2017**, *127*, 346–351. [CrossRef]

49. Cheng, J.; Gussev, M.; Allen, J.; Hu, X.; Moustafa, A.R.; Splitter, D.A.; Shyam, A. Deformation and failure of PrintCast A356/316 L composites: Digital image correlation and finite element modeling. *Mater. Des.* **2020**, *195*, 109061.

50. Xu, W.; Yu, A.; Lu, X.; Tamaddon, M.; Wang, M.; Zhang, J.; Zhang, J.; Qu, X.; Liu, C.; Su, B. Design and performance evaluation of additively manufactured composite lattice structures of commercially pure Ti (CP–Ti). *Bioact. Mater.* **2021**, *6*, 1215–1222. [CrossRef] [PubMed]

51. McDonald-Wharry, J.; Amirpour, M.; Pickering, K.L.; Battley, M.; Fu, Y. Moisture sensitivity and compressive performance of 3D-printed cellulose-biopolyester foam lattices. *Addit. Manuf.* **2021**, *40*, 101918. [CrossRef]

52. Mahmoud, D.; Al-Rubaie, K.S.; Elbestawi, M.A. The influence of selective laser melting defects on the fatigue properties of Ti6Al4V porosity graded gyroids for bone implants. *Int. J. Mech. Sci.* **2021**, *193*. [CrossRef]

53. Feng, J.; Liu, B.; Lin, Z.; Fu, J. Isotropic octet-truss lattice structure design and anisotropy control strategies for implant application. *Mater. Des.* **2021**, *203*, 109595. [CrossRef]

54. Traxel, K.D.; Groden, C.; Valladares, J.; Bandyopadhyay, A. Mechanical properties of additively manufactured variable lattice structures of Ti6Al4V. *Mater. Sci. Eng. A* **2021**, *809*, 140925. [CrossRef]

55. Hajjari, M.; Jafari Nedoushan, R.; Dastan, T.; Sheikhzadeh, M.; Yu, W.R. Lightweight weft-knitted tubular lattice composite for energy absorption applications: An experimental and numerical study. *Int. J. Solids Struct.* **2021**, *213*, 77–92. [CrossRef]

56. Li, P.Y.; Ma, Y.E.; Sun, W.B.; Qian, X.; Zhang, W.; Wang, Z.H. Fracture and failure behavior of additive manufactured Ti6Al4V lattice structures under compressive load. *Eng. Fract. Mech.* **2021**, *244*, 107537. [CrossRef]

57. Cao, X.; Xiao, D.; Li, Y.; Wen, W.; Zhao, T.; Chen, Z.; Jiang, Y.; Fang, D. Dynamic compressive behavior of a modified additively manufactured rhombic dodecahedron 316L stainless steel lattice structure. *Thin-Walled Struct.* **2020**, *148*, 106586. [CrossRef]

58. Xiao, L.; Song, W.; Xu, X. Experimental study on the collapse behavior of graded Ti-6Al-4V micro-lattice structures printed by selective laser melting under high speed impact. *Thin-Walled Struct.* **2020**, *155*, 106970. [CrossRef]

59. Goodall, R.; Hernandez-Nava, E.; Jenkins, S.N.M.; Sinclair, L.; Tyrwhitt-Jones, E.; Khodadadi, M.A.; Ip, D.H.; Ghadbeigi, H. The effects of defects and damage in the mechanical behavior of ti6al4v lattices. *Front. Mater.* **2019**, *6*, 1–11. [CrossRef]

60. Liu, F.; Zhang, D.Z.; Zhang, P.; Zhao, M.; Jafar, S. Mechanical properties of optimized diamond lattice structure for bone scaffolds fabricated via selective laser melting. *Materials* **2018**, *11*, 374. [CrossRef]

61. Xiao, L.; Song, W. Additively-manufactured functionally graded Ti-6Al-4V lattice structures with high strength under static and dynamic loading: Experiments. *Int. J. Impact Eng.* **2018**, *111*, 255–272. [CrossRef]

62. Montanini, R.; Rossi, G.; Quattrocchi, A.; Alizzio, D.; Capponi, L.; Marsili, R.; Di Giacomo, A.; Tocci, T. Structural characterization of complex lattice parts by means of optical non-contact measurements. In Proceedings of the I2MTC 2020 IEEE International Instrumentation and Measurement Technology Conference, Dubrovnik, Croatia, 25–28 May 2020; pp. 1–6. [CrossRef]

63. Hao, W.; Liu, Y.; Wang, T.; Guo, G.; Chen, H.; Fang, D. Failure analysis of 3D printed glass fiber/PA12 composite lattice structures using DIC. *Compos. Struct.* **2019**, *225*, 111192. [CrossRef]

64. Fíla, T.; Koudelka, P.; Falta, J.; Zlámal, P.; Rada, V.; Adorna, M.; Bronder, S.; Jiroušek, O. Dynamic impact testing of cellular solids and lattice structures: Application of two-sided direct impact Hopkinson bar. *Int. J. Impact Eng.* **2021**, *148*, 103767. [CrossRef]

65. Li, S.; Hu, M.; Xiao, L.; Song, W. Compressive properties and collapse behavior of additively-manufactured layered-hybrid lattice structures under static and dynamic loadings. *Thin-Walled Struct.* **2020**, *157*, 107153. [CrossRef]

66. *ASTM E8: ASTM E8/E8M standard test methods for tension testing of metallic materials 1*; ASTM International: West Conshohocken, PA, USA, 2010; Volume 4, pp. 1–27. [CrossRef]

67. Ghasri-Khouzani, M.; Li, X.; Bogno, A.A.; Liu, J.; Henein, H.; Chen, Z.; Qureshi, A.J. Investigation of compressive and tensile behavior of stainless steel/dissolvable aluminum bimetallic composites by finite element modelling and digital image correlation. *J. Manufact. Proc.* submitted.

68. Abaqus. *Abaqus 6.14. Abaqus 6.14 Anal. User's Guide*; Dassault Systèmes Simulia Corp.: Providence, RI, USA, 2014.

69. Lombard, M. *Introducing Solid Works*; Wiley: Hoboken, NJ, USA, 2013; pp. 1–35. [CrossRef]

70. Mower, T.M.; Long, M.J. Mechanical behavior of additive manufactured, powder-bed laser-fused materials. *Mater. Sci. Eng. A* **2016**, *651*, 198–213. [CrossRef]

Article

Thermo-Mechanical Modeling of Wire-Fed Electron Beam Additive Manufacturing

Fatih Sikan [1,2], Priti Wanjara [2], Javad Gholipour [2], Amit Kumar [1] and Mathieu Brochu [1,*]

[1] Department of Materials Engineering, McGill University, Montreal, QC H3A 0C5, Canada; fatih.sikan@mail.mcgill.ca (F.S.); amit.kumar2@mail.mcgill.ca (A.K.)
[2] National Research Council Canada, Aerospace Research Center, Montreal, QC H3T 2B2, Canada; priti.wanjara@cnrc-nrc.gc.ca (P.W.); javad.gholipourbaradari@cnrc-nrc.gc.ca (J.G.)
* Correspondence: mathieu.brochu@mcgill.ca

Abstract: The primary objective of this research was to develop a finite element model specifically designed for electron beam additive manufacturing (EBAM) of Ti-6Al-4V to understand metallurgical and mechanical aspects of the process. Multiple single-layer and 10-layer build Ti-6Al-4V samples were fabricated to validate the simulation results and ensure the reliability of the developed model. Thin wall plates of 3 mm thickness were used as substrates. Thermocouple measurements were recorded to validate the simulated thermal cycles. Predicted and measured temperatures, residual stresses, and distortion profiles showed that the model is quite reliable. The thermal predictions of the model, when validated experimentally, gave a low average error of 3.7%. The model proved to be extremely successful for predicting the cooling rates, grain morphology, and the microstructure. The maximum deviations observed in the mechanical predictions of the model were as low as 100 MPa in residual stresses and 0.05 mm in distortion. Tensile residual stresses were observed in the deposit and the heat-affected zone, while compressive stresses were observed in the core of the substrate. The highest tensile residual stress observed in the deposit was approximately 1.0 σ_{ys} (yield strength). The highest distortion on the substrate was approximately 0.2 mm.

Keywords: thermo-mechanical modelling; finite element analysis; residual stresses; microstructure; Ti-6Al-4V

Citation: Sikan, F.; Wanjara, P.; Gholipour, J.; Kumar, A.; Brochu, M. Thermo-Mechanical Modeling of Wire-Fed Electron Beam Additive Manufacturing. *Materials* **2021**, *14*, 911. https://doi.org/10.3390/ma14040911

Academic Editor: Mika Salmi

Received: 8 January 2021
Accepted: 3 February 2021
Published: 15 February 2021

Publisher's Note: MDPI stays neutral with regard to jurisdictional claims in published maps and institutional affiliations.

1. Introduction

In recent years, additive manufacturing (AM) technologies have revolutionized the manufacturing industry due to their original layer-by-layer processing nature and known advantages over conventional forming, forging, casting, and/or machining technologies [1]. Currently, AM is used to produce custom parts from a vast selection of materials in many industries, such as aerospace [2], automotive [3], satellite and space [3], and biomedical [4]. The main reasons for AM attracting such attention are (1) the potential of complex shape fabrication, design freedom to highlight creativity and part uniqueness [5], and (2) promising sustainability through cost effectiveness by decreasing lead time, production waste of high cost materials, energy consumption, and, thus, overall cost per part [6].

Despite several advantages, AM technology also has limitations that need to be addressed prior to becoming a mainstream production method, especially for critically loaded structures [7]. Production, geometry, and microstructural issues—such as porosity, lack of fusion between layers [8], cracking and distortion due to thermal and residual stresses [9], low spatial resolution in the final shape [10], high surface roughness [11], or need for post-processing (such as heat treatments, surface treatments, thermo-mechanical treatments or machining) [12]—hinder the advancement of AM in many fields. Such engineering challenges cause massive issues for strategic and critical industries such as aerospace where part requirements are very stringent for safety and reliability assurance. In addition, current materials used to build metal parts with AM are quite limited, and the readily available alloys for AM are inherited from welding and powder metallurgy fields. These include a few titanium [13] and aluminum [14] alloys, stainless steels [15], and nickel-based superalloys [16]. Increasing the inventory of additively manufacturable materials by developing new alloys or solving issues of current alloys are required. Thus, AM also attracts interest amongst materials and manufacturing researchers. The main efforts in AM literature are focused on several sub-groups, such as process parameter optimization to minimize defects [1], macro and microstructure control to tailor advanced properties [17], remedies for cracking problems in Ni-base superalloys [18], and developments of AM models to understand and predict thermal and mechanical phenomenon occurring during the process [19].

Electron beam additive manufacturing (EBAM) is a technique that belongs to the direct energy deposition (DED) sub-group of AM technologies [20]. EBAM involves the melting and solidification of a metal wire feedstock using an electron beam source to form the part geometry in a layer-by-layer manner. Although research on laser-based techniques has been more widely reported for AM processing of metal components, EBAM provides valuable advantages such as higher energy efficiency, faster build rate, material versatility (conductive, reflective, refractive), higher quality due to the reducing vacuum environment, and low residual stresses and part distortion [20,21]. However, in general, EBAM and DED processes are still not broadly applied and remain limited to cost-intensive parts. The reason DED often lacks cost-competitiveness against traditional production methods is due to process development complexity associated with parametric optimization and toolpath planning, which arises usually because the layer thickness during building requires corrective actions due to slight deviation in the process environment from layer to layer (e.g., heat build-up). Thus, for individual applications, the necessary trial-and-error iterative process developments become cost-/time-intensive as they involve considerable manual operator/engineer interventions. So, reliable and efficient models are of utmost importance for furthering development, and especially in the case of EBAM, they are still deficient for maturing and diversifying process applications.

Due to the complexity of AM processing, thermal and mechanical models considering heat and mass transfer are essential to calculate temperature fields to predict melt pool shape and size, cooling rates, residual stresses, and distortion. Finite element modeling is extensively used in AM processes for such purposes [21–25]. Some models focus on heat and mass transfer during the process and neglect mechanical phenomenon [19]. These models solve two- or three-dimensional steady-state or transient energy conservation equations with convective and radiative boundary conditions. The primary objective of such models is to monitor temperature variations within the built part in order to predict and control metallurgical aspects, such as segregation and microstructure of the material. Few of these heat transfer models also consider fluid dynamics in order to precisely calculate melt-pool dimensions and temperature distributions within the melt [26]. Such models solve conservation of mass, momentum, and energy equations and are generally computation intensive. Some other models also consider mechanical phenomena, such as yielding, thermal expansion, and elastic modulus with mechanical boundary conditions to predict residual stresses and distortion of the built geometry [27,28]. Most of these thermal and/or mechanical models neglect fluid flow for the sake of computational efficiency and focus on the prediction of residual stresses or their remedies. However, accurate calculation of the residual stresses generally requires improved features. Using appropriate material properties that consider strain rate, temperature, and phase dependence are critical for these models. In some cases, adopting hardening and creep models [24,29] reduce the error significantly [30]. However, limited material property databases force researchers to adjust their thermo-mechanical models.

This study focuses on Ti-6Al-4V, which is the workhorse alloy for the aerospace industry due to its high specific strength, high corrosion resistance, and high temperature properties [31,32]. This research targets exploring a deposition model for Ti-6Al-4V using EBAM to progress the understanding of the key factors and their effects on final deposition success, which are gauged by both reliability and performance. A thermo-mechanical finite element model (FEM) was constructed to predict residual stresses and distortion for a specified geometry, which would be purposeful for greatly reducing the amount of experimental iterations needed to optimize deposition procedures to meet the requirements (e.g., geometry, distortion tolerances) of manufactured and repaired parts. In order to validate this model, temperature variations on the substrate were recorded using thermocouples. In addition, microstructural characterization, residual stress, and distortion measurements were conducted. An important milestone in the present study was to advance the development of a part-scale thermo-mechanical model and confirm its prediction accuracy at levels similar to other reported works in the literature [21,26]. Here, it is worth mentioning that EBAM models present in the literature have mostly focused on mechanical model validation using distortion magnitudes from a singular or a few measurement points: however, in this study, residual stresses were thoroughly analyzed from nearly 100 unique measurement points. In addition, for the first time, the specific microstructural features of Ti6Al4V built by EBAM were used to validate thermal aspects of the model. In this regard, the model validation approach presented in this study gains another level of significance.

2. Materials and Methods

A series of single bead thin wall structures were fabricated using a wire-fed 42 kW Sciaky EBAM and welding system (Sciaky, Chicago, IL, USA) under 5×10^{-3} Pa (5×10^{-5} mbar) vacuum at the Aerospace Manufacturing Technologies Centre of the National Research Council of Canada in Montreal, Canada to emulate additive processing. Timetal Ti64 (AMS4911M) wrought plates (Timet, Warrensville Heights, OH, USA) were used as the substrate in the deposition experiments. Substrate plates were cut with dimensions of 85 mm in width, 65 mm in height, and 3 mm in thickness. The height of the substrate was oriented parallel to the rolling direction (RD) of the plate. The top (deposition) surface of the substrate was ground with 120 grit SiC paper and cleaned with acetone prior to deposition. Temperature measurements were undertaken using a 0.5 mm diameter K-type thermocouple attached to the side of the substrate. To attach the thermocouple, a hole, approximately 1 mm in diameter, was drilled roughly 3 mm away from the top surface (Figure 1). Timetal Ti64 (AMS 4954) filler wire with a 0.9 mm diameter was used for deposition. The chemical compositions of the wire and plate are given in Table 1. In the experiment, a bi-directional scan strategy was used where subsequent layers have opposite scanning directions. Ten successive layers were deposited to produce a 5 mm build height (500 μm layer thickness). One of the experiments was stopped after depositing a single pass/layer to have a snapshot of the microstructure and residual stresses for the first layer as a point of comparison. The photograph and the schematic in Figure 1 show the substrate geometry and experimental setup. The travel speed and wire feed rate were set at 3.81 mm/s and 8.5 mm/s, respectively. Samples were cooled down to room temperature under vacuum.

Table 1. Chemical composition (wt.%) of Timetal Ti64 wire and substrate used in the study.

Component	Al	V	Fe	C	O	N	H	Y	Ti
AMS 4911M Plate	6.21	4.00	0.18	0.006	0.18	0.005	-	0.005	Balance
AMS 4954 Wire	6.66	4.18	-	0.03	0.18	0.007	0.003	0.005	Balance

A three-dimensional (3D), transient fully coupled temperature-displacement thermo-mechanical model for EBAM was developed using the commercial finite element analysis (FEA) software CAE/2018 ABAQUS™ (Dassault Systems, Detroit, MI, USA). Analysis was run on a system with 8 cores AMD Ryzen 7-1700 CPU at 3.00 GHz with 64 GB physical memory (RAM). Simulations took approximately 20 h to complete. The transient temperature fields over the part geometry in all directions were calculated from the 3D heat conduction equation, as shown in Equation (1):

$$\frac{\partial}{\partial x}\left(k\frac{\partial T}{\partial x}\right) + \frac{\partial}{\partial y}\left(k\frac{\partial T}{\partial y}\right) + \frac{\partial}{\partial z}\left(k\frac{\partial T}{\partial z}\right) + Q = \frac{\partial\left(\rho C_p T\right)}{\partial t} \tag{1}$$

where T is the temperature, k is the thermal conductivity, C_p is the specific heat, ρ is the density, t is the time, Q is the power generated per unit volume, and x, y, and z are local coordinates of a point on part geometry.

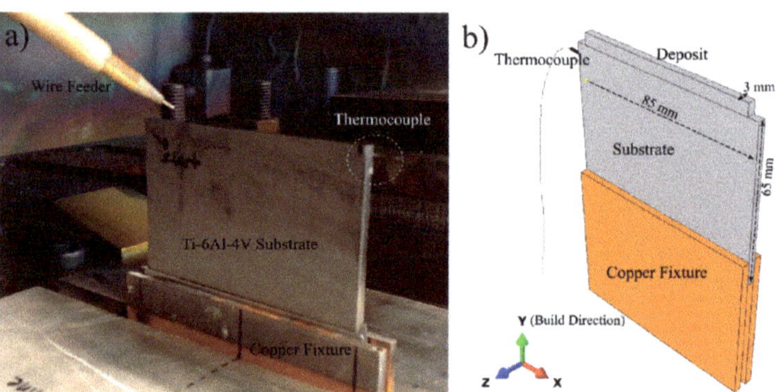

Figure 1. (**a**) Photograph of the experimental setup and (**b**) schematic drawing of the sample geometry.

The high-density electron beam heat source was modeled as a conical volumetric heat source [33] with a Gaussian distribution, as illustrated in Equation (2). The intensity distribution profile was adopted and modified from the equation modeled by Rouquette et al. [34]:

$$Q = \frac{2}{h}\left(1 - \frac{z}{h}\right)\frac{3 \times \eta \times V_a \times I_b}{\pi \times \varnothing_e^2}\exp\left(-\frac{2\left(x^2 + y^2\right)}{\varnothing_e^2}\right) \tag{2}$$

where Q is the power generated per unit volume, h is the penetration depth, η is the heat source efficiency, V_a is the accelerating voltage, I_b is the beam current, $\varnothing$ is the beam diameter, and x, y, and z are local coordinates of a point on part geometry. Heat source parameters such as beam diameter and penetration depth were determined according to the process parameters. Figure 2 shows the radial power intensity distribution profile for the heat source model at the surface. Circular contour plots on the right side of the figure illustrate the change in intensity with respect to increasing penetration depth. According to Equation (2), linear reduction in power intensity with increasing penetration depth was adopted and is shown in Figure 2. All other process parameters were the same as for the actual EBAM experiment. Heat source energy efficiency, η, was assumed to be 0.9 based on multiple literature studies [21,27,35]. In order to simulate thermal heat losses on the part, thermal boundary conditions were implemented. Forced surface convective heat losses were neglected since the EBAM process is carried out under vacuum. Radiation heat losses were considered from all outer free surfaces according to the Stefan–Boltzmann law—please see Equation (3):

$$q_{rad} = \varepsilon\sigma\left(T_{srf}^4 - T_\infty^4\right) \tag{3}$$

where ε is the emissivity, σ is the Stefan–Boltzmann constant, T_{srf} is the surface temperature of the part, and T_∞ is the ambient temperature (298 K). Thermal parameters used in the analysis are given in Table 2. Temperature-dependent emissivity values were used. The effect of the melting/solidification on the temperature calculations was considered in the model. In order to do this, the solidus temperature, liquidus temperature, and latent heat of fusion were defined and are tabulated in Table 2. Marangoni flow (liquid mass transfer due to temperature variations and capillary forces) within the melt pool was not implemented in the model; however, to compensate this significant effect, thermal conductivity was multiplied with a correction factor of 3 for temperatures above the liquidus temperature based on the approach proposed by Lampa et al. [36] and commonly used in the literature [21].

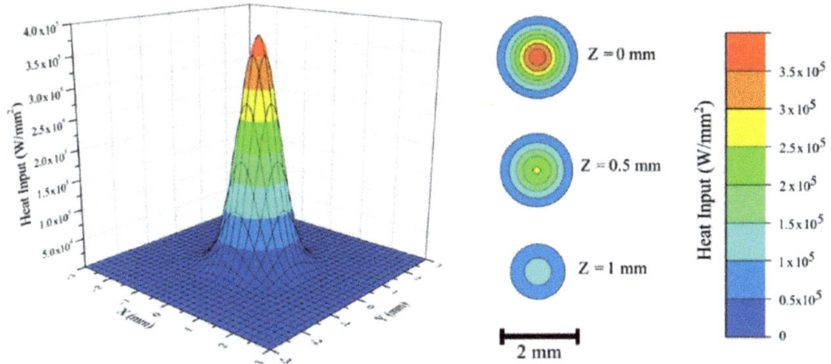

Figure 2. Power intensity distribution profile for an electron beam heat source model at the surface. Circular contour plots illustrate intensity variations at different penetration depths.

Table 2. Parameters used in FEA-electron beam additive manufacturing (EBAM) simulation.

Heat Source Efficiency, η	0.9 [35]	**Ambient Temperature, T_∞**	298 K
Initial Temperature, T_0	298 K	**Liquidus Temperature, $T_{liquidus}$**	1933 K [19]
Inter-pass Waiting Time	35 s	**Stefan–Boltzmann Constant, σ**	5.67×10^{-8} W/m^{2}K^4
Solidus Temperature	1877 K [19]	**Latent Heat of Fusion**	360 kJ/kg [19]

Figure 3 shows the temperature-dependent material properties of the Ti-6Al-4V alloy that were implemented in ABAQUS™ for both the substrate and the deposit. Thermal properties such as density, specific heat, thermal conductivity, and emissivity with respect to temperature were defined. Similarly, yield strength, elastic modulus, and thermal expansion coefficient with respect to temperature were defined as mechanical properties of Ti-6Al-4V [30,37]. For simplicity, the mechanical properties were assumed to be strain rate independent. In addition, no hardening profile was adopted in the model, i.e., perfect plasticity during deformation after the yield stress was assumed in order to reduce the computation effort and in consideration of previous high temperature tensile test studies that showed a hardening behavior which was close to perfect plasticity above 723 K for Ti-6Al-4V [38]. Anisotropy in the mechanical properties of the alloy due to microstructural texture was neglected. To simulate copper fixtures that are holding the substrate plate, mechanical boundary conditions were applied. According to the actual experimental condition, all nodes that were within a 10 mm range from the bottom of the substrate were fully fixed, meaning no displacement and rotation were allowed in any direction.

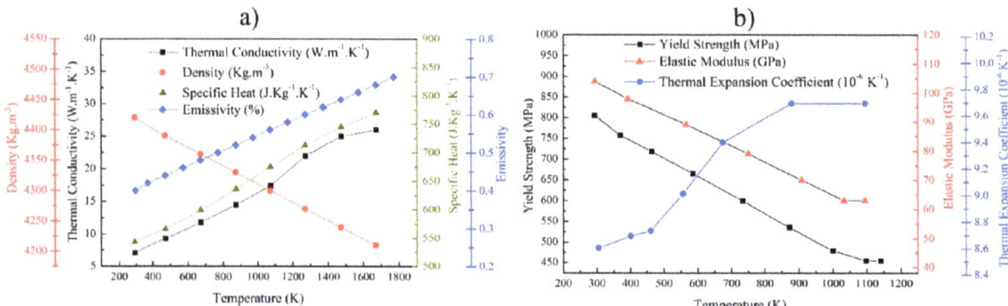

Figure 3. Temperature-dependent material properties used in FEA simulations: (**a**) thermal and (**b**) mechanical properties [30,37].

Standard solid 8-node (C3D8T) coupled displacement–temperature elements were used in the analysis. The overview of the mesh layout is shown in Figure 4a. In addition, a magnified version of the interface and the first layer's mesh is depicted in Figure 4b. Mesh refinement was done for the elements at the interface and in the deposit to ensure accurate spatial resolution in the analysis. For substrate elements further away from the interface, a gradual coarsening in the mesh size was applied. Another mesh refinement was also applied on two side ends of the substrate to ensure the presence of a node at the thermocouple location. The mesh size on the deposit was kept constant, as shown in the inset of Figure 4b. This mesh size was selected according to the layer thickness and to ensure the presence of multiple elements within the melt pool. In order to simulate the deposition of the molten material, the "element birth technique" [21] was used. In this technique, all the elements related to the deposited material pre-exist in the model, but they only become activated with the stepwise movement of the heat source. The heat source was applied as a body heat flux on newly activated elements and previous elements within the beam diameter and penetration depth. As illustrated in Figure 4b, 4 new elements are activated in each step. Deposited elements are activated at the liquidus temperature for Ti-6Al-4V, while all other elements were set at an initial temperature, T_0 at t = 0 s. Step time was accurately designed to imitate the process time and travel speed of the heat source in the real experiment. Step time was calculated by dividing the total length of the deposit with the multiplication of the total number of steps in 1 layer and experimental travel speed. In the simulation, the deposition of 1 layer and 10 layers takes 19.5 and 511 s, respectively. After each layer was deposited, an inter-pass waiting time of 35 s was applied to reduce heat accumulation on the part. Once full deposition is completed, another 7200 s (2 h) long step was applied for cooling to room temperature. The total deposition time was simulated as 7710 s. Finally, after the full sample geometry cools down to room temperature, the mechanical boundary conditions on the bottom of the substrate were deactivated to simulate fixture removal. A mesh sensitivity analysis was done to ensure the repeatability and reliability of the simulated results. No significant differences were observed in the residual stresses and distortion for different mesh sizes. The shown mesh size was selected for further experimental validation considering computational efficiency.

To validate the residual stresses on the part, experimental residual stresses for both single layer and 10-layer samples were measured with the X-ray diffraction technique (cosα method) using a Pulstec μ-X360s portable X-ray residual stress analyzer (Pulstec, Hamamatsu, Japan) with a Vanadium X-ray tube. Residual stresses of a few plates were measured before deposition to record the as-received residual stresses. As-received residual stresses were as low as ±15 MPa. Samples after deposition were not cleaned, ground, or polished before residual stress measurements to avoid the development of artificial stresses. Images of the substrate before and after deposition can be seen in Figure 5a,b, respectively. To create a measured residual stress contour map of the substrate, a region near the interface with an area of 85×20 mm^2 was analyzed. This area can be seen in Figure 5a. In this area of the substrate, 5 mm apart evenly spaced grid points were demarcated with a marker pen. Residual stresses along the X axis were measured from each of these points and repeated 3 times for statistical accuracy. Measured residual stress results were used for validation of the predicted residual stress results. Distortion measurements were carried out using an ATOS Core optical 3D scanner system (GOM-Trilion Quality Systems, Seattle, WA, USA). Substrate plates were scanned before and after deposition to compare dimensional differences. Then, samples were cut roughly 20 mm away from the interface through the X-axis for investigation of the microstructure, melt pool size, and heat-affected zone (HAZ). Grinding and polishing was performed by automated techniques using a Struers grinding/polishing system. Samples were initially ground using 240 and 1200 grit SiC grinding papers; then, they were polished with a 9 μm diamond suspension on a rigid composite disk followed by a 0.02 μm silica suspension on a synthetic cloth. A few drops of H$_2$O$_2$ solution was added into the suspension for a better polish. Then, samples were chemically etched for microstructural examination using Kroll's Etchant with 2 mL of HF, 10 mL of HNO$_2$, and 88 mL of H$_2$O. All metallographic investigation was undertaken using a Keyence VK-X laser scanning confocal/optical microscope (OM) (Keyence Canada, Mississauga, Canada).

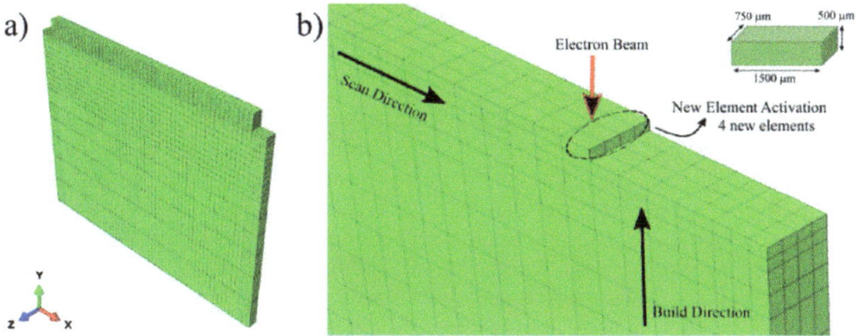

Figure 4. (**a**) Overview of the mesh layout for the substrate and 10-layer sample and (**b**) Mesh layout of the interface showing element activation routine and moving heat source. Inset depicts size of each elements on the deposit.

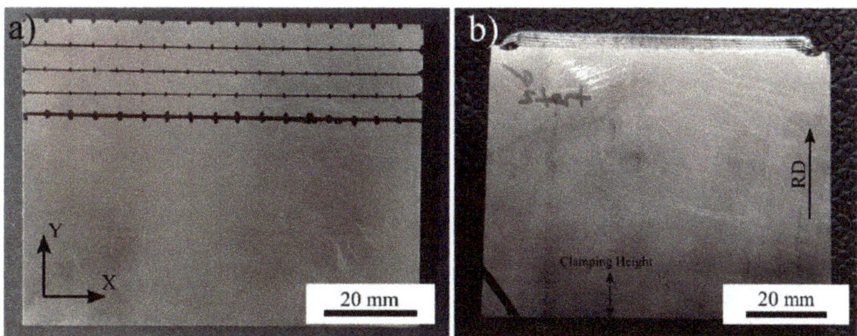

Figure 5. Images of the Ti-6Al-4V substrate (**a**) before and (**b**) after deposition experiment. Grid points used for XRD residual stress measurements are shown.

3. Results and Discussion

Transient thermo-mechanical simulations can yield valuable information such as temperature fields, distortion, and residual stresses as a function of process time and position.

However, the reliability of such information needs to be evaluated with a series of validation experiments. Contour plots exported from ABAQUS™ in Figure 6 show the predicted temperature distributions during deposition of the first and the last layers. The thermal history of some nodes was used for validation of the model. The thermocouple node, which was selected according to the position of the real thermocouple in the experiment, is shown in Figure 6a. Similarly, nodes A and B, which are on the mid-width of the first and the last layers, are illustrated in Figure 6a,b, respectively.

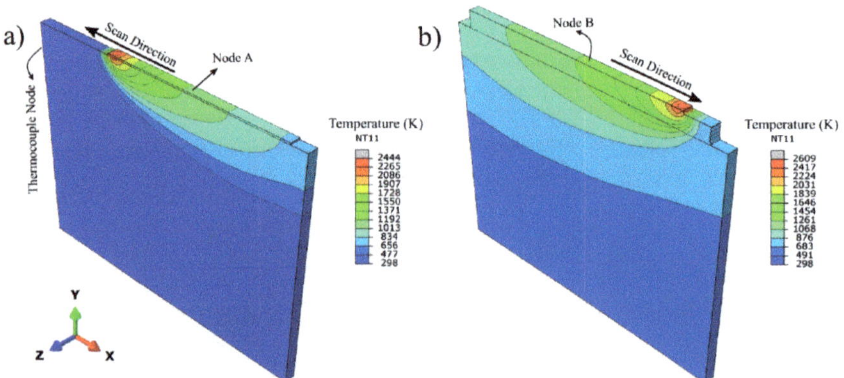

Figure 6. FEA predicted temperature profiles during deposition: (**a**) the first and (**b**) the last layer.

Figure 7 depicts the comparison of the simulated and experimental fusion zone (FZ). In Figure 7a, the temperature distributions during the deposition of the first layer in two dimensions (2D) are shown. Since the sample geometry is a single bead thin wall, the width of the melt pool is fixed and equals the width of the substrate plate; thus, the melt pool dimensions were only analyzed in 2D. The gray zone indicates the regions where the temperature exceeds $T_{liquidus}$; in other words, the size of the melt pool. The depth of the gray zone was simulated to be 1.6 mm throughout the deposition of the first layer. This depth increased up to 2 mm for the last layer due to heat accumulation on the substrate. Figure 7b shows the cross-sectional OM micrograph of the single layer sample. The microstructure reveals that the melt pool depth varies between 1.5 and 2 mm for the single layer sample, which suggests good agreement with the model. Figure 7b also shows that the initially solidified β grains are relatively fine with an equiaxed morphology and have an approximate grain size of 67 ± 13 μm. On the other hand, the microstructure of the FZ shows gradual coarsening with further distance from the fusion interface toward columnar grains parallel to the build direction in the deposit microstructure. Detailed analysis of the microstructure in the FZ, HAZ, and calculated cooling rates will be discussed later in this paper.

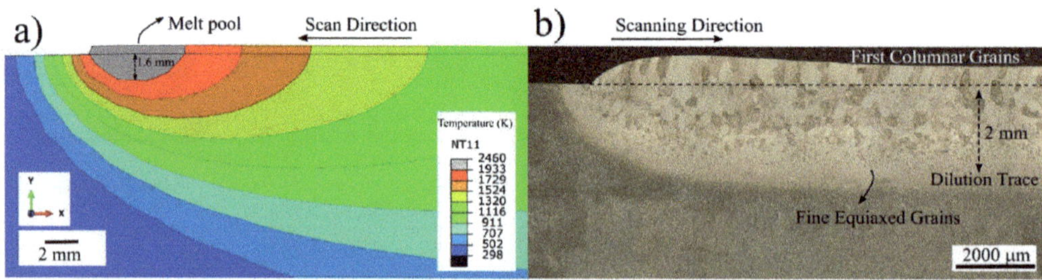

Figure 7. (**a**) Predicted temperature profiles showing melt pool depth during first layer in 2D and (**b**) optical microscope (OM) micrograph of the single layer sample that reveals dilution trace.

For node A and B (as identified in Figure 6), the temperature sequences were experimentally recorded and simulated as a function of processing time. Figure 8a compares the predicted results from the thermocouple node in the FEM model against the measured temperature data from actual experimentation. This figure shows that the predicted and experimental temperature profiles are in good agreement. The temperature profile plots gathered from nodes A and B were used to comment on the thermal history of the sample and to calculate the local cooling rates for the first and last layer of the deposit. Figure 8b depicts the thermal history of nodes A and B along with the key temperature thresholds of significant microstructural changes ($T_{liquidus}$, 1933 K and $T_{\beta\text{-transus}}$, 1273 K) for Ti-6Al-4V alloy. Each temperature peak in Figure 8 corresponds to a layer pass. After each peak, a cooling period can be seen. This period corresponds to the summation of cooling time due to the heat source moving away from the node of interest and inter-pass waiting time. Comparison of the predicted and experimental temperature peaks for the first few layers showed that the model does an over-estimation of approximately 100 K. Further analysis revealed that the mismatch is the highest for the first 3 layers. The calculated mismatches in the peak temperatures and the corresponding layer number can be listed as 6.53% for the 1st layer, 7.92% for the 2nd layer, and 5.57% for the 3rd layer. The average mismatch for the rest of the layers was 2.3%. The agreement indicates that the computed temperature fields can be used for microstructural predictions and the residual stress calculations. From the thermal history of the sample shown in Figure 8b, it is possible to state that every point in the first layer melts and solidifies with each pass at least 5 more times after deposition. Considering the position of the node A, the melt pool depth, and the layer thickness, this finding is plausible. It should be noted that the melt pool depth (2 mm) is at least 4 times the layer thickness (500 μm ascertained as explained below). Another important feature of the thermal history is the number of events where the local temperature exceeds $T_{\beta\text{-transus}}$, the allotropic transformation. According to Figure 8b, it is clear that the temperature of the first layer exceeds $T_{\beta\text{-transus}}$ for each layer throughout the deposition process for 10 layers. This means that every local point in the deposit experiences an α to β phase transformation during each heating cycle and a β to α phase transformation during each cooling cycle. This phenomenon is expected to cause homogenization of the microstructure in the deposit. The cooling rates were calculated by taking the first derivatives of the plots in Figure 8b. Calculated cooling rates for the first layer at node A can be listed as: 264 K/s at $T_{liquidus}$ (during first solidification), 89 K/s at $T_{\beta\text{-transus}}$ (during first pass), 52 K/s at $T_{\beta\text{-transus}}$ (during last pass). Similarly, the cooling rate of the last layer at node B was calculated as 49 K/s at $T_{\beta\text{-transus}}$. This result indicates that the final cooling rate at $T_{\beta\text{-transus}}$ is quite similar everywhere over the build height in the deposit and approximately equal to 50 K/s.

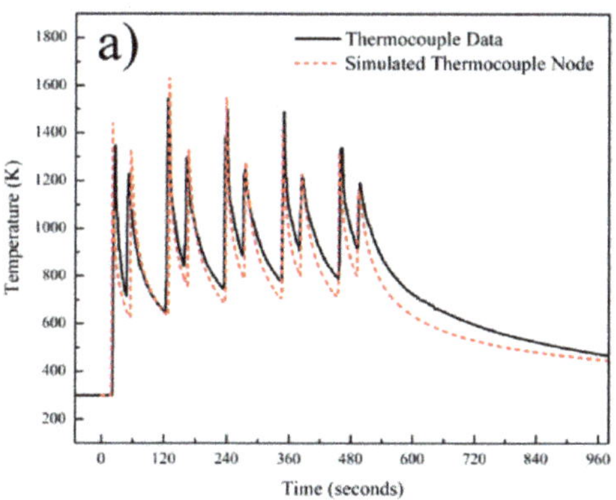

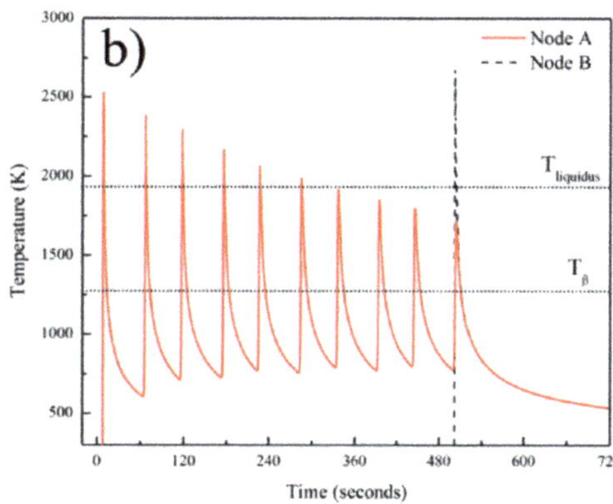

Figure 8. (a) Plot showing comparison between predicted and experimentally measured temperature profiles and (b) predicted thermal history of the node A and node B.

Figure 9 shows the OM micrograph of the 10-layer sample macrostructure that indicates a wider HAZ (extending approximately 1.2 mm) compared to the single layer sample. The main reason for this phenomenon is the longer exposure of the substrate to the elevated temperatures during the process. The temperature values recorded by the thermocouple in Figure 8a can provide an idea of the thermal cycle experienced in the HAZ. It can be stated that the temperatures in that zone varied between 800 and 1550 K for most of the deposition time. This resulted in the coarsening of the α laths in the rolled plate. It has been reported that the HAZ is observed when the temperature exceeds 708 °C, where α dissolution starts to affect microstructure [39]. Figure 9 gives a detailed description of the thermal events and their corresponding effects on the microstructure of the sample. The region shown in the figure is the end-side of the odd numbered layers. The yellow trace and arrow illustrated in the figure reveal the depth of the melt pool in the first layer. Similar to the single layer results, the melt pool depth is approximately 1.5 mm. Note that a wide curved melt pool trace is visible at the end of the scan paths. In addition, the melt pool depths appear to be slightly shorter at the beginning of the scan paths. These features were created due to the acceleration and deceleration routines of the EBAM machine during processing and do not reflect the stable moving melt pool dimensions. After deposition of the first layer, the second layer was started with an opposite scanning direction, as shown with the red arrow. During the deposition of the second layer, another dilution trace was created, which is shown with the red dashed line. The second dilution trace can also be used to validate the melt pool depth. The height of the first layer was marked on the micrograph, 500 µm above the substrate level. The distance between the first layer surface and 2nd dilution trace was measured as 1.5 mm, revealing the melt pool depth once again. The macrostructure of the 10-layer sample showed equiaxed prior β grains at the FZ of the substrate and columnar prior β grains parallel to the build direction at the deposit. This type of grain structure is common in the literature and formed due to the directional heat flux during cooling [40,41]. Cooling in the FZ of the substrate occurs faster and in multiple directions since the substrate acts as a heat sink [42]. This results in relatively finer irregular or equiaxed prior β grains [43]. However, cooling in the deposit is strongly directional, and the heat flux is opposite to the build direction. Then, the columnar grains grow parallel to the build direction along the maximum temperature gradient with epitaxial growth [44]. In order to reinforce these discussions, thermal gradients, G, and solidification rates, R, from various regions of the sample were calculated using the thermo-mechanical model. Calculated G and R values were compared with solidification maps of Ti-6Al-4V from the literature [45] to predict solidification microstructures. Figure 10 shows the Ti-6Al-4V solidification map along with the calculated G and R values from these regions. The plot shows that calculated G and R from the FZ is in the equiaxed grain morphology map field. This result agrees well with the microstructural observations in Figure 9. On the other hand, calculated G and R values from the top and mid-height of the deposit are located at the mixed grain morphology map field. Although the majority of the grains in these regions are in columnar nature, there are few grains with equiaxed morphology as well. It is safe to report that equiaxed grains could still form under these solidification conditions. The final note from the solidification map is the difference between predicted cooling rates from the FZ and the deposit. The higher predicted cooling rate in the FZ explains the finer grain structure of the corresponding region shown in Figure 9. Another structural feature of the Ti-6Al-4V wire-feed deposits in the literature are the layer bands observed in the microstructure. These bands are shown to be in curved shapes for multi-bead deposits [19,40] or straight lines for single-bead deposits [42,46]. These bands are known to originate from α/β allotropic phase transformations during multiple heating and cooling cycles [19,40]. Brandl et al. stated that the zones that experience temperatures above $T_{\beta\text{-transus}}$ during the deposition of the last layer become free of this band structure [40]. This band-free region is called the "transient region" in the literature [47] because of its incomplete thermal history. On the other hand, zones that are no longer experiencing temperatures above $T_{\beta\text{-transus}}$ are called the "steady-state region", since their thermal history is completed and is no longer undergoing phase transformations. In this study, the microstructure of the 10-layer sample illustrated in Figure 9 shows no evidence of a band structure. Considering the predicted thermal history of the sample shown in Figure 8a, the whole sample was heated above $T_{\beta\text{-transus}}$ even during the last pass. This explains the lack of bands in the microstructure for the 10-layer sample. It can be stated that the 10-layer sample is fully in the "transient region", and more layers are required to observe the band structure for this alloy under these processing conditions. Finally, the dilution trace of the last layer can be observed in the microstructure approximately 1.5 mm away from the top of the sample. Dilution traces for the intermediate layers were not observed probably due to the homogenization of the microstructure during the process. The last trace is visible since the sample was never heated to homogenize the microstructure after that layer.

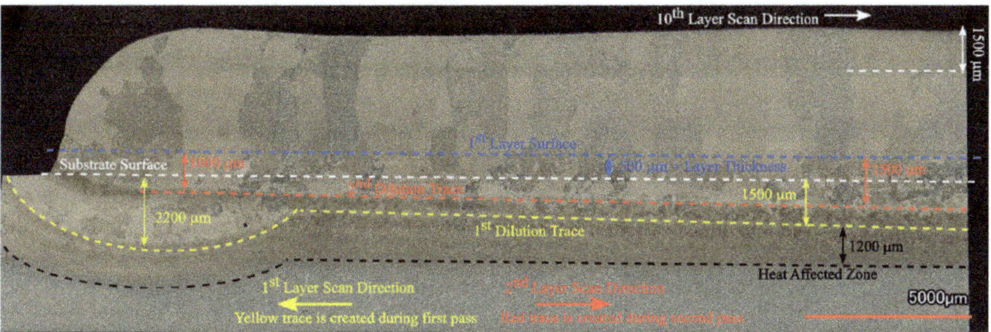

Figure 9. Low magnification OM micrograph of the 10-layer Ti-6Al-4V sample showing macrostructure overview of the heat-affected zone (HAZ), fusion zone (FZ) and the deposit.

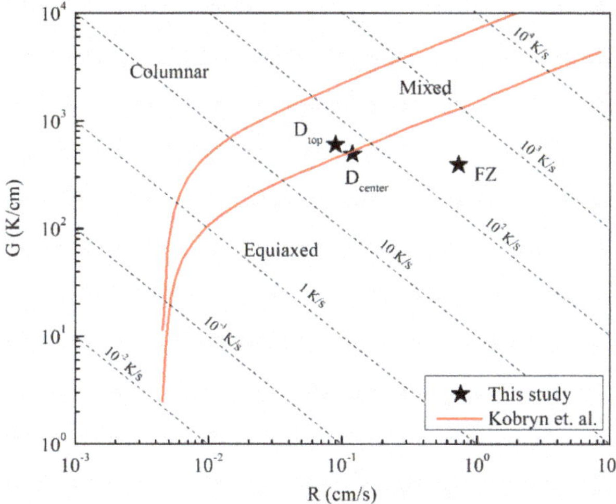

Figure 10. The Ti-6Al-4V solidification map [45] showing the calculated G and R values from various regions of the sample: D_{top} (top of the deposit), D_{center} (mid-height of the deposit) and FZ (fusion zone).

Higher magnification OM micrographs of the 10-layer Ti-6Al-4V sample are shown in Figure 11. All micrographs were taken from the deposit side, while Figure 11a,c are showing the upper portion and Figure 11b,d are showing the lower portion of the sample near the FZ. A basket-weave α/β microstructure was observed within the prior β grains as seen in

Figure 11c,d. A small amount of allotriomorphic (grain boundary) α or Widmanstätten α phase was observed in the lower side of the sample. It is important to recall that the calculated cooling rates (52–49 K/s) were below the critical cooling rate (410 K/s) for α' martensite formation [48]. Observation of the basket-weave α/β microstructure further validates the predicted thermal history. In addition to the micrographs, the size of the microstructural features from various zones of the single layer and ten-layer samples are reported in Figure 12. Prior β grain widths were measured as 923 $\pm$ 183 μm and 517 $\pm$ 196 μm for the deposit zone and FZ, respectively. Similarly, prior β grain lengths were measured as 3.2 $\pm$ 0.6 mm and 0.6 $\pm$ 0.1 mm for the deposit zone and FZ, respectively. The average length of the columnar prior β grains in the deposit extend up to 6-7 layers, further highlighting the epitaxial growth of β grains over subsequent layers. On the other hand, equiaxed prior β grains within the FZ had an aspect ratio close to one, as shown in Figure 12b. Another key microstructural feature for Ti-6Al-4V alloy are the needle-shaped α laths. During cooling through $T_{\beta\text{-transus}}$, the high-temperature body centered cubic (BCC)-β phase undergoes allotropic phase transformation to the hexagonal closed packed (HCP)-α phase. This HCP-α phase begins to nucleate at the β grain boundaries according to the crystallographic relationship known as Burgers Orientation Relationship (BOR). This relationship dictates that from the —{110} $_\beta$ | | (0001) $_\alpha$, <11$\bar{2}$0> α | | <1$\bar{1}\bar{1}$> β –, 12 possible α variants can nucleate at the boundary of the β grains. Specifically, {110} planes of the BCC-β phase set the basal planes of the HCP-α phase [32,43,49–51]. Once the α laths nucleate, they grow and form a basket-weave microstructure. The size of the individual α laths also reflect the effect of thermal history and the cooling rate on the final microstructure [31,52]. In the observed microstructure, no distinct change in α lath thickness was observed throughout the 10-layer deposit. This is also illustrated in Figure 11c,d. This is an expected result considering the predicted thermal history, since all layers undergo a final allotropic phase transformation and cool down with a similar cooling rate after the last pass. However, the measured α lath thicknesses were coarser in the FZ and HAZ of the substrate. Measured α lath thicknesses were 0.7 $\pm$ 0.2 μm, 1.2 $\pm$ 0.4 μm, and 1.6 $\pm$ 0.4 μm for the deposit, FZ, and HAZ, respectively. The reason for the coarser microstructure in the FZ and HAZ is that at some point, they stop experiencing allotropic phase transformations as new layers are added, due to the increasing distance from the heat source. However, the microstructure is still exposed to sub-transus temperatures due to the heat diffusion. Thus, a gradual coarsening occurs with every new layer. The measured α lath thicknesses for the single layer sample are 0.24 $\pm$ 0.04 μm and 0.21 $\pm$ 0.04 μm from the deposit and FZ, respectively. Laths were approximately 0.4 μm thinner in the single layer sample compared to the ten-layer sample. This can be explained with the difference between the final cooling rates of the single layer sample (89 K/s) and the ten-layer sample (49 K/s), as the simulations predicted. In the literature, the dimensions of the microstructural features for DED fabricated Ti-6Al-4V parts depend on the build height and heat source types [53]. For example, the length of the prior β grains reported in the literature for 10 mm long samples are 5.1 $\pm$ 3.2 mm and 1.2 $\pm$ 0.6 mm for the electron beam and laser, respectively [53]. The prior β lengths in this study are slightly shorter, but this can be due to the smaller sample height. Further layers would cause longer prior β grains. On the other hand, α lath thickness for electron beam, laser, and tungsten inert gas torch (TIG) DED deposits usually vary between 0.5 and 2 μm [42,49,53–55]. Although general understanding in the literature suggests that increasing the cooling rate results in a reduction in α lath thickness [40,48,50], this decrease does not occur linearly with the cooling rate change [31]. Lütjering stated that the α lath thickness decreased drastically from 5 to 0.5 μm due to the changes in cooling rate in the range of 0.02 to 2 K/s. A further increase in the cooling rate up to 150 K/s resulted in only a 0.3 μm decrease [31]. In another study, Kelly studied the effect of cooling rate on the α laths thickness and showed micrographs having α laths with approximately 1, 0.7, and 0.3 μm thickness for 0.6, 10, and 94 K/s, respectively [47]. Therefore, it can be concluded that the calculated cooling rates and the corresponding microstructural features agree well with the previous literature studies with similar cooling rates.

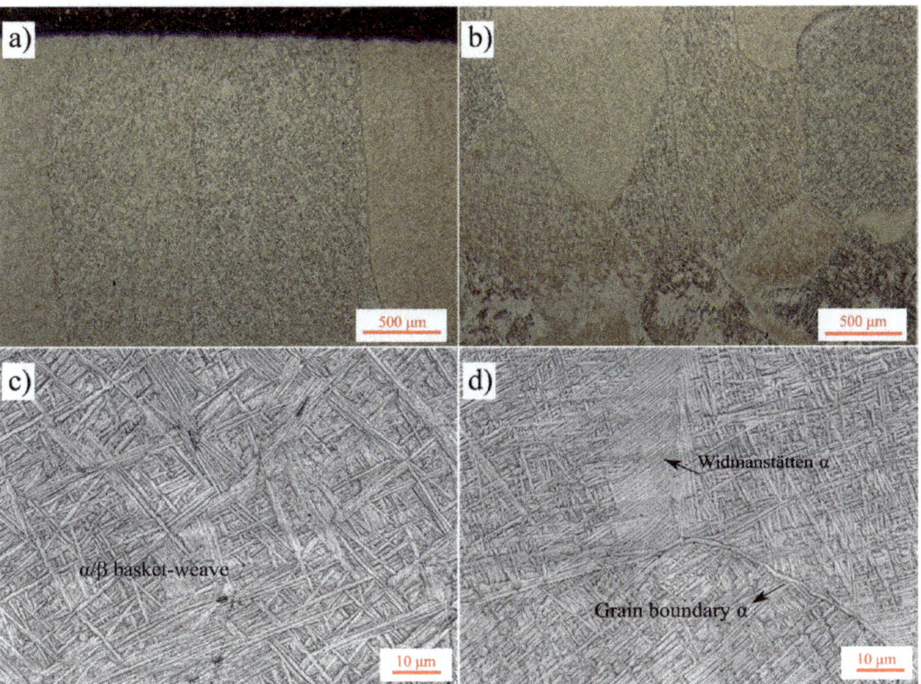

Figure 11. Higher magnification OM micrographs of the 10-layer Ti-6Al-4V sample from (**a,c**) first layers and (**b,d**) last layers of the deposit.

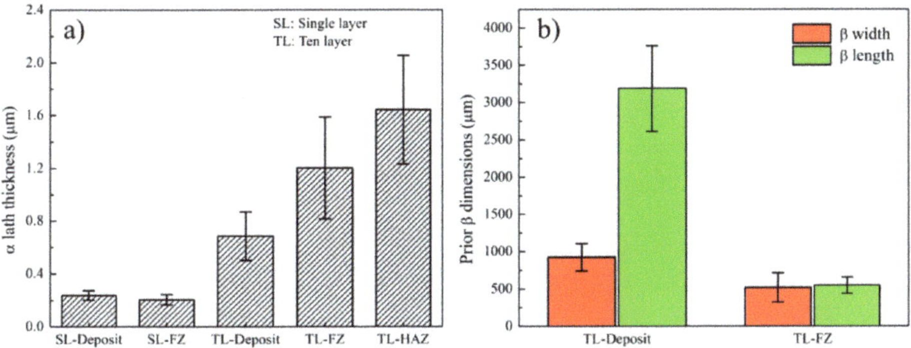

Figure 12. Plots showing dimensions of microstructural features such as (**a**) α lath thickness and (**b**) prior β dimensions for the single layer and 10-layer Ti-6Al-4V samples.

Residual stresses in AM processes mainly develop because of the thermal cycles experienced with the addition of new layers. Local differences in temperatures and their corresponding non-uniform volume expansions and contractions result in their development [1,21]. Figure 13 shows the measured and predicted residual stresses along the scan direction (x) for the single layer sample. Residual stresses were compared along the scan direction (x) since they were the highest and dominant residual stress on the substrate.

Residual stresses beyond 20 mm distance from the deposit/substrate interface are not shown since they were relatively low (0–10 MPa) and considered negligible. Experimental residual stresses developed in the substrate shown in Figure 13a were measured using XRD from each of the pen marks depicted in Figure 5a. Corresponding predicted residual stresses are given in Figure 13b with the exported image from ABAQUS™. A residual stress color scale was adjusted to be similar to the one in Figure 13a for visual comparison. Residual stresses above 400 MPa, as shown in the gray zone, were only calculated on the deposit. Thus, the following observations and discussions can be derived from the results of the single layer sample. First, predicted and measured residual stress contour plots are in good agreement. The residual stresses are highest along the scan direction, concurring with previously reported findings in the literature for AM parts [1,56]. The deposit and HAZ on the substrate are under tension, while the core of the substrate is under compression for the single-layer sample. The highest tensile residual stress on the deposit and the substrate are approximately 800 ($1.0\sigma_{ys}$) and 400 ($0.5\sigma_{ys}$) MPa, respectively. Similarly, in the literature, the ratio of the maximum residual stress to the alloy yield strength at room temperature for Ti-6Al-4V deposits were reported to be around 1.0 σ_{ys} [25]. The results also showed that the depth of the tensile residual stress region is approximately 4 mm from the substrate surface, which is similar to the HAZ depth. Due to the heating, during the new layer deposition, the deposited material and HAZ expand as its yield strength decreases. However, this expansion is hindered by the cooler substrate material. Thus, compressive stress is observed on the substrate material. Due to subsequent cooling, the deposited material and HAZ try to contract. Contraction is again obstructed by the surrounding material. Although contraction is easier initially due to the lower yield strength at elevated temperatures, it becomes harder as the part cools down. This results in a tensile stress in the deposited material and HAZ [21]. Figure 13c shows the measured and predicted residual stresses along Path 1, as shown in Figure 13b. The predicted residual stress profile has a decent fit with the experimental data points. The highest error was observed to be at the deposit/substrate interface, near x = 0 along Path 1. The measured residual stress at this point is approximately 100 MPa higher. One possible reason for this error could be the overflow of the deposited material to the substrate in the experiment, which was not simulated in the model.

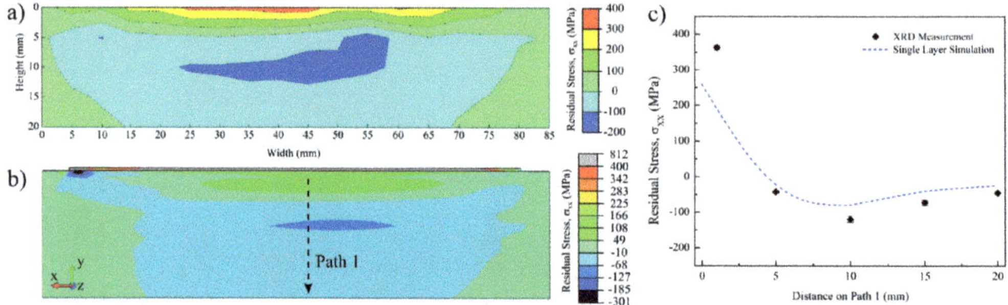

Figure 13. Contour plots showing the residual stresses along the scan direction (XX) for the single layer Ti-6Al-4V sample and substrate: (**a**) measured by XRD, (**b**) predicted from the FEM and (**c**) compared along Path 1.

The predicted residual stresses along the scan direction (x) for the 10-layer Ti-6Al-4V sample and pre-defined three paths are shown in Figure 14. A residual stress profile similar to the single layer sample was calculated for the 10-layer sample. A compressive stress region is located at the core of the substrate just below the HAZ, while most of the deposit is under tension. Tensile residual stresses were calculated to be higher at the interface or at the top and mid-width of the deposit. The highest calculated tensile residual stress was approximately 400 MPa in the deposit. Predicted residual stresses were considerably lower for the 10-layer sample compared to the single layer sample. The major reason of this phenomenon is the stress relaxation and softening in the 10-layer sample due to subsequent heating cycles and longer heat exposure. Another affecting factor for such phenomenon can be that the single-layer sample was deposited on the cold substrate at the room temperature, while the last layer of the 10-layer sample was deposited on a surface, which was already relatively hot. This could work as a pre-heat treatment before processing [57]. It was previously reported in the literature that the elevated temperatures and reduced temperature gradients experienced during EBAM processing are the main basis of the low residual stresses on the substrates [49]. The second affecting factor that could influence the residual stresses are the solid-state phase transformations occurring due to the multiple thermal cycles [30,58]. Denlinger et al. stated that volumetric changes due to allotropic phase transformations cause transformation strains within the alloy. These strains could potentially oppose contraction strains during cooling and relieve residual stresses [27,30]. Similarly, Elmer et al. studied the lattice expansions of Ti-6Al-4V during allotropic transformations via in situ high-energy synchrotron X-ray radiation and found that α and β phases have dramatically different changes in their lattice parameters during allotropic transformation. These differences in lattice parameters and thermal expansion behaviors are considered to be the origin of the stress relief in Ti-6Al-4V during allotropic transformations [58]. The residual stress profiles encountered in this study for the thin wall substrate were found to be similar to the ones in the literature for the bulky substrates [25]. Mukherjee et al. showed the evolutions of the residual stresses progressively in each layer. Their study revealed a compressive stress region on the substrate below the HAZ and a tensile stress region on the deposit. This tensile region was most pronounced at the mid-width of the deposit. Each layer deposition resulted in the tensile region shifting to an upper layer on the deposit. Residual stresses on the previous layers were stress relieved partially because of the heating during the process [25]. These findings are in good agreement with the residual stress evolution observed from the single layer to the 10-layer sample of this study. Figure 14 shows a gradual increase in the predicted tensile residual stresses with each layer, further highlighting the shift in residual stresses with each pass. Residual stresses in the 10-layer sample were further compared with the experimental measurements using the pre-defined paths shown in Figure 14. Path 1 extends 20 mm from the deposit/substrate interface to the substrate at the mid-width of the substrate. Path 2 and Path 3 are located approximately 1 mm and 20 mm away from the interface, respectively.

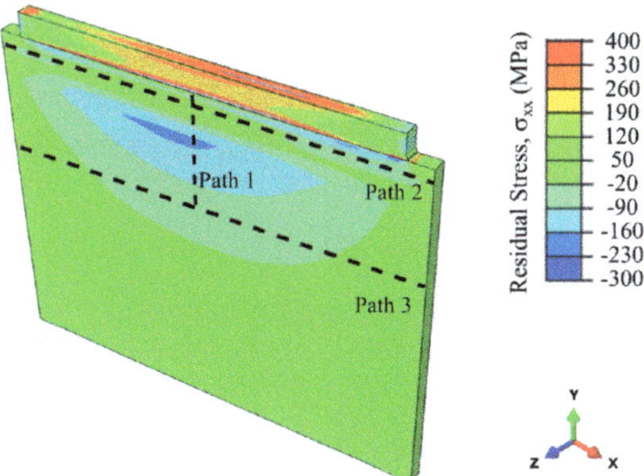

Figure 14. Predicted residual stresses along the scan direction for 10-layer Ti-6Al-4V deposit and substrate. Figure also illustrates pre-defined paths for residual stress plots.

Calculated and measured residual stresses along the scan direction on Path 1 and Paths 2 and 3 are shown in Figure 15a,b, respectively. Both figures show that acceptable agreement was achieved between the residual stress predictions and experimental values. Calculated and measured profiles fit well with each other for the majority of the path lengths. The highest error was observed on Path 3 near the side ends of the plate. This might be due to the fact that the experimental deposit is slightly longer than the simulated geometry. In addition, metal overflow at ends is more effective because of the acceleration/deceleration routine of the EBAM machine, as previously mentioned. Calculation error and residual stresses were low on Path 2, as it was relatively far from the effective thermal gradient and high residual stress region of the substrate. Figure 15a better illustrates the residual stress variations along the substrate height. Residual stresses on the substrate decrease from approximately 125 MPa down to −30 MPa with increasing distance from the deposition interface. When compared with the single layer sample, it is clear that residual stresses below the HAZ were relieved from approximately −100 to −30 MPa. This is probably the result of stress relieving on the substrate due to elevated temperatures experienced with each layer deposition. Although calculation and measurement follow a similar trend with increasing distance along Path 1, the predicted values are approximately 50 MPa toward compressive. The possible reason of this discrepancy between simulation and the experiment could be the defined material properties in the model. Lu et al. studied the effect of different material properties on the calculated residual stresses for Ti-6Al-4V and found that small differences in yield strength, coefficient of thermal expansion, and elastic modulus could significantly change the final residual stress profiles [59]. Considering that these properties depend on the microstructure/texture that are constantly changing with the temperature during AM, they are locally different even in a single part. So, it is relatively hard to predict the exact properties at all times.

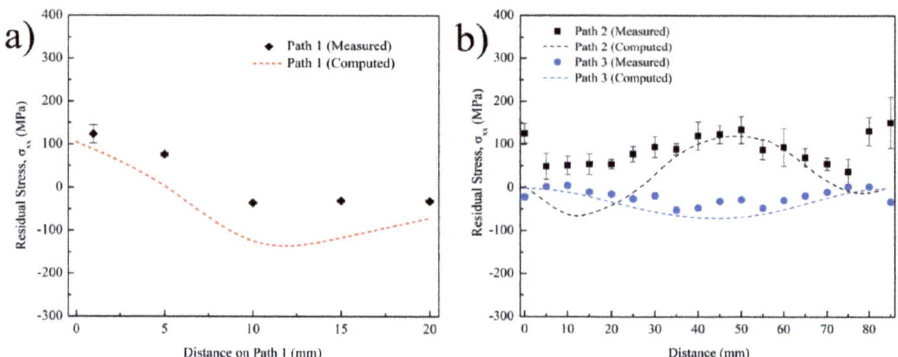

Figure 15. Predicted and measured (XRD) residual stress comparison plots along scan direction on (**a**) Path 1 and (**b**) Paths 2 and 3 for 10-layer Ti-6Al-4V substrate.

Due to the thermal gradients and uneven thermal expansion throughout the part, the initial geometry is continuously experiencing distortion during the process. Figure 16 depicts the contour plots showing the comparison of predicted and measured out-of-plane distortion maps of the 10-layer Ti-6Al-4V deposit and the substrate. Figure 16a shows that the calculated distortion is relatively low, and the magnitude varies between 50 and 250 µm. The highest distortion is localized at the end side of the first layer pass. The measured out-of-plane distortion map is in good agreement with the predicted values, as shown in Figure 16b. Similar to the model predictions, the highest distortion was measured on the end side of the first layer pass. In addition, the deposit region is shown to have high distortion as an experimental artifact. During the measurement procedure via the ATOS Core Optical 3D scanner, the geometry was scanned before and after deposition. Since the deposit was not present in the initial scan (before deposition), the contour map shows a high distortion region on the top. Measured out-of-plane distortion magnitudes also vary between 90 and 200 μm. Measured values are shown to have negative magnitude; this reveals that the substrate plane is distorted inwards, meaning the plate is thinner after deposition. Such deformation is plausible, since the highest residual stresses are tensile along the scan direction (x) and are experienced at the top of the substrate. In our previous study, highest substrate distortion was observed at the same corner with the magnitude of approximately 300–400 μm for 50 mm build height under the same conditions [49]. In this study, similar or slightly lower distortion was observed for a 5 mm build height. Previous studies in the literature showed that the substrate distortion evolution rate during DED processing is much more pronounced during the initial few layer deposition and gradually diminishes throughout the process for Ti-6Al-4V components [27,59]. A simple explanation for this is the shifting of high thermal gradients and residual stresses away from the deposit with increasing build height. In addition, the decrease in experienced cooling rates would lower the distortion rate. These may reasonably account for the similar distortion observed for two different build heights.

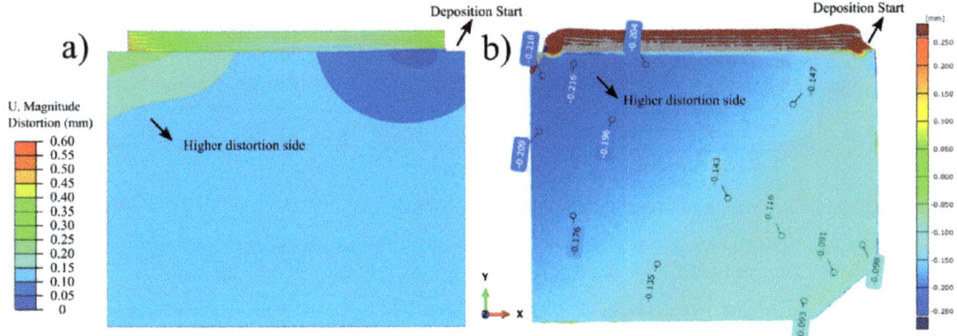

Figure 16. Contour plots showing (**a**) predicted and (**b**) measured out-of-plane distortion maps of the 10-layer Ti-6Al-4V deposit and the substrate.

In future work, the developed model will be utilized to investigate EBAM process planning as well as parametric optimization for both manufacturing and repair methodology processing in which the substrate/part integrity is also critical for return to service. Acquired knowledge from the model will be used to facilitate process development to assure suitable heat dissipation and minimize distortion for complex part geometries, such as fan blades that have twisted double curvature surface profiles.

4. Summary and Conclusions

In this study, a 3D transient fully coupled thermo-mechanical model for EBAM was developed using the commercial FEA software ABAQUS™ in order to investigate and plan Ti-6Al-4V deposition methodology and process. The developed model was calibrated for computational efficiency through a series of mesh sensitivity studies. The developed model was also validated using multiple EBAM and characterization experiments. Thermal validation through melt-pool size analysis, thermal history analysis, and microstructure analysis showed that the developed model is extremely reliable to predict local temperatures and cooling rates, melt-pool and dilution depths, and the resulting macro and microstructure over the deposit and the substrate. The microstructure of the EBAM deposit was observed as an α/β basket-weave structure. Initial β grains within the FZ and first layers were observed to have equiaxed morphology. On the other hand, β grains in the lateral layers of the deposit were observed to have mostly columnar morphology parallel to the build direction with few exceptions. Predicted G and R values were validated using solidification maps [45] and microstructure. No banding in the structure was observed due to the low build height and thin wall geometry of the samples. Mechanical validation was undertaken via comprehensive residual stress and distortion measurements. Predicted and measured residual stresses were in good agreement, further highlighting the reliability of the developed model. Residual stresses were more pronounced along the scan direction. Tensile residual stresses were dominant over the deposit and HAZ, while compressive residual stresses were observed at the core of the substrate. Higher residual stresses along the substrate were observed for the single layer sample compared to the 10-layer sample. Increasing the build height and thus the longer exposure to elevated temperatures are thought to be the origin of this stress relaxation on the substrate. ATOS Core 3D optical measurements revealed that the model predictions were very accurate on calculating the distortion profile and magnitude on the substrate. The highest distortion observed on the substrate was approximately 250 µm, revealing success on maintaining initial geometry.

Author Contributions: Conceptualization, F.S., P.W., J.G., A.K. and M.B.; methodology, F.S., P.W., J.G., A.K. and M.B.; software, F.S., A.K., J.G. and M.B.; validation, F.S., P.W. and M.B.; formal analysis, F.S.; investigation, F.S.; resources, P.W., J.G. and M.B.; data curation, F.S.; writing—original draft preparation, F.S.; writing—review and editing, P.W., J.G. and M.B.; visualization, F.S.; supervision, P.W., J.G. and M.B.; project administration, P.W., J.G. and M.B.; funding acquisition, M.B., P.W., J.G. All authors have read and agreed to the published version of the manuscript.

Funding: This research received financial support from the Natural Sciences and Engineering Research Council of Canada (NSERC) network for Holistic Innovation in Additive Manufacturing (HI-AM) (NSERC Project Number: NETGP 494158-16) and the National Research Council of Canada.

Institutional Review Board Statement: Not applicable.

Informed Consent Statement: Not applicable.

Data Availability Statement: The authors confirm that the data supporting the findings of this study are available within the article.

Acknowledgments: The authors are thankful to Xavier Pelletier and Maxime Guérin at the National Research Council of Canada for technical support on EBAM experimentation, metallography, and ATOS Core 3D measurements.

Conflicts of Interest: The authors declare no conflict of interest.

References

1. Debroy, T.; Wei, H.L.; Zuback, J.S.; Mukherjee, T.; Elmer, J.W.; Milewski, J.O.; Beese, A.M.; Wilson-Heid, A.; De, A.; Zhang, W. Additive manufacturing of metallic components—Process, structure and properties. *Prog. Mater. Sci.* **2018**, *92*, 112–224. [CrossRef]
2. Boyer, R. An overview on the use of titanium in the aerospace industry. *Mater. Sci. Eng. A* **1996**, *213*, 103–114. [CrossRef]
3. Liu, S.; Shin, Y.C. Additive manufacturing of Ti6Al4V alloy: A review. *Mater. Des.* **2019**, *164*, 107552. [CrossRef]
4. Emmelmann, C.; Scheinemann, P.; Münsch, M.; Seyda, V. Laser Additive Manufacturing of Modified Implant Surfaces with Osseointegrative Characteristics. *Phys. Procedia* **2011**, *12*, 375–384. [CrossRef]
5. Uhlmann, E.; Kersting, R.; Klein, T.B.; Cruz, M.F.; Borille, A.V. Additive Manufacturing of Titanium Alloy for Aircraft Components. *Procedia CIRP* **2015**, *35*, 55–60. [CrossRef]
6. Huang, R.; Riddle, M.; Graziano, D.; Warren, J.; Das, S.; Nimbalkar, S.; Cresko, J.; Masanet, E. Energy and emissions saving potential of additive manufacturing: The case of lightweight aircraft components. *J. Clean. Prod.* **2016**, *135*, 1559–1570. [CrossRef]
7. Wanjara, P.; Gholipour, J.; Watanabe, E.; Sugino, T.; Patnaik, P.; Sikan, F.; Brochu, M. High Frequency Vibration Fatigue Behavior of Ti6Al4V Fabricated by Wire-Fed Electron Beam Additive Manufacturing Technology. *Adv. Mater. Sci. Eng.* **2020**, *2020*, 1–14. [CrossRef]
8. King, W.E.; Barth, H.D.; Castillo, V.M.; Gallegos, G.F.; Gibbs, J.W.; Hahn, D.E.; Kamath, C.; Rubenchik, A.M. Observation of keyhole-mode laser melting in laser powder-bed fusion additive manufacturing. *J. Mater. Process. Technol.* **2014**, *214*, 2915–2925. [CrossRef]
9. Mercelis, P.; Kruth, J. Residual stresses in selective laser sintering and selective laser melting. *Rapid Prototyp. J.* **2006**, *12*, 254–265. [CrossRef]
10. Thompson, S.M.; Bian, L.; Shamsaei, N.; Yadollahi, A. An overview of Direct Laser Deposition for additive manufacturing; Part I: Transport phenomena, modeling and diagnostics. *Addit. Manuf.* **2015**, *8*, 36–62. [CrossRef]
11. Mower, T.M.; Long, M.J. Mechanical behavior of additive manufactured, powder-bed laser-fused materials. *Mater. Sci. Eng. A* **2016**, *651*, 198–213. [CrossRef]

12. Shui, X.; Yamanaka, K.; Mori, M.; Nagata, Y.; Kurita, K.; Chiba, A. Effects of post-processing on cyclic fatigue response of a titanium alloy additively manufactured by electron beam melting. *Mater. Sci. Eng. A* **2017**, *680*, 239–248. [CrossRef]

13. Xu, W.; Brandt, M.S.; Sun, S.; Elambasseril, J.; Liu, Q.; Latham, K.; Xia, K.; Qian, M. Additive manufacturing of strong and ductile Ti–6Al–4V by selective laser melting via in situ martensite decomposition. *Acta Mater.* **2015**, *85*, 74–84. [CrossRef]

14. Javidani, M.; Arreguin-Zavala, J.; Danovitch, J.; Tian, Y.; Brochu, M. Additive Manufacturing of AlSi10Mg Alloy Using Direct Energy Deposition: Microstructure and Hardness Characterization. *J. Therm. Spray Technol.* **2017**, *26*, 587–597. [CrossRef]

15. Murr, L.E.; Martinez, E.; Amato, K.N.; Gaytan, S.M.; Hernandez, J.; Ramirez, D.A.; Shindo, P.W.; Medina, F.; Wicker, R.B. Fabrication of Metal and Alloy Components by Additive Manufacturing: Examples of 3D Materials Science. *J. Mater. Res. Technol.* **2012**, *1*, 42–54. [CrossRef]

16. Tian, Y.; Gontcharov, A.; Gauvin, R.; Lowden, P.; Brochu, M. Effect of heat treatment on microstructure evolution and mechanical properties of Inconel 625 with 0.4 wt% boron modification fabricated by gas tungsten arc deposition. *Mater. Sci. Eng. A* **2017**, *684*, 275–283. [CrossRef]

17. Geiger, F.; Kunze, K.; Etter, T. Tailoring the texture of IN738LC processed by selective laser melting (SLM) by specific scanning strategies. *Mater. Sci. Eng. A* **2016**, *661*, 240–246. [CrossRef]

18. Chauvet, E.; Kontis, P.; Jägle, E.A.; Gault, B.; Raabe, D.; Tassin, C.; Blandin, J.-J.; Dendievel, R.; Vayre, B.; Abed, S.; et al. Hot cracking mechanism affecting a non-weldable Ni-based superalloy produced by selective electron Beam Melting. *Acta Mater.* **2018**, *142*, 82–94. [CrossRef]

19. Chekir, N.; Tian, Y.; Gauvin, R.; Brodusch, N.; Sixsmith, J.J.; Brochu, M. Laser Wire Deposition of Thick Ti-6Al-4V Buildups: Heat Transfer Model, Microstructure, and Mechanical Properties Evaluations. *Met. Mater. Trans. A* **2018**, *49*, 6490–6508. [CrossRef]

20. Negi, S.; Nambolan, A.A.; Kapil, S.; Joshi, P.S.; Manivannan, R.; Karunakaran, K.; Bhargava, P. Review on electron beam based additive manufacturing. *Rapid Prototyp. J.* **2019**, *26*, 485–498. [CrossRef]

21. Cao, J.; Gharghouri, M.A.; Nash, P. Finite-element analysis and experimental validation of thermal residual stress and distortion in electron beam additive manufactured Ti-6Al-4V build plates. *J. Mater. Process. Technol.* **2016**, *237*, 409–419. [CrossRef]

22. Parry, L.; Ashcroft, I.; Wildman, R.D. Understanding the effect of laser scan strategy on residual stress in selective laser melting through thermo-mechanical simulation. *Addit. Manuf.* **2016**, *12*, 1–15. [CrossRef]

23. Ahn, J.; He, E.; Chen, L.; Wimpory, R.; Dear, J.; Davies, C. Prediction and measurement of residual stresses and distortions in fibre laser welded Ti-6Al-4V considering phase transformation. *Mater. Des.* **2017**, *115*, 441–457. [CrossRef]

24. Rae, W. Thermo-metallo-mechanical modelling of heat treatment induced residual stress in Ti–6Al–4V alloy. *Mater. Sci. Technol.* **2019**, *35*, 747–766. [CrossRef]

25. Mukherjee, T.; Zhang, W.; Debroy, T. An improved prediction of residual stresses and distortion in additive manufacturing. *Comput. Mater. Sci.* **2017**, *126*, 360–372. [CrossRef]

26. Manvatkar, V.D.; De, A.; Debroy, T. Heat transfer and material flow during laser assisted multi-layer additive manufacturing. *J. Appl. Phys.* **2014**, *116*, 124905. [CrossRef]

27. Denlinger, E.R. Residual Stress and Distortion Modeling of Electron Beam Direct Manufacturing Ti-6Al-4V ✶ ✶This chapter is based upon the original work: Denlinger, Erik R., Jarred C. Heigel, and Pan Michaleris. "Residual stress and distortion modeling of electron beam direct manufacturing Ti-6Al-4V.". *Proc. Inst. Mech. Eng. Part B J. Eng. Manuf.* **2015**, *229*, 1803–1813.

28. Heigel, J.; Michaleris, P.; Reutzel, E.W. Thermo-mechanical model development and validation of directed energy deposition additive manufacturing of Ti-6Al-4V. *Addit. Manuf.* **2015**, *5*, 9–19. [CrossRef]

29. Muransky, O.; Hamelin, C.J.; Smith, M.; Bendeich, P.J.; Edwards, L. The effect of plasticity theory on predicted residual stress fields in numerical weld analyses. *Comput. Mater. Sci.* **2012**, *54*, 125–134. [CrossRef]

30. Denlinger, E.R.; Michaleris, P. Effect of stress relaxation on distortion in additive manufacturing process modeling. *Addit. Manuf.* **2016**, *12*, 51–59. [CrossRef]

31. Lütjering, G.; Williams, J.C. Titanium. In *Engineering Materials and Processes*, 2nd ed.; Derby, B., Ed.; Spinger: Berlin/Heidelberg, Germany, 2007; pp. 1–442.

32. Boyer, R.; Collings, E.W.; Welsch, G. *Materials Properties Handbook: Titanium Alloys*; ASM International: Cleveland, OH, USA, 1994.

33. Zhang, Z.; Huang, Y.; Kasinathan, A.R.; Shahabad, S.I.; Ali, U.; Mahmoodkhani, Y.; Toyserkani, E. 3-Dimensional heat transfer modeling for laser powder-bed fusion additive manufacturing with volumetric heat sources based on varied thermal conductivity and absorptivity. *Opt. Laser Technol.* **2019**, *109*, 297–312. [CrossRef]

34. Rouquette, S.; Guo, J.; Le Masson, P. Estimation of the parameters of a Gaussian heat source by the Levenberg–Marquardt method: Application to the electron beam welding. *Int. J. Therm. Sci.* **2007**, *46*, 128–138. [CrossRef]

35. Kou, S. *Welding Metallurgy*; Wiley: Hoboken, NJ, USA, 2002.

36. Lampa, C.; Kaplan, A.F.H.; Powell, J.; Magnusson, C. An analytical thermodynamic model of laser welding. *J. Phys. D Appl. Phys.* **1997**, *30*, 1293–1299. [CrossRef]

37. Zhao, X.; Iyer, A.; Promoppatum, P.; Yao, S.-C. Numerical modeling of the thermal behavior and residual stress in the direct metal laser sintering process of titanium alloy products. *Addit. Manuf.* **2017**, *14*, 126–136. [CrossRef]

38. Tao, Z.; Yang, H.; Li, H.; Fan, X. Quasi-static tensile behavior of large-diameter thin-walled Ti-6Al-4V tubes at elevated temperature. *Chin. J. Aeronaut.* **2016**, *29*, 542–553. [CrossRef]

39. Brandl, E.; Michailov, V.; Viehweger, B.; Leyens, C. Deposition of Ti–6Al–4V using laser and wire, part I: Microstructural properties of single beads. *Surf. Coat. Technol.* **2011**, *206*, 1120–1129. [CrossRef]

40. Brandl, E.; Schoberth, A.; Leyens, C. Morphology, microstructure, and hardness of titanium (Ti-6Al-4V) blocks deposited by wire-feed additive layer manufacturing (ALM). *Mater. Sci. Eng. A* **2012**, *532*, 295–307. [CrossRef]

41. Brandl, E.; Greitemeier, D. Microstructure of additive layer manufactured Ti–6Al–4V after exceptional post heat treatments. *Mater. Lett.* **2012**, *81*, 84–87. [CrossRef]

42. Baufeld, B.; Van Der Biest, O. Mechanical properties of Ti-6Al-4V specimens produced by shaped metal deposition. *Sci. Technol. Adv. Mater.* **2009**, *10*, 015008. [CrossRef] [PubMed]

43. Chekir, N.; Tian, Y.; Sixsmith, J.; Brochu, M. Effect of travel speed and sub-β transus post deposition heat treatments on thin Ti-6Al-4V laser wire deposits. *Mater. Sci. Eng. A* **2018**, *724*, 376–384. [CrossRef]

44. Karimzadeh, F.; Ebnonnasir, A.; Foroughi, A. Artificial neural network modeling for evaluating of epitaxial growth of Ti6Al4V weldment. *Mater. Sci. Eng. A* **2006**, *432*, 184–190. [CrossRef]

45. Kobryn, P.; Semiatin, S. Microstructure and texture evolution during solidification processing of Ti–6Al–4V. *J. Mater. Process. Technol.* **2003**, *135*, 330–339. [CrossRef]

46. Åkerfeldt, P.; Antti, M.-L.; Pederson, R. Influence of microstructure on mechanical properties of laser metal wire-deposited Ti-6Al-4V. *Mater. Sci. Eng. A* **2016**, *674*, 428–437. [CrossRef]

47. Kelly, S.M. Thermal and microstructure modeling of metal deposition processes with application to Ti-6Al-4V. Ph.D. Thesis, Virginia Polytechnic Institute and State University, Blacksburg, VA, USA, 2004.

48. Ahmed, T.T.; Rack, H. Phase transformations during cooling in α+β titanium alloys. *Mater. Sci. Eng. A* **1998**, *243*, 206–211. [CrossRef]

49. Wanjara, P.; Watanabe, K.; De Formanoir, C.; Yang, Q.; Bescond, C.; Godet, S.; Brochu, M.; Nezaki, K.; Gholipour, J.; Patnaik, P. Titanium Alloy Repair with Wire-Feed Electron Beam Additive Manufacturing Technology. *Adv. Mater. Sci. Eng.* **2019**, *2019*, 1–23. [CrossRef]

50. Pederson, R. Microstructure and Phase transformation of Ti–6Al–4V. Licentiate Thesis, Lulea University of Technology, Lulea, Sweden, 2002; pp. 4–62.

51. De Formanoir, C.; Martin, G.; Prima, F.; Allain, S.Y.; Dessolier, T.; Sun, F.; Vivès, S.; Hary, B.; Bréchet, Y.; Godet, S. Micromechanical behavior and thermal stability of a dual-phase α+α' titanium alloy produced by additive manufacturing. *Acta Mater.* **2019**, *162*, 149–162. [CrossRef]

52. Lütjering, G. Influence of processing on microstructure and mechanical properties of (α+β) titanium alloys. *Mater. Sci. Eng. A* **1998**, *243*, 32–45. [CrossRef]

53. Waryoba, D.; Keist, J.; Ranger, C.; Palmer, T. Microtexture in additively manufactured Ti-6Al-4V fabricated using directed energy deposition. *Mater. Sci. Eng. A* **2018**, *734*, 149–163. [CrossRef]

54. Baufeld, B.; Brandl, E.; Van Der Biest, O. Wire based additive layer manufacturing: Comparison of microstructure and mechanical properties of Ti–6Al–4V components fabricated by laser-beam deposition and shaped metal deposition. *J. Mater. Process. Technol.* **2011**, *211*, 1146–1158. [CrossRef]

55. Gong, X.; Lydon, J.; Cooper, K.; Chou, K. Beam speed effects on Ti–6Al–4V microstructures in electron beam additive manufacturing. *J. Mater. Res.* **2014**, *29*, 1951–1959. [CrossRef]

56. Liu, Y.; Yang, Y.; Wang, D. A study on the residual stress during selective laser melting (SLM) of metallic powder. *Int. J. Adv. Manuf. Technol.* **2016**, *87*, 647–656. [CrossRef]

57. Leung, C.L.A.; Tosi, R.; Muzangaza, E.; Nonni, S.; Withers, P.J.; Lee, P.D. Effect of preheating on the thermal, microstructural and mechanical properties of selective electron beam melted Ti-6Al-4V components. *Mater. Des.* **2019**, *174*, 107792. [CrossRef]

58. Elmer, J.; Palmer, T.; Babu, S.; Specht, E. In situ observations of lattice expansion and transformation rates of α and β phases in Ti–6Al–4V. *Mater. Sci. Eng. A* **2005**, *391*, 104–113. [CrossRef]

59. Lu, X.; Lin, X.; Chiumenti, M.; Cervera, M.; Li, J.; Ma, L.; Wei, L.; Hu, Y.; Huang, W. Finite element analysis and experimental validation of the thermomechanical behavior in laser solid forming of Ti-6Al-4V. *Addit. Manuf.* **2018**, *21*, 30–40. [CrossRef]

materials

Article

A Comparative Analysis of Laser Additive Manufacturing of High Layer Thickness Pure Ti and Inconel 718 Alloy Materials Using Finite Element Method

Sapam Ningthemba Singh [1,2], Sohini Chowdhury [1], Yadaiah Nirsanametla [1,*], Anil Kumar Deepati [3], Chander Prakash [4,5,*], Sunpreet Singh [6], Linda Yongling Wu [5], Hongyu Y. Zheng [5] and Catalin Pruncu [7,8,*]

1. Department of Mechanical Engineering, North Eastern Regional Institute of Science and Technology, Nirjuli 791109, India; sapamthemba@gmail.com (S.N.S.); sohinidme@gmail.com (S.C.)
2. Department of Mechanical Engineering, National Institute of Technology Silchar, Silchar 788010, India
3. College of Applied Industrial Technology, Jazan University, Jizan 45142, Saudi Arabia; flytoanil@gmail.com
4. School of Mechanical Engineering, Lovely Professional University, Phagwara 144411, India
5. Department of Mechanical Engineering, Shandong University of Technology, Zibo 255000, China; ylwu06@sdut.edu.cn (L.Y.W.), zhenghongyu@sdut.edu.cn (H.Y.Z.)
6. Mechanical Engineering, National University of Singapore, Singapore 119077, Singapore; snprt.singh@gmail.com
7. Mechanical Engineering, Imperial College, Exhibition Rd., London SW7 2AZ, UK
8. Design, Manufacturing & Engineering Management, University of Strathclyde, Glasgow G1 1XJ, UK
* Correspondence: ny@nerist.ac.in (Y.N.); chander.mechengg@gmail.com (C.P.); c.pruncu@imperial.ac.uk or catalin.pruncu@strath.ac.uk (C.P.)

Citation: Singh, S.N.; Chowdhury, S.; Nirsanametla, Y.; Deepati, A.K.; Prakash, C.; Singh, S.; Wu, L.Y.; Zheng, H.Y.; Pruncu, C. A Comparative Analysis of Laser Additive Manufacturing of High Layer Thickness Pure Ti and Inconel 718 Alloy Materials Using Finite Element Method. *Materials* **2021**, 14, 876. https://doi.org/10.3390/ma14040876

Academic Editor: Mika Salmi
Received: 23 December 2020
Accepted: 8 February 2021
Published: 12 February 2021

Publisher's Note: MDPI stays neutral with regard to jurisdictional claims in published maps and institutional affiliations.

Abstract: Investigation of the selective laser melting (SLM) process, using finite element method, to understand the influences of laser power and scanning speed on the heat flow and melt-pool dimensions is a challenging task. Most of the existing studies are focused on the study of thin layer thickness and comparative study of same materials under different manufacturing conditions. The present work is focused on comparative analysis of thermal cycles and complex melt-pool behavior of a high layer thickness multi-layer laser additive manufacturing (LAM) of pure Titanium (Ti) and Inconel 718. A transient 3D finite-element model is developed to perform a quantitative comparative study on two materials to examine the temperature distribution and disparities in melt-pool behaviours under similar processing conditions. It is observed that the layers are properly melted and sintered for the considered process parameters. The temperature and melt-pool increases as laser power move in the same layer and when new layers are added. The same is observed when the laser power increases, and opposite is observed for increasing scanning speed while keeping other parameters constant. It is also found that Inconel 718 alloy has a higher maximum temperature than Ti material for the same process parameter and hence higher melt-pool dimensions.

Keywords: laser additive manufacturing; Inconel 718; pure Ti; finite element modeling; melt-pool formation

1. Introduction

Additive manufacturing (AM) has sustained and advanced as a manufacturing process from rapid prototyping (RP). Earlier, AM was mainly used for prototyping, but with due time and thanks to the contributions of many researchers, now, it can be employed in the commercial manufacturing of aircraft parts, prosthesis, spacecraft parts, high strength military equipment, automobile parts, etc. Against the conventional machining principle where the product is manufactured by removing unwanted materials from the blank, AM manufactures the components from scratch by adding materials in a layer by layer order directly from the sliced CAD model. AM offers various advantages such as printing multi-material parts, manufacturing of functionally graded materials, the ability to print highly customized biomedical parts, a little change in tools and machine layout for new designs, etc. The ability to manufacture complex shapes using lightweight materials by AM is another leverage that AM has over the conventional machining processes in the aerospace industry, as lightweight materials can be manufactured into complex shapes without losing strength using AM. Despite these recent advances, it remains very costly, low efficiency in terms of material handling, process costs, overall machine acquisition, and maintenance costs. One of the most critical factors in optimizing these bottlenecks is the slow build time. Slow manufacturing time can be improved if the AM of high layer thickness materials is achieved. Hence, a clear understanding of the process is required.

In the aerospace industries, safety and capability are the two main concerns besides the cost involved. In such cases, Titanium (Ti) is the preferred material because of its desired properties being relatively lightweight yet high in strength, higher operating temperature range, ability to resist corrosion, and the ability to adapt to other composites [1]. However, in traditional manufacturing processes, the usage of Ti is limited as the cost is high as compared to other materials. With AM technology's advancement, AM of Ti is drawing more tractions [2–5]. Apart from the material-specific problems such as high surface roughness, irregular and concentrated residual stresses; long manufacturing lead time and slow build rate are also significant hindrances in adopting laser additive manufacturing (LAM) processes in aerospace industries. Inconel 718 alloy material has its applications in aerospace industries as well as in biomedical applications. Ti and Inconel 718 alloys are used in manufacturing aircraft fan blades, turbine discs, pressure compressors, etc. Some of the major applications in biomedical industries are surgical, dental, orthopaedic devices, and stents [6,7]. With the ever-increasing applications of these two materials in aerospace and biomedical applications, both face unique yet identical problems while fabricating. There are many process parameters in a laser-based AM process. Laser power, scanning speed, hatching distance, layer thickness, and scanning pattern are important process parameters in a laser-based AM process that can be adjusted to optimize the manufactured parts' properties [8–11]. Jang et al. reported that laser power is the most influential parameter on surface roughness, hardness, and density [12]. However, Calignano et al. [13] reported that scanning speed has the highest impact on the surface roughness of the fabricated components. The contradicting reports may be due to the difference in materials and the use of different machines. The Taguchi method, response surface method (RSM), and artificial neural networks (ANNs) are the popular optimization methods used to optimize the process parameters in a laser-based AM process. The Taguchi method has been applied to optimize the AM process for various materials including Ti, Inconel, SS316L, AlSi10Mg, etc. [12–18]. Similarly, the RSM method has been used to optimize the major process parameters i.e., laser power, scanning speed, scanning pattern, etc. [19–21]. Fotovvati et al. reported that layer thickness is one of the important factors regarding surface roughness of Ti-6Al-4V alloy in a laser-based powder bed fusion AM process. Using the Taguchi method, the authors also recommended using the highest energy density i.e., high laser power and low scanning speed. Laser power and hatching space are the most significant parameters for micro-hardness and density of the fabricated parts [22]. Khorasani et al. implemented ANN-based models and found that the heat treatment conditions play a major role in the final surface roughness of the manufactured parts [23].

Several research works have been reported with regards to the numerical modeling and finite element-based analysis of the AM process. The basic model consists of a heat source applied to the layers where analysis of the heat interaction is observed based on the finite element modeling [24,25]. A single layer model is very helpful in terms of initial modeling and validation of the model, extending into a multi-layer model to replicate the real-world scenario. From the analytical point of view, the thermal cycle repeats after a certain number of layers. Keeping the computational cost and time in mind, a model should be restricted to a limit where there is a significant influence on the thermal cycle.

A simple thermal model was analysed by Alexander et al. that considered the heat source model, thermal conductivity, and other parameters [26]. Other researchers have also proposed their own models based on different heat source models [27–29]. There are different heat source models such as the Gaussian distribution heat source model, cone type heat source models, Goldak's ellipsoidal model, logarithmic decay model, etc. [8,30–38]. Li et al. analysed the residual stress during the selective laser melting (SLM) of 355 steel using the Gaussian volumetric model and found that the model can predict the stress distribution accurately [31]. The use of the Gaussian distributed heat source model is further supported by Teixeira et al. in their numerical simulation of the TIG welding process using the Gaussian distributed heat source model [34]. Wei et al. performed a thermos-mechanical analysis on the hot cracking of laser welding of lap joints by adopting the Gaussian model, but they extended the model by taking the linear decay in the Z direction [36]. Soldner et al. extended numerical models to account for crystallization of the printed parts in the selective laser sintering (SLS) of PA12 materials and found that the crystallization occurs even during the manufacturing process. They also found that the temperature distribution is dependent on the geometry orientation of the parts [39].

Wu et al. found that the heat flow and melt-pool dimensions have a great impact on the final product's residual stress as well as on other mechanical properties during the SLM of AlSi10Mg [40]. The melt-pool size increases as the laser scanning continues and reaches a steady state. Higher beam intensity and slower beam speed have faster melt-pool evolution, while lower intensity with faster beam speed has the opposite effect. A larger melt pool is observed in the subsequent layers. Similar patterns are also observed for other materials. Another study was carried out by Hu et al. on multi-layer SLM of AlSi10Mg material. The gradual increase of temperature and melt-pool were reported with the addition of new layers, which helped in obtaining fine grains for low energy input [41]. A similar numerical study was carried out by Khan et al. [42]. An experimental investigation was performed to measure the residual stress of SLM-manufactured AISI 316L samples. On the top surfaces, higher residual stresses in the scan direction were reported than in the downward direction. However, in the lateral surface, the residual stresses are higher in the perpendicular direction than that of the scan direction. This is due to the temperature gradient and cooling down of the substrate [43]. In another study, it was found that Ti-6A1-4V experiences more residual stress than Inconel 718 material [44]. However, little research is being pursued on a high layer thickness material. As opposed to the general concept, Jingjing et al. found that layer thickness has very little effect on the melt-pool dimensions for a Ti-6Al-4V material by the SLM process [45]. Most of the reported studies concentrated on the very thin layer thickness, which is a limiting factor on the final material addition rate and hence the manufacturing time.

Song et al. presented the effects of scanning patterns on residual stress in an SLM process of Ti-6Al-4V. Even though there is an influence of scanning patterns on the residual stress, no significant change in the melt-pool size was observed [46]. Yang and Wang presented a case study of a repair process by direct laser fabrication (DLF) in ABAQUS software to study the influence of non-linear temperature-dependent properties in a pure nickel material on temperature distribution [47]. It is obvious that the highest peak temperature value is at the point or region where the laser is applied while the second peak value is at the border of the current layer and the previous layer, and the third peak value is at the border of the previous layer and second previous layer. There is a sudden rise in the thermal conductivity and heat transfer at the point where the phase change takes place. Direct metal laser fabrication (DMLF) is designed specifically for metal parts. However, it can be used for non-metal and polymer materials as well [48].

A high layer thickness of an AM process comes with some inherent problems such as an increase in surface roughness, a higher tendency to dimensional distortions and, improper melting. Martínez et al. found that the scan strategy, thermal cycle, and residual stresses affect, in part, distortion, especially on parts where a high strength to weight ratio is considered [49]. Nadammal et al. found that the rotational and alternate scanning pattern results in the least residual stress and distortion [50]. A study by Ventola et al. observed that additively manufactured rough surfaces have enhanced heat transfer and can be used in electronics cooling [51]. Yang et al. investigated the vertical surface roughness of AlSi10Mg parts manufactured by the SLM process instead of the common research on horizontal surface roughness. The authors determined that the surface roughness can be reduced by about 70% from 15 μm to 4 μm by using the energy density approach [52]. Such inherent problems like the surface roughness can be used to our advantage in applications where enhanced heat transfer is required without much concern regarding the surface roughness of the materials. In such areas, the use of a high layer thickness will improve the manufacturing time.

Ding et al. performed a numerical analysis of heat transfer along with the fluid flow analysis for different laser scanning patterns in an SLM process. In a point laser exposure, a wide melt-pool was observed. In addition to this, there was a longer processing time with multiple thermal cycles, which resulted in finer grains [53]. Another advantage that AM offers is the ability to provide onsite demand and manufacturing of spare parts, which will reduce the manufacturing lead time (MLT), inventory, and transportation cost [54]. By adopting onsite demand, the US Navy was able to reduce the cost by 30% and increase on-demand part production [55,56]. A comparative study of samples is reported based on different manufacturing processes. Chastand et al. investigated different fatigues of Ti-6Al-4V samples produced by electron beam melting (EBM) and the SLM process. It is observed that surface defects have a high impact, followed by un-melted zones and small internal defects [57]. A recent study found that the short-time creep resistance of Inconel 718 manufactured by the SLM process is comparatively the same as that of manufactured by forging process. However, it is far below when compared to a sample manufactured by the casting method [58]. Without denying the importance of these investigations, most of the research concentrates on the comparison of the same material manufactured by different manufacturing processes. Limited information is available that shows a comparative study of two materials viz. Ti and Inconel 718 in terms of temperature distribution and different melt-pool dimensions under the same processing conditions manufactured by the LAM process. Such a study will help in determining which exact material to use in specific conditions or even use both the materials.

However, the AM process faces some major problems that are hindering the mass adoption of this technology on commercial scales. Another major issue is the limited availability of high performance yet light in weight materials. Some of the materials that are available are difficult to process using the AM technique, or it exhibits poor mechanical properties even if it is possible for AM to process those materials. Given the importance of LAM in aerospace and biomedical applications combined with the ever-expanding use of Ti material and Inconel 718 alloy, these two materials are considered in the present study. Another major hindrance is the high manufacturing lead time, which is largely because of the very thin layer thickness. The layer thickness of the most SLM process mentioned in this section is below 100 μm. Manufacturing a whole part with this scale of layer thickness will take a long time to complete. Adopting a higher layer thickness will substantially reduce the MLT of the item. Additionally, limited information is available on the interaction and process information for different materials for additively manufactured parts of high layer thickness materials. Therefore, the present study is focused on the investigation of thermal interaction and change in melt-pool dimensions due to thermal interactions in between the layers and substrates in a high layer thickness multi-layer LAM process of pure Ti and Inconel 718 materials and comparatively analyze the results.

The present paper shows the comparative numerical investigation of laser-based additive manufacturing of high layer thickness materials. A Gaussian distributed disc heat source model is used as the laser heat source model. To simulate the layer addition, the element death and birth feature is employed. It also investigates the effects of laser power and laser scanning speed on the heat flow and melt-pool dimensions. Based on the observed results, it is found that LAM of high layer thickness materials is possible with sufficient process parameters, well under the capability of currently available LAM machines.

2. Numerical Modeling

A finite element-based numerical model is developed for transient thermal analysis using ANSYS (APDL) 14.5 developed by the ANSYS Inc. Canonsburg, PA, USA. The SOLID70 element having a single degree of freedom (DOF) is chosen during finite element (FE) thermal modeling. Figure 1 shows the FE model of Ti and Inconel 718 materials. The substrate size is 10 mm × 31 mm × 4 mm with a layer size of 1 mm × 31 mm × 0.5 mm. Figure 1b shows the meshed FE model along with a magnified view. Most of the heat transfer and interaction will take place in and around the layers. With an aim to reduce the computational time and cost, fine-mesh is adopted for the layers, and coarse-mesh is adopted for the substrates. To mimic the addition of new layers on top of existing layers, the element birth and death feature is employed. Even though five layers are already modelled, the upper four layers are marked as death elements and do not take part in any physical and thermal interactions. After scanning of the first layer, practically a new powder layer is added, and the 2nd layer is marked as active while the other three upper layers are still marked as death elements. This process is continued until the scanning of the last layer (5th).

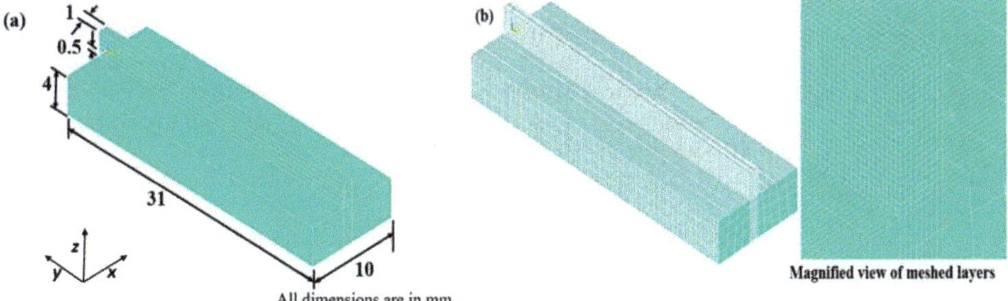

Figure 1. (a) Geometry of the FE model and (b) FE model of a five-layer selective laser melting (SLM) for Ti and Inconel 718 alloy material.

The process parameters employed during the LAM process for Inconel 718 and pure Ti are given in Table 1. The layer thickness is kept constant at 0.5 mm for both materials. Temperature-dependent physical material properties of Ti and Inconel 718 are given in Figure 2. The other material properties of the two materials are given in Table 2.

Table 1. LAM processing parameters for pure Ti Inconel 718 material.

Data Set No.	1	2	3	4	5	6
Speed (mm/s)	150	150	200	200	300	300
Power (W)	200	250	200	250	200	250

Table 2. Material properties of Ti and Inconel 718 used in the present work [61–63].

Parameter	Material		Unit
	Inconel 718	**Pure Ti**	
Density	8190	4.51×10^3	$Kg{\cdot}m^{-3}$
Thermal conductivity	11.4	20	$W{\cdot}m^{-1}{\cdot}K^{-1}$
Melting point	1609–1700	1941	K
Specific heat	435	518	$J{\cdot}kg^{-1}{\cdot}K^{-1}$
Latent heat	152×10^3	292×10^3	$J{\cdot}kg^{-1}$
Co-efficient of thermal expansion	1.3×10^{-5}	2.09×10^{-5}	K^{-1}

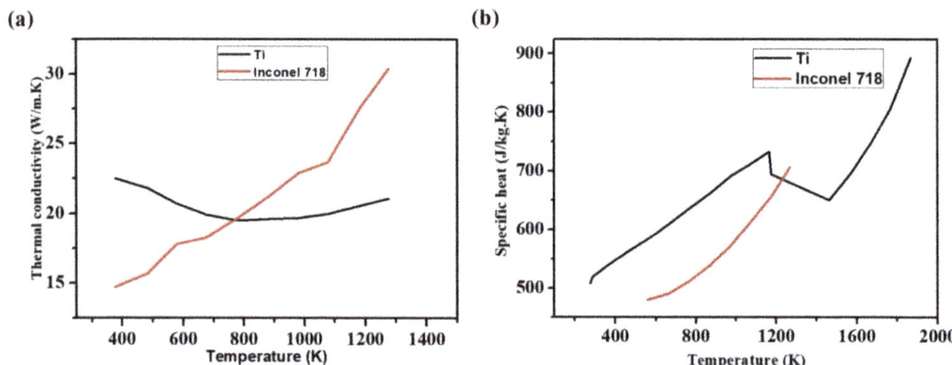

Figure 2. Temperature-dependent (**a**) thermal conductivity and (**b**) specific heats of Ti and Inconel 718 [44,59–61].

During the SLM process, the energy from the laser beam strikes the metal powders, which melts, and the energy is transferred to subsequent layers. By repeating this process, some layers are subjected to thermal changes several times, thus resulting in phase transformation, grain growth, a final product with non-uniform grain size, shape, and thermal distribution. Moreover, during the SLM process, the object is subjected to convection and radiation heat losses. Figure 3 presents a schematic of thermal boundary constraints of a multi-layered SLM process.

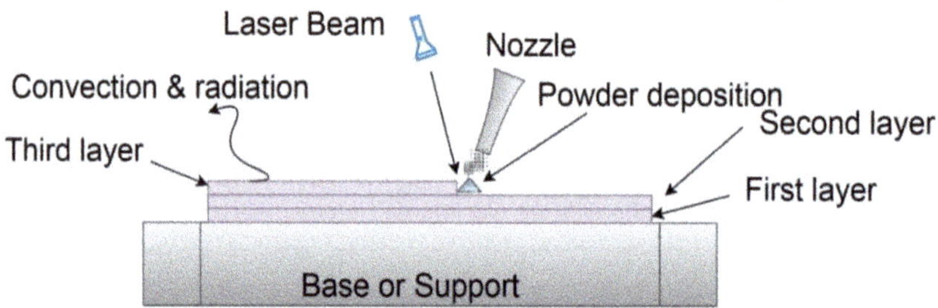

Figure 3. Schematic of boundary conditions of a multi-layered SLM process.

A finite 3D model is developed for the laser additive manufacturing process using ANSYS software to determine the non-linearity and temperature variations with respect to time. The 3D heat conduction equation (transient) is represented by [34]

$$k\left(\frac{\partial^2 T}{\partial x^2} + \frac{\partial^2 T}{\partial y^2} + \frac{\partial^2 T}{\partial z^2}\right) + \dot{Q} = \rho c_p\left(\frac{\partial T}{\partial t}\right) \tag{1}$$

where k, $\dot{Q}$, v, C_p, T, and ρ are thermal conductivity, rate of internal heat generation per unit volume, velocity vector, specific heat, temperature, and density of the material, respectively. The terms on equation 1 represent heat transfers in X, Y, and Z directions with laser heat power supplied for melting of the powder layers. The initial condition at the start of the operation is given by

$$T(x, y, z)|_{t=0} = T_0 \tag{2}$$

where T_0 represents ambient temperature (298 K i.e., 25 °C).

Most of the heat transfer takes place due to phase change, latent heat of conduction in and around the area where laser heat is applied, and melt-pool occurs. Convection and radiation heat transfer are assumed in all the exposed surfaces of the substrate. The natural boundary condition can be represented mathematically, which involves convective heat transfer and radiative heat transfer [64–70]:

$$k\frac{\partial T}{\partial n} - q + q_c + q_r = 0 \tag{3}$$

where q is the input heat flux, q_c and q_r are heat loss due to convection and radiation respectively. The basic equation of convective and radiative heat transfer can be represented as:

$$q_c = h(T - T_0) \tag{4}$$

$$q_r = \sigma\epsilon\left(T^4 - T_0^4\right) \tag{5}$$

where h is the convective heat transfer co-efficient, σ is the Stephan–Boltzmann constant and ϵ is the emissivity, T_0 and T are the initial temperature and instant temperature.

The Gaussian heat distribution model is the most common model used in a laser-based additive manufacturing process as the heat distribution, which follows Gaussian distribution. Gaussian distributed disc model is implemented as the heat source to mimic the laser heat source onto the powder layers from the laser source, which is mathematically represented as [32,33,71].

$$q = \frac{3\eta P}{\pi r^2}exp(\frac{-3(x^2 + y^2)}{r^2}), \tag{6}$$

where q, η, P and r represents the rate of laser heat flux, the efficiency of the laser beam absorption, absolute laser beam power, effective radius of the laser beam at the surface of the layer. x and y depict the position of the centre of the laser beam at a specific time and at the respective layer.

3. Results and Discussion

The numerical simulation is performed for Ti and Inconel 718, corresponding to the dataset in Table 1, and a detailed analysis of the LAM process is studied. The temperature behavior against time and nature of the melt-pool of Ti and Inconel 718 materials are studied separately. For different datasets, the comparative study of the two materials under the same process parameters are studied and analysed to determine the nature of transient heat transfer and temperature over time. The developed model has been validated with the experimental data available in the literature [45]. First, a single layer FE is developed for Ti-6Al-4V material and is compared with the experimental data available in the literature. As shown in Figure 4, the melt-pool dimensions of computed value are ~0.12 and ~0.13 mm for the experimental data. The pattern and data agree with the computed value and experimental value. The percentage difference in the melt-pool depth is estimated to be 5.38% which is under the acceptable range. This model is further extended into a multi-layer model with the material properties considered in the present paper.

3.1. Melt-Pool Evolution and Temperature Distribution

The melt-pool dimension increases as the laser power move along the direction of the laser scan. Moreover, the melt-pool size and length increased as the new layers are added. However, there are un-melted powders in some parts of the layer when laser power is 150 W and the laser velocity is 150 mm·s^{-1}. This is possibly due to the applied laser power is not enough to melt the whole layer thickness. Due to this, the specific dataset is not mentioned in Table 1. Figure 5 represents the 3D view of laser movement along the different layers and evolution of the melt pool. Figure 5a shows the laser movement along the layer at the beginning of the 1st layer. Similarly, Figure 5b–e shows laser movement along the 2nd layer, at the middle of the 3rd layer, at the 4th layer, and at the end of the 5th layer. Figure 5f displays cooling after laser scanning is over for dataset#6 of Table 1. The cross-sectional view of melt-pool alteration is shown in Figure 5g, corresponding to different layers. Moreover, the cumulative effect of the increasing size and length of the melt-pool can be visually recognized in Figure 5g. The zone above 1941 K denotes part of the layer which is fully melted. The 1400–1941 K zone indicates the sintered area. The other zones do not take part either in melting or sintering.

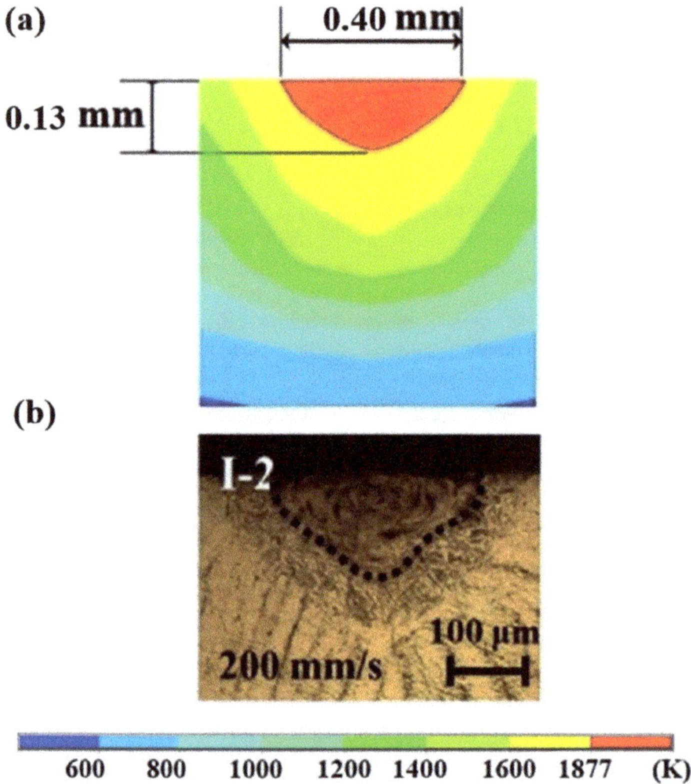

Figure 4. Comparison of (**a**) computed and (**b**) experimental melt-pool shape and size.

In Figure 5g, it is observed that the melt-pool depth and width increased as the process continue and as layers are deposited on top of one another. The width increases from ~0.42 to ~0.62 mm from 1st layer to the 2nd layer and then to ~0.65 mm for the third, while, fourth and fifth layers have a width of 1 mm or higher. It was observed that the melt pool depth increased from ~0.088 to ~0.41 mm from the 1st layer to the 5th layer. The melt pool depth corresponding to the 2nd layer, 3rd layer, and 4th layer was determined to be ~0.21, ~0.24, and ~0.38 mm, respectively. At the beginning of the 1st layer, it is observed that the whole depth of the layer is not melted. This can be attributed to the fact that at the beginning of the process, the substrate's temperature and layers are at ambient temperature, which prevents it from reaching the melting point. Preheating the powder materials may reduce the improper melting of layers at the beginning of the process. However, as the laser supply and heat transmission from melt pool to substrate continue, the overall temperature of the substrate and the layers increases that enables the laser focus area to attain higher maximum temperatures. With temperature rise, the melt-pool depth increases. This shows that the manufacturing of products by the LAM process is possible. From the 4th to 5th layer, there is a decrease in the rate of increasing melt-pool dimensions when compared to the other layers. This indicates the stagnation point where a further increase in the layers will not affect the melt-pool dimensions significantly. This phenomenon can be attributed to the fact that the temperature of the initial layers is nearing the approximate substrate temperature. However, the overall temperature of the substrate is increased and remains the principal heat transfer agent. Considering the high layer thickness, it will be of great interest if further research is undertaken for a sloped surface as performed by Xiang et al., where the slope's quality is analysed. Such an analysis by considering different oblique angles under different process parameters will be of great interest as a higher layer thickness may produce more extreme results in increasing or decreasing the quality of slope on the surfaces [72].

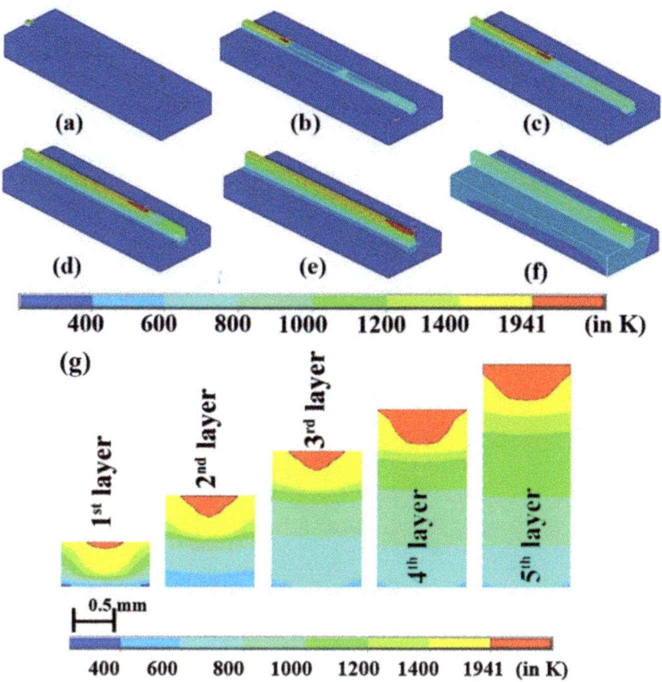

Figure 5. Laser movement and cooling over different layers (**a–f**) and changing melt-pool dimensions of laser scanning during a laser AM process of 5 layers for pure Ti for dataset#6 (**g**).

Figure 6 represents the time-temperature graph at the midpoint of each layer when power is 250 W and speed is 300 mm·s^{-1}. The maximum temperature obtained at the middle of the 1st layer is 2195 K, 2342 K for the 2nd layer, 2475 K for the 3rd layer, 2552 K for the 4th layer, and 2622 K for the 5th layer. This confirms the general prediction that the maximum peak temperature increases with each increasing layer number. Thereby, the melt-pool size is increased, and it is above the melting temperature of the Ti material.

Theoretically, there should be five and four decreasing peak temperatures corresponding to the 1st and 2nd layer. Similarly, three, two, and one peak temperature corresponding to the 3rd layer, 4th layer and 5th layer are observed. The same phenomenon is also observed in the present case, and it can be identified from Figure 7a,b. However, Figure 6 has only 4 visible peaks due to low laser power. This shows that when the laser is in the 5th layer, the temperature at the 1st layer hardly has any effect. However, there is a rapid decrease in temperature after the laser is applied. After the rapid decline in the temperature, it is gradually reduced as cooling takes place.

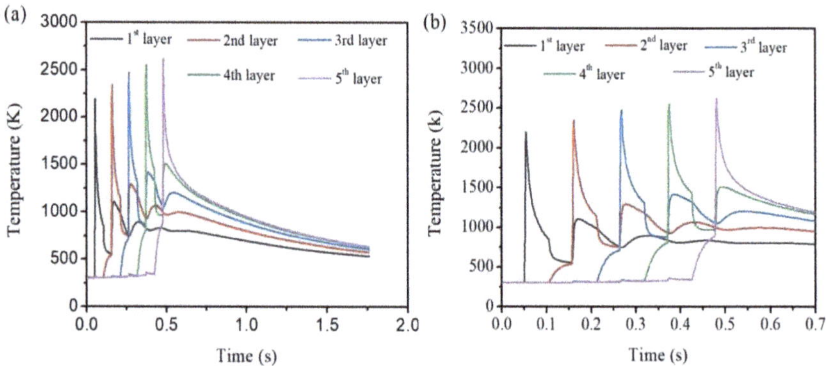

Figure 6. Time-temperature history (**a**) full view, (**b**) zoomed view considering only up to 0.7 s of scanning time, at the centre of each layer for dataset#6.

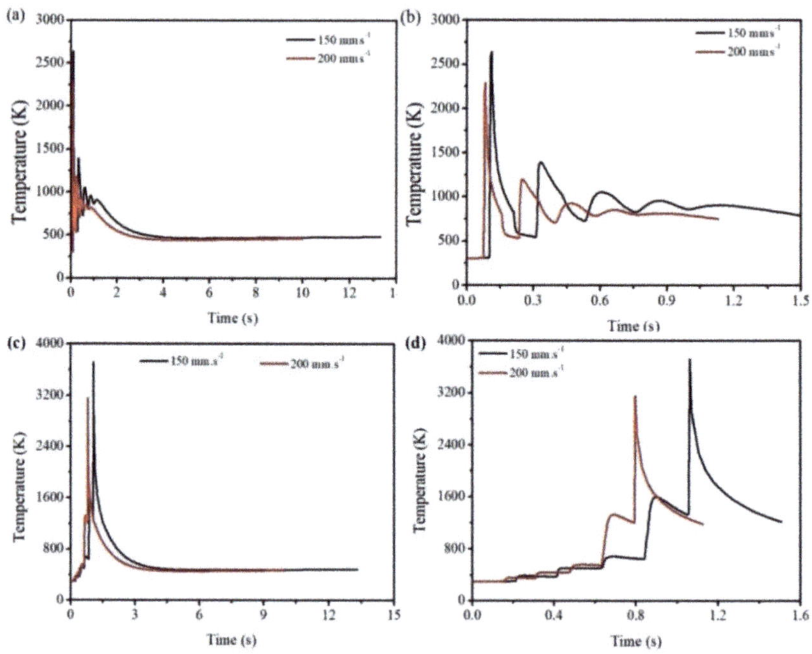

Figure 7. Time-temperature history (**a**) full view, (**b**) zoomed view at the centre of 1st layer, (**c**) full view, (**d**) zoomed view at the end of 5th layer at 200 W laser power and at a scan speed of 150 and 200 mm·s^{-1}.

3.2. Effects of Scanning Speed on the Melt-Pool and Temperature Distribution

As the laser scan speed increases with constant laser power, the depth as well as the width of the melt-pool decreases. Time-temperature history in the middle of the 1st layer is shown in Figure 7a,b. It can be identified that at a scan speed of 150 mm·s^{-1}, a maximum temperature of 2640 K is attained, while at a scan speed of 200 mm·s^{-1}, the highest temperature is 2287 K. Similar trend is also observed in the case of 5th layer (Figure 7c,d). At a scan speed of 150 mm·s^{-1}, the maximum peak temperature is 3722 K, while at a scan speed of 200 mm·s^{-1}, the peak temperature is 3154 K. This indicates that the current LAM machines have the capability to produce high layer thickness parts.

Figure 8 shows the decreasing depth of melt-pool corresponding to the 1st layer, 3rd layer, and 5th layer at different scan speeds. The depth of the melt-pool decreases from ~0.31 to ~0.12 mm for 1st layer, from ~0.45 mm to ~0.36 mm for the 3rd layer, and for the 5th layer, it decreases from ~0.55 to ~0.41 mm. Similarly, the width of the melt-pool decreases from ~0.87 to ~0.5 mm for the 1st layer, from ~1 to ~0.7 mm, for the 5th layer. It is evident that for both the processing conditions, melting of the powder material over the whole layer thickness is incomplete in the 1st layer. However, subsequent layers are fully melted. Moreover, a further study on increasing layer thickness will provide extensive information to determine the optimal parameters.

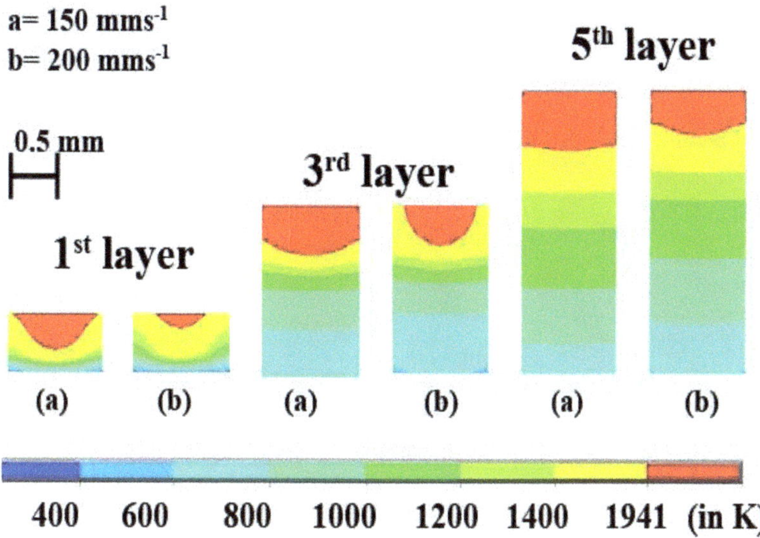

Figure 8. Changing melt-pool shapes for the changing scanning speed for the constant power source of 200 W and scanning speed of (**a**) 150 mm·s^{-1} and (**b**) 200 mm·s^{-1}.

3.3. Effects of Laser Power on the Melt-Pool and Temperature Distribution for Pure Ti

Laser power is directly proportional to the temperature at any point during the LAM process, provided all other parameters are kept constant. Figure 9 shows the increasing final maximum temperature for increasing laser power at constant laser scan speed. The temperature variation at different laser power and at a constant scan speed of 200 mm·s^{-1} is shown in Figure 9. As discussed above, during the process, there are five major peaks in the 1st layer. The maximum temperature obtained at the centre of the 1st layer is 2288 K when laser power is 200 W and 2685 K when laser power is 250 W. Figure 10 represents the difference in melt-pool profiles corresponding to different laser power and at a constant scanning speed of 200 mm·s^{-1}. A similar trend of Figure 7 is observed in Figure 9 wherein, 1st layer is not fully melted. However, as the process continues with new layers are being deposited, the temperature keeps increasing. Thereby, a fully melted pool is obtained across the whole cross-section of the layer.

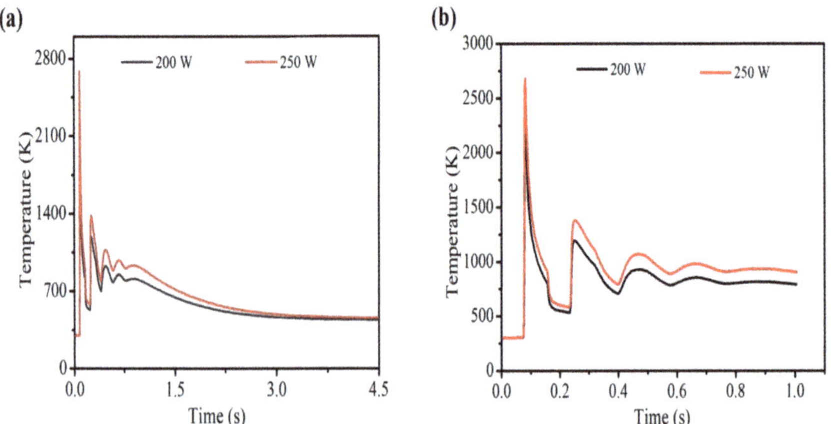

Figure 9. Effect of laser power in the middle of the 1st layer when laser power is 200, 250 W and scan speed is 200 mm·s^{-1} (**a**) full view and (**b**) magnified view.

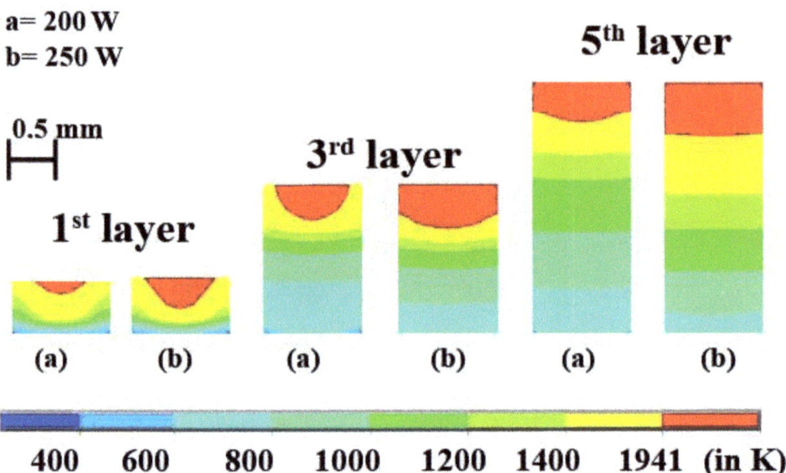

Figure 10. Evolution of melt pool dimensions for (**a**) P = 200, and (**b**) 250 W, v = 200 mm·s^{-1} in the 1st, 3rd and 5th layer.

The decreasing nature of peaks represents the gradual decrease of the impacts of laser power on the preceding layers as new layers are being deposited. In Figure 9b, the 5th peak has the lowest temperature in the 1st layer, and it is just above the temperature before cooling starts. This shows the negligible significance of laser in the 5th layer on the maximum temperature in the 1st layer. Figure 9 has overlapping peaks, but their peak values are different since both have the same scanning speed and different laser power. Figure 10 displays the evolution of melt-pool shape as new layers are added. It is evident that the width of the melt-pool increases faster than the depth of the melt-pool. This is due to the larger surface width area interaction with the laser. The increase in depth of melt-pool is significant and should not be discarded.

3.4. Comparative Study of Pure Ti and Inconel 718 Alloy

The pure Ti and Inconel 718 have different mechanical and thermal properties. It is self-evident that the melt-pool dimensions and temperature distribution of the two materials will be different for the same process parameters. Since the main influencing factors viz. laser power and laser scanning speed are the same for the two materials, a quantitative analysis is performed. Such an analysis will give more details about the intensity of temperature distribution and the melt-pool nature. This helps in choosing the right material under different circumstances where cost-saving, or safety may be the driving factor. It will be quite beneficial when multi-material manufacturing is being carried out and the difference in temperature and melt-pool becomes very important for the two materials to be fused in a homogenous way. This is especially very interesting concerning the printing of functionally graded materials. In such cases, there is a need to know if the different materials will melt and interact at the time at a given set of process parameters.

It is observed that the pure Ti material has a lower temperature and melt-pool dimensions than the Inconel 718 alloy for the same process parameters. It indicates that Inconel 718 requires lesser laser power to attain the same temperature as pure Ti material. Figure 11a shows the cross-sectional view of the variation of melt-pool of the two materials at 1st layer, 3rd layer, and 5th layer when the laser power is 250 W and laser scanning speed is 200 mm·s^{-1}. The difference in the melt-pool dimensions can be attributed to the difference in material properties. From Figure 2, it is observed that the Inconel 718 has higher thermal conductivity as temperature increases while pure Ti has higher specific heat as the temperature increases. This shows that the specific heat outweighs the thermal conductivity whenever the maximum temperature of a layer is concerned. Higher specific heat means a longer time to attain a higher temperature. The melt-pool depths for the first layer are ~0.28 and ~0.38 mm for pure Ti and Inconel 718 alloy, respectively. For the third layer, the melt-pool depths are ~0.46 and ~0.54 mm for pure Ti and Inconel 718 alloy, respectively. The melt-pool shape of the fifth layer is almost the same as flat for both cases. However, there is a difference in the melt-pool thickness, which is ~0.56 and ~0.87 mm for pure Ti and Inconel 718.

Figure 11b represents the difference of melt-pool dimensions between pure Ti and Inconel 718 alloy when laser intensity is 200 W, and scan speed is 200 mm·s^{-1}. The depth of melt-pool in the first layer is ~0.14 and ~0.28 mm for pure Ti and Inconel 718 alloy, respectively. For the third layer, the depth of the melt-pool is ~0.38 and ~0.47 mm for Ti and Inconel 718, respectively. In the 5th layer, the depth of melt-pool is ~0.42 and ~0.59 mm for Ti and Inconel 718 alloy, respectively.

Figure 12a,b shows the corresponding time-temperature history. Figure 12a is the time-temperature history considering the whole manufacturing time, while Figure 12b is the magnified view where peak temperature is represented. The maximum temperatures achieved by the laser power at the centre of the 1st layer are 2684 and 2732 K for pure Ti and Inconel 718, respectively. Figure 12c,d shows the temperature profile at the middle point of the 3rd layer over time. A similar trend is found in this case also. The peak temperature attained at the centre of the third layer in the case of pure Ti is 2599 K, while for Inconel 718 alloy, the peak temperature is 2644 K. Another interesting observation is the nature of the changing shape of melt-pool for both materials as the process continues. At first, the cross-sectional melt-pool shape is semi-circular, but it gradually changes to the shape of a ship's hull. This shows that the melt-pool width has a higher sensitivity of change than that of the melt-pool depth.

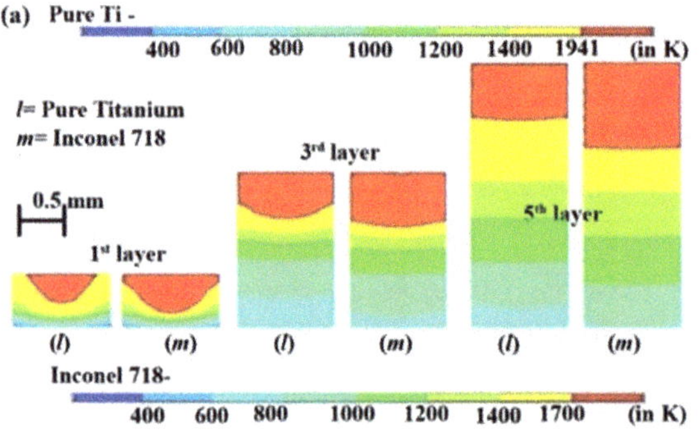

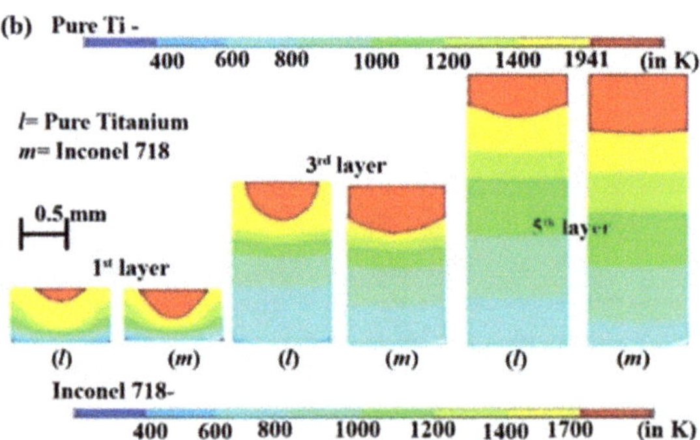

Figure 11. The difference of melt-pool dimensions for pure Ti and Inconel 718 alloy for the (**a**) dataset#4 and (**b**) dataset#3.

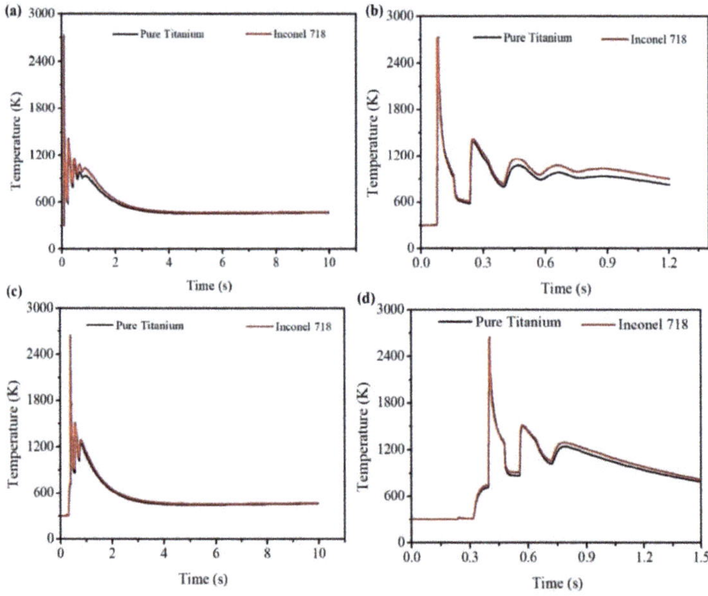

Figure 12. Time-temperature history of pure Ti and Inconel 718 alloy (**a**,**b**) at the centre of the 1st layer for dataset#4 and (**c**,**d**) at the centre of the 3rd layer for dataset#3.

4. Conclusions

A 3D finite-element model of laser-based additive manufacturing is developed to investigate melt-pool complexity and temperature behavior during the laser application throughout the model. One of the peculiar features of the model is the consideration of a high layer thickness as opposed to the general consideration of very thin layer thickness. The five-layer FEM simulation is performed to understand the effects of laser power and laser scan velocity on heat flow, melt-pool shape, and temperature distribution of the Ti and Inconel 718 materials. With the current study, the following conclusions can be drawn:

- For enough laser power and laser scan speed, LAM for a high layer thickness is possible for both Ti and Inconel 718 alloy materials. The current technology specifications have the capability to perform it.
- The temperature of subsequent layers increases as the process continues, and the number of layers increases. During the initial layers, the substrate is the primary agent for the heat transfer away from the layers.
- At the initial stage, there are parts of the layer that are not fully melted. However, this changes as the process continues and the number of layers increases.
- For low laser power, the impacts of laser power become very low in the 1st layer as the laser is acting on the 5th layer. For high laser power, there is a significant rise in overall temperature in the 1st layer even if the laser is scanning at the top layer.
- The melt-pool and maximum peak temperature are increased when the laser power is increased while the laser scanning velocity is kept constant for each material. However, the melt-pool dimensions and temperature are decreased when the scan velocity is increased while keeping a constant laser power.
- It is also found that the pure Ti has a lower peak temperature and melt-pool dimensions than the Inconel 718 alloy for the same influencing parameters, such as laser power and laser scan velocity, due to the difference in the material properties.
- There is a rapid change in the melt-pool shape of Inconel 718 from a semi-circle to a convex hull shape as the process continues.
- Further research of the current model can be implemented with the mechanical analysis to study the mechanical properties and residual stress analysis. Additionally, this model can be used as a base model to extend the research to a multi-track, multi-layer model during a laser-based additive manufacturing process.

Author Contributions: Conceptualization, S.N.S.; S.C.; and Y.N.; Software, S.N.S.; S.C.; and Y.N.; validation, S.N.S.; S.C.; and Y.N.; formal analysis, S.N.S.; S.C.; Y.N.; A.K.D. and C.P. (Chander Prakash); investigation, S.N.S.; S.C.; Y.N.; H.Y.Z.; and S.S.; resources, Y.N.; C.P. (Chander Prakash); C.P. (Catalin Pruncu), and S.S.; data curation, S.N.S.; S.C.; A.K.D.; and L.Y.W.; writing—original draft preparation, S.N.S.; S.C.; and Y.N.; writing—review and editing, S.N.S.; S.C.; Y.N.; A.K.D., C.P. (Chander Prakash); L.Y.W.; H.Y.Z.; S.S. and C.P. (Catalin Pruncu); visualization, Y.N.; L.Y.W. and H.Y.Z.; supervision, Y.N.; S.S.; and C.P. (Catalin Pruncu); project administration, Y.N.; H.Y.Z.; and S.S.; funding acquisition, Y.N.; C.P. (Chander Prakash) and C.P. (Catalin Pruncu). All authors have read and agreed to the published version of the manuscript.

Funding: This work was supported by TEQUIP II grant at the North Eastern Regional Institute of Science and Technology, Itanagar, Arunachal Pradesh, India.

Institutional Review Board Statement: Not applicable.

Informed Consent Statement: Not applicable.

Data Availability Statement: The data presented in this study are available on request from the corresponding author.

Conflicts of Interest: There exists no conflict of interest.

References

1. Facchini, L.; Magalini, E.; Robotti, P.; Molinari, A.; Hoges, S.; Wissenbach, K. Ductility of a Ti-6Al-4V alloy produced by selective laser melting of pre-alloyed powders. *Rapid Prototyp. J.* **2010**, *16*, 450–459. [CrossRef]

2. Qiu, C.; Adkins, N.J.E.; Attallah, M.M. Microstructure and tensile properties of selectively laser-melted and of HIPed laser-melted Ti–6Al–4V. *Mater. Sci. Eng. A* **2013**, *578*, 230–239. [CrossRef]

3. Murr, L.E.; Esquivela, E.V.; Quinonesb, S.A.; Gaytana, S.M.; Lopeza, M.I.; Martineza, E.Y.; Medinac, F.; Hernandeza, D.H.; Martineza, E.; Martineza, J.L.; et al. Microstructures and mechanical properties of electron beam-rapid manufactured Ti–6Al–4V biomedical prototypes compared to wrought Ti–6Al–4V. *Mater. Charact.* **2009**, *60*, 96–105. [CrossRef]

4. Edwards, P.; Ramulu, M. Fatigue performance evaluation of selective laser melted Ti–6Al–4V. *Mater. Sci. Eng. A* **2014**, *598*, 327–337. [CrossRef]

5. Balichakra, M.; Krishna, P.; Balla, V.K.; Das, M. Understanding Thermal Behavior in Laser Processing of Titanium Aluminide Alloys. *Proc. Int. India Manuf. Technol. Des. Res. Conf.* **2016**, 73–77.

6. Liang, X.; Liu, Z.; Wang, B. State-of-the-art of surface integrity induced by tool wear effects in machining process of titanium and nickel alloys: A review. *Measurement* **2019**, *132*, 150–181. [CrossRef]

7. Ngo, T.D.; Kashani, A.; Imbalzano, G.; Nguyen, K.T.Q.; Hui, D. Additive manufacturing (3D printing): A review of materials, methods, applications and challenges. *Compos. Part B Eng.* **2018**, *143*, 172–196. [CrossRef]

8. Ökten, K.; Biyikoğlu, A. Development of thermal model for the determination of SLM process parameters. *Opt. Laser Technol.* **2021**, *137*, 106825. [CrossRef]

9. Koutiri, I.; Pessard, E.; Peyre, P.; Amlou, O.; Terris, D.T. Influence of SLM process parameters on the surface finish, porosity rate and fatigue behavior of as-built Inconel 625 parts. *J. Mater. Process. Technol.* **2018**, *255*, 536–546. [CrossRef]

10. Wang, Z.; Xiao, Z.; Tse, Y.; Huang, C.; Zhang, W. Optimization of processing parameters and establishment of a relationship between microstructure and mechanical properties of SLM titanium alloy. *Opt. Laser Technol.* **2019**, *112*, 159–167. [CrossRef]

11. Khorasani, A.M.; Gibson, I.; Awan, U.S.; Ghaderi, A. The effect of SLM process parameters on density, hardness, tensile strength and surface quality of Ti-6Al-4V. *Addit. Manuf.* **2019**, *25*, 176–186. [CrossRef]

12. Jiang, H.Z.; Li, Z.Y.; Feng, T.; Wu, P.-Y.; Chen, Q.-S.; Feng, Y.-L.; Li, S.-W.; Gao, H.; Xu, H.-J. Factor Analysis of Selective Laser Melting Process Parameters with Normalised Quantities and Taguchi Method. *Opt. Laser Technol.* **2019**, *119*, 105592. [CrossRef]

13. Calignano, F.; Manfredi, D.; Ambrosio, E.P.; Iuliano, L.; Fino, P. Influence of Process Parameters on Surface Roughness of Aluminum Parts Produced by DMLS. *Int. J. Adv. Manuf. Technol.* **2013**, *67*, 2743–2751. [CrossRef]

14. Campanelli, S.; Casalino, G.; Contuzzi, N.; Ludovico, A. Taguchi Optimization of the Surface Finish Obtained by Laser Ablation on Selective Laser Molten Steel Parts. *Procedia CIRP* **2013**, *12*, 462–467. [CrossRef]

15. Rathod, M.K.R.; Karia, M.M.C. Experimental Study for Effects of Process Parameters of Selective Laser Sintering for AlSi10mg. *Int. J. Technol. Res. Eng.* **2020**, *7*, 6957–6960.

16. Joguet, D.; Costil, S.; Liao, H.; Danlos, Y. Porosity Content Control of CoCrMo and Titanium Parts by Taguchi Method Applied to Selective Laser Melting Process Parameter. *Rapid Prototyp. J.* **2016**, *22*, 20–30. [CrossRef]

17. Sathish, S.; Anandakrishnan, V.; Dillibabu, V.; Muthukannan, D.; Balamuralikrishnan, N. Optimization of Coefficient of Friction for Direct Metal Laser Sintered Inconel 718. *Adv. Manuf. Technol.* **2019**, 371–379.

18. Dong, G.; Marleau-Finley, J.; Zhao, Y.F. Investigation of Electrochemical Post-Processing Procedure for Ti-6AL-4V Lattice Structure Manufactured by Direct Metal Laser Sintering (DMLS). *Int. J. Adv. Manuf. Technol.* **2019**, *104*, 3401–3417. [CrossRef]

19. Dada, M.; Popoola, P.; Mathe, N.; Pityana, S.; Adeosun, S. Parametric Optimization of Laser Deposited High Entropy Alloys Using Response Surface Methodology (RSM). *Int. J. Adv. Manuf. Technol.* **2020**, *109*, 2719–2732. [CrossRef]

20. Read, N.; Wang, W.; Essa, K.; Attallah, M.M. Selective Laser Melting of ALSi10MgAlloy: Process Optimisation and Mechanical Properties Development. *Mater. Des.* **2015**, *65*, 417–424. [CrossRef]

21. El-Sayed, M.A.; Ghazy, M.; Youssef, Y.; Essa, K. Optimization of SLM Process Parameters for Ti6Al4V Medical Implants. *Rapid Prototyp. J.* **2019**, *25*, 433–447. [CrossRef]

22. Fotovvati, B.; Balasubramanian, M.; Asadi, E. Modeling and Optimization Approaches of Laser-Based Powder-Bed Fusion Process for Ti-6Al-4V Alloy. *Coatings* **2020**, *10*, 1104. [CrossRef]

23. Khorasani, A.M.; Gibson, I.; Ghasemi, A.; Ghaderi, A. Modelling of Laser Powder Bed Fusion Process and Analysing the Effective Parameters on Surface Characteristics of Ti-6Al-4V. *Int. J. Mech. Sci.* **2020**, *168*, 105299. [CrossRef]

24. Alimardani, M.; Paul, C.P.; Toyserkani, E.; Khajepour, A. Multiphysics Modelling of Laser Solid Freeform Fabrication Techniques. In *Woodhead Publishing Series in Welding and Other Joining Technologies, Advances in Laser Materials Processing*, 2nd ed.; Woodhead Publishing: Cambridge, UK, 2018; pp. 665–691.

25. Huang, Y.; Behrad, M.; Ehsan, K. A comprehensive analytical model for laser powder-fed additive manufacturing. *Addit. Manuf.* **2016**, *12*, 90–99. [CrossRef]

26. Rubenchik, A.M.; King, W.E.; Wu, S.S. Scaling laws for the additive manufacturing. *J. Mater. Process. Technol.* **2018**, *257*, 234–243. [CrossRef]

27. Lu, X.; Lin, X.; Chiumenti, M.; Cervera, M.; Li, J.; Ma, L.; Wei, L.; Hu, Y.; Huang, W. Finite element analysis and experimental validation of the thermomechanical behavior in laser solid forming of Ti-6Al-4V. *Addit. Manuf.* **2018**, *21*, 30–40. [CrossRef]

28. Huang, W.; Zhang, Y. Finite element simulation of thermal behavior in single-track multiple-layers thin wall without-support during selective laser melting. *J. Manuf. Process.* **2019**, *42*, 139–148. [CrossRef]

29. Liu, B.; Li, B.; Li, Z.; Bai, P.; Wang, Y.; Kuai, Z. Numerical investigation on heat transfer of multi-laser processing during selective laser melting of AlSi10Mg. *Results Phys.* **2019**, *12*, 454–459. [CrossRef]

30. Cao, L. Workpiece-scale numerical simulations of SLM molten pool dynamic behavior of 316L stainless steel. *Comput. Math. Appl.* **2020**. [CrossRef]
31. Li, T.; Zhang, L.; Chang, C.; Wei, L. A Uniform-Gaussian distributed heat source model for analysis of residual stress field of S355 steel T welding. *Adv. Eng. Softw.* **2018**, *126*, 1–8. [CrossRef]
32. Ning, J.; Sievers, D.E.; Garmestani, H.; Liang, S.Y. Analytical modeling of in-process temperature in powder bed additive manufacturing considering laser power absorption, latent heat, scanning strategy, and powder packing. *Materials* **2019**, *12*, 808. [CrossRef]
33. Cao, L.; Yuan, X. Study on the numerical simulation of the SLM molten pool dynamic behavior of a nickel-based superalloy on the workpiece scale. *Materials* **2019**, *12*, 2272. [CrossRef] [PubMed]
34. Teixeira, P.R.F.; Araújo, D.B.; Cunha, L.A.B. Study of the Gaussian distribution heat source model applied to numerical thermal simulations of TIG welding processes. *Cienc. Eng. Sci. Eng. J.* **2014**, *23*, 115–122.
35. Xu, G.X.; Wu, C.S.; Qin, G.L.; Wang, X.Y.; Lin, S.Y. Adaptive volumetric heat source models for laser beam and laser + pulsed GMAW hybrid welding processes. *Int. J. Adv. Manuf. Technol.* **2011**, *57*, 245–255. [CrossRef]
36. Wei, H.; He, Q.; Chen, J.S.; Wang, H.P.; Carlson, B.E. Coupled thermal-mechanical-contact analysis of hot cracking in laser welded lap joints. *J. Laser Appl.* **2017**, *29*, 022412. [CrossRef]
37. Zain-ul-abdein, M.; Nélias, D.; Jullien, J.F.; Deloison, D. Experimental investigation and finite element simulation of laser beam welding induced residual stresses and distortions in thin sheets of AA 6056-T4. *Mater. Sci. Eng. A* **2010**, *527*, 3025–3039. [CrossRef]
38. He, Q.; Wei, H.; Chen, J.S.; Wang, H.P.; Carlson, B.E. Analysis of hot cracking during lap joint laser welding processes using the melting state-based thermomechanical modeling approach. *Int. J. Adv. Manuf. Technol.* **2018**, *94*, 4373–4386. [CrossRef]
39. Soldner, D.; Greiner, S.; Burkhardt, C.; Drummer, D.; Steinmann, P.; Mergheim, J. Numerical and experimental investigation of the isothermal assumption in selective laser sintering of PA12. *Addit. Manuf.* **2020**, 101676. [CrossRef]
40. Wu, J.; Wang, L.; An, X. Numerical analysis of residual stress evolution of AlSi10Mg manufactured by selective laser melting. *Optik* **2017**, *137*, 65–78. [CrossRef]
41. Hu, H.; Ding, X.; Wang, L. Numerical analysis of heat transfer during multi-layer selective laser melting of AlSi10Mg. *Optik* **2016**, *127*, 8883–8891. [CrossRef]
42. Khan, H.M.; Dirikolu, M.H.; Koç, E.; Oter, Z.C. Numerical investigation of heat current study across different platforms in SLM processed multi-layer AlSi10Mg. *Optik* **2018**, *170*, 82–89. [CrossRef]
43. Simson, T.; Emmel, A.; Dwars, A.; Bohm, J. Residual stress measurements on AISI 316L samples manufactured by selective laser melting. *Addit. Manuf.* **2017**, *17*, 183–189. [CrossRef]
44. Mukherjee, T.; Zhang, W.; DebRoy, T. An improved prediction of residual stresses and distortion in additive manufacturing. *Comput. Mater. Sci.* **2017**, *126*, 360–372. [CrossRef]
45. Yang, J.; Han, J.; Yu, H.; Yin, J.; Gao, M.; Wang, Z.; Zeng, X. Role of molten pool mode on formability, microstructure and mechanical properties of selective laser melted Ti-6Al-4V alloy. *Mater. Des.* **2016**, *110*, 558–570.
46. Song, J.; Wu, W.; Zhang, L.; He, B.; Lu, L.; Ni, X.; Long, Q.; Zhu, G. Role of scanning strategy on residual stress distribution in Ti-6Al-4V alloy prepared by selective laser melting. *Optik* **2018**, *170*, 342–352. [CrossRef]
47. Yan, J.; Wang, F. 3D finite element temperature field modelling for direct laser fabrication. *Int. J. Adv. Manuf. Technol.* **2009**, *43*, 1060–1068.
48. Yang, J.; Ouyang, H.; Wang, Y. Direct metal laser fabrication: Machine development and experimental work. *Int. J. Adv. Manuf. Technol.* **2010**, *46*, 1133–1143. [CrossRef]
49. Martínez, S.; Ortega, N.; Celentano, D.; Sánchez Egea, A.J.; Ukar, E.; Lamikiz, A. Analysis of the part distortions for inconel 718 SLM: A case study on the NIST test artifact. *Materials* **2020**, *13*, 5087. [CrossRef]
50. Nadammal, N.; Mishurova, T.; Fritsch, T.; Muñoz, I.S.; Kromm, A.; Haberland, C.; Bruno, G. Critical role of scan strategies on the development of microstructure, texture, and residual stresses during laser powder bed fusion additive manufacturing. *Addit. Manuf.* **2021**, *38*, 101792.
51. Ventola, L.; Robotti, F.; Dialameh, M.; Calignano, F.; Manfredi, D.; Chiavazzo, E.; Asinari, P. Rough surfaces with enhanced heat transfer for electronics cooling by direct metal laser sintering. *Int. J. Heat Mass Transf.* **2014**, *75*, 58–74. [CrossRef]
52. Yang, T.; Liu, T.; Liao, W.; MacDonald, E.; Wei, H.; Chen, X.; Jiang, L. The influence of process parameters on vertical surface roughness of the AlSi10Mg parts fabricated by selective laser melting. *J. Mater. Process. Technol.* **2019**, *266*, 26–36. [CrossRef]
53. Ding, X.; Wang, L.; Wang, S. Comparison study of numerical analysis for heat transfer and fluid flow under two different laser scan pattern during selective laser melting. *Optik* **2016**, *127*, 10898–10907. [CrossRef]
54. Khajavi, S.H.; Partanen, J.; Holmström, J. Additive manufacturing in the spare parts supply chain. *Comput. Ind.* **2014**, *65*, 50–63. [CrossRef]
55. Frazier, W.E. Navy workshop aims to cut costs. *Adv. Mater. Process.* **2008**, *166*, 43–46.
56. Frazier, W.E. Direct digital manufacturing of metallic components: Vision and roadmap. *Annu. Int. Solid Free. Fabr. Symp.* **2010**, 717–732.
57. Chastand, V.; Quaegebeur, P.; Maia, W.; Charkaluk, E. Comparative study of fatigue properties of Ti-6Al-4V specimens built by electron beam melting (EBM) and selective laser melting (SLM). *Mater. Charact.* **2018**, *143*, 76–81. [CrossRef]
58. Wang, L.Y.; Zhou, Z.J.; Li, C.P.; Chen, G.F.; Zhang, G.P. Comparative investigation of small punch creep resistance of Inconel 718 fabricated by selective laser melting. *Mater. Sci. Eng. A* **2019**, *745*, 31–38. [CrossRef]

59. Li, Y.; Gu, D. Parametric analysis of thermal behaviour during selective laser melting Additive manufacturing of Aluminium alloy powder. *Mater. Des.* **2014**, *63*, 856–867. [CrossRef]

60. Zhang, D.; Zhang, P.; Liu, Z.; Feng, Z.; Wang, C.; Gu, Y. Thermofluid field of molten pool and its effects during selective laser melting (SLM) of Inconel 718 alloy. *Addit. Manuf.* **2018**, *21*, 567–578. [CrossRef]

61. Mills, K.C. *Ni-IN 718, Recommended Values of Thermophysical Properties for Selected Commercial Alloys*; Woodhead Publishing: Cambridge, UK, 2002; pp. 181–190.

62. Yadaiah, N.; Bag, S. Role of oxygen as surface active element in linear GTA welding process. *J. Mater. Eng. Perform.* **2013**, *22*, 3199–3209. [CrossRef]

63. Mills, K.C. *Ti Pure Titanium, Recommended Values of Thermophysical Properties for Selected Commercial Alloys*; Woodhead Publishing: Cambridge, UK, 2002; pp. 205–210.

64. Hua, T.; Jing, C.; Xin, L.; Fengying, Z.; Weidong, H. Research on molten pool temperature in the process of laser rapid forming. *J. Mater. Process. Technol.* **2008**, *198*, 454–462. [CrossRef]

65. Griffith, M.L.; Keicher, D.M.; Atwood, C.L.; Romero, J.A.; Smugeresky, J.E.; Harwell, L.D.; Greene, D.L. Free Form Fabrication of Metallic Components Using Laser Engineered Net Shaping (LENSTM). In Proceeding of 7th Solid Freeform Fabrication Symposium, Austin, TX, USA, 12–14 August 1996; pp. 125–132.

66. Wen, S.; Shin, Y.C. Modeling of Transport Phenomena during the Coaxial Laser Direct Deposition Process. *J. Appl. Phys.* **2010**, *108*, 044908-1–044908-9. [CrossRef]

67. Griffith, M.L.; Ensz, M.T.; Puskar, J.D.; Robino, C.V.; Brooks, J.A.; Philliber, J.A.; Smugeresky, J.E.; Hofmeister, W.H. Understanding the Microstructure and Properties of Components Fabricated by Laser Engineered Net Shaping (LENS). In *MRS Proceeding, 625s*; Cambridge University Press: Cambridge, UK, 2000.

68. Kelly, S.M.; Kampe, S.L. Microstructural Evolution in Laser-Deposited Multilayer Ti-6Al-4V Builds: Part II. Thermal Modeling. *Metall. Mater. Trans. A* **2004**, *35*, 1869–1879. [CrossRef]

69. Baufeld, B.; Biest, O.V.D.; Gault, R.; Ridgway, K. Manufacturing Ti-6Al-4V Components by Shaped Metal Deposition: Microstructure and Mechanical Properties. *IOP Conf. Ser. Mater. Sci. Eng.* **2011**, *26*, 1–8. [CrossRef]

70. Bontha, S.; Klingbeil, N.W.; Kobryn, P.A.; Fraser, H.L. Thermal Process Maps for Predicting Solidification Microstructure in Laser Fabrication of Thin-wall Structures. *J. Mater. Process. Technol.* **2006**, *178*, 135–142. [CrossRef]

71. Yadaiah, N.; Bag, S. Development of egg-configuration heat source model in numerical simulation of autogenous fusion welding process. *Int. J. Therm. Sci.* **2014**, *86*, 125–138. [CrossRef]

72. Xiang, Z.; Wang, L.; Yang, C.; Yin, M.; Yin, G. Analysis of the quality of slope surface in selective laser melting process by simulation and experiments. *Optik* **2019**, *176*, 68–77. [CrossRef]

Article

Numerical Study on Thermodynamic Behavior during Selective Laser Melting of 24CrNiMo Alloy Steel

Xiangpeng Luo **, Minghuang Zhao, Jiayi Li and Chenghong Duan ***

College of Mechanical and Electrical Engineering, Beijing University of Chemical Technology, North Third Ring Road 15, Chaoyang District, Beijing 100029, China; xpluo@mail.buct.edu.cn (X.L.); zhaomh26@163.com (M.Z.); 18811409192@163.com (J.L.)
* Correspondence: duanch@mail.buct.edu.cn; Tel.: +86-010-6443-6678

Received: 4 December 2019; Accepted: 18 December 2019; Published: 20 December 2019

Abstract: In this paper, a multi-layer and multi-track finite element model of 24CrNiMo alloy steel by selective laser melting (SLM) is established by using the ABAQUS software. The distribution and evolution of temperature field and stress field and the influence of process parameters on them are systematically studied. The results show that the peak temperature increases from 2153 °C to 3105 °C and the residual stress increases from 335 MPa to 364 MPa with increasing laser power from 200 W to 300 W; the peak temperature decreases from 2905 °C to 2405 °C and the residual stress increases from 327 MPa to 363 MPa with increasing scanning speed from 150 mm/s to 250 mm/s; the peak temperature increases from 2621 °C to 2914 °C and the residual stress decreases from 354 MPa to 300 MPa with increasing preheating temperature from 25 °C to 400 °C. Far away from scanning area, far away from starting point, and the adjacent areas with vertical scanning direction, resulting in a uniform temperature distribution, help to reduce the residual stress. Due to the remelting effect, the interlayer scanning angle changing helps to release the residual stress of the former layer causing a smaller residual stress after redistribution.

Keywords: selective laser melting; finite element analysis; thermodynamic behavior; substrate preheating; scanning strategy; 24CrNiMo alloy steel

---

1. Introduction

As an academic term for 3D printing, rapid prototyping, and layer manufacturing, additive manufacturing (AM) combines materials processing and forming technology, computer-aided design, and so on. Based on the principle of discrete-stacking, this technology stacks metal materials or non-metal materials by layers to directly produce three-dimensional solid parts through software and control systems, by means of sintering, melting, spraying, photocuring, etc. Compared with the traditional subtractive manufacturing of cutting raw materials, AM is a manufacturing method that is able to process three-dimensional complex structural parts by adding materials point by point, line by line, and surface by surface [1–3]. Moreover, AM technology has the advantages of flexible processing, short production cycle, and high material utilization, and can manufacture any complex shape structure parts in theory [4–6]. Selective laser melting (SLM) is a technique, which utilizes a high-energy laser beam with a planned scanning strategy, used to scan pre-placed metal powders to make them melt rapidly. The powders are then cooled and solidified rapidly, and finally formed into solid components layer by layer [7,8].

SLM, one of the most important means of metal laser additive manufacturing, has attracted considerable attention in recent years, and its numerical simulation research has also become a hot issue in the industry. Due to the rapid movement of the laser heat source, the temperature of the material rapidly rises and then shows an extremely fast material solidification cooling rate. The temperature history of a locally fabricated area is extremely complicated, and the molten pool has a very short lifetime in a SLM process [9]. It is difficult to observe and record the transient temperature field, stress field, and molten pool morphology through experimental methods. Numerical simulation is an effective way to solve this problem.

To date, many scholars have carried out numerical simulation research on a metal powder SLM process. Chen et al. [10] established a multilayer finite element (FE) model to study the temperature field during a SLM process of TiB2/Ti6Al4V multi-material parts under various processing parameters. The simulation results showed that the maximum temperature gradient increased from 24.920 °C/μm to 37.754 °C/μm with increasing the laser power from 300 W to 450 W, but the maximum temperature gradient decreased from 33.884 °C/μm to 31.478 °C/μm with increasing the scanning speed from 400 mm/s to 1000 mm/s. Woo et al. [11] developed a three-dimensional FE model to investigate the thermal behavior during the SLM process of WC-reinforced H13 steel composite powder. The molten pool geometry under different parameters was analyzed based on the temperature field obtained by numerical simulation. The results showed that the distribution factor, packing efficiency, and absorption coefficient are the main factors affecting the molten pool structure. Ali et al. [12] developed a modeling approach to simulate the temperature field during a SLM process of Ti6Al4V powder. The validity of the model was validated by comparing the width and depth of the molten pool obtained by finite element analysis and experimental measurement. Based on this model, the effects of process parameters on cooling rates and temperature gradients inducing the accumulation of residual stress build-up was studied. Foroozmehr et al. [13] used a three-dimensional FE model to simulate the size of the molten pool in the SLM process. The model considered the penetration characteristics of laser beam on powder bed, and the depth depended on the thickness of powder. The temperature distribution, depth, width, and length of the molten pool of each track were studied, and the influence of scanning speed on them was analyzed. Wang et al. [14] developed a three-dimensional FE model to research the thermal behavior and residual stress during the SLM process of AlSi10Mg powder. The simulation results showed that the stress gradually increased during the SLM process due to the effect of heat accumulation, and the temperature gradient affected the final distribution of the residual stress. However, numerical studies have not yet been reported on the temperature field and stress field of 24CrNiMo alloy steel, an alloy steel material commonly used for processing brake discs for high-speed rail, fabricated by SLM, and the application of SLM technology in high-speed train and rail industry is rarely studied, which greatly restrict a wider application of this technology in the high-speed train and rail industry. Moreover, there are few researches on the influence of SLM scanning strategy on the temperature field and stress field.

In this paper, the distribution and evolution of temperature field and stress field of 24CrNiMo alloy steel by SLM are studied by finite element analysis method. The influence of different process parameters on temperature field distribution, temperature change rate, thermal stress evolution, and residual stress distribution is clarified. The influence rules and mechanisms of laser power, scanning speed, preheating temperature of substrate, subarea scanning strategy, and interlayer scanning angle changing on temperature field and stress field during a SLM process are revealed.

2. Modeling

2.1. Mathematical Model

According to the basic theory of heat transfer, heat conduction, heat convection, and heat radiation are the basic modes of heat transfer. In a SLM process, metal powders melt rapidly and then solidify rapidly. The analysis of temperature field in this process is a non-linear transient thermal analysis, as shown in Figure 1.

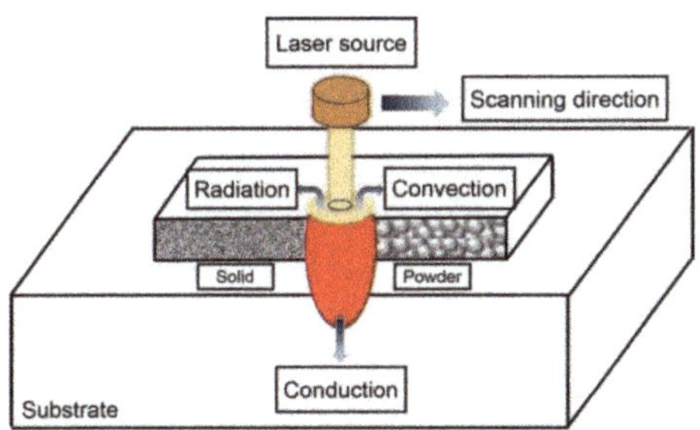

Figure 1. Schematic diagram of heat transfer during selective laser melting (SLM) process.

For transient heat issues, based on Fourier heat transfer law and energy conservation law, and taking material anisotropic into account, the three-dimensional non-linear transient heat transfer control equation is

$$\rho c \frac{\partial T}{\partial t} = \frac{\partial}{\partial x}(k_x \frac{\partial T}{\partial x}) + \frac{\partial}{\partial y}(k_y \frac{\partial T}{\partial y}) + \frac{\partial}{\partial z}(k_z \frac{\partial T}{\partial z}) + Q, \tag{1}$$

where k_x, k_y, k_z are the thermal conductivity in x, y, z direction, T is the powder temperature, Q is the heat flux density, ρ is the density, and c is the specific heat capacity.

Heat transfer between powder surface and ambient environment is heat convection and radiation. It can be described as follows:

$$-k \frac{\partial T}{\partial n} = h(T_a - T_s) + \sigma \varepsilon (T_a^{4} - T_s^{4}), \tag{2}$$

where k is the thermal conductivity, n is the normal vector to the powder surface, T_a is the powder temperature, T_s is the ambient temperature, h is the convection heat transfer coefficient, σ is the Stephen-Boltzmann constant (5.67×10^{-8} W/m^2·K^4), and ε is the thermal emissivity.

2.2. Boundary Conditions

Considering the radiation impact of the laser on the powder, it is used as the correction coefficient of the boundary heat transfer. Moreover, the Equation (3) can be obtained by mathematical transformation of the Equation (2).

$$-k \frac{\partial T}{\partial n} = \beta \left[h + \sigma \varepsilon (T_a^{3} + T_a^{2} T_s + T_a T_s^{2} + T_s^{3}) \right] \times (T_a - T_s). \tag{3}$$

Total heat transfer coefficient is defined by:

$$H = \beta \left[h + \sigma \varepsilon (T_a^{3} + T_a^{2} T_s + T_a T_s^{2} + T_s^{3}) \right]. \tag{4}$$

According to the experimental results of the temperature field from Reference [14], H at lower temperature is set to be thousands of times of that at steady state, and H at higher temperature is set to 80 times of that at steady state in this paper. Furthermore, at steady state, $\beta = 1$.

The boundary condition is set to be three-point constraint [15], that is the three vertex constraints on the bottom of the substrate are $U_x = U_y = U_z = 0$, $U_y = U_z = 0$, $U_z = 0$, respectively.

2.3. Heat Source Model

The double ellipsoidal heat source model used in this paper can be described as follows [16]:

$$q(x, y, z) = \frac{6\sqrt{3}P\eta}{\pi\sqrt{\pi}abc} e^{(-3\frac{x^2}{a^2} - 3\frac{y^2}{b^2} - 3\frac{z^2}{c^2})}, \tag{5}$$

where P is the laser power, η is the laser absorptivity, a, b, and c are the shape parameters of the heat source, x, y, and z are the local coordinates.

2.4. FE Model

In this paper, two models are used for analysis in ABAQUS (version 6.14.4, Dassault Systèmes, Paris, France), both of them include the substrate and the powder bed. As shown in Figure 2a, the dimensions of the substrate in FE model 1 are 2.0 mm × 1.2 mm × 0.3 mm, and the dimensions of the powder bed are 1.4 mm × 0.6 mm × 0.1 mm with two layers, the height of each layer is 0.05 mm. As shown in Figure 2b, the dimensions of the substrate in FE model 2 are 1.80 mm × 1.80 mm × 0.5 mm, and the dimensions of the powder bed are 1.2 mm × 1.2 mm × 0.10 mm with two layers, the height of each layer is 0.05 mm. Considering the computational efficiency and accuracy, the powder bed is meshed by DC3D8 element, which is an eight-node linear heat transfer brick element, with dimensions of 0.02 mm × 0.02 mm × 0.02 mm, and the substrate is gradually coarsely meshed away from the powder layer. The C3D8R element, an eight-node linear brick element, is used to calculate the stress field by applying the indirect coupling method based on the temperature field.

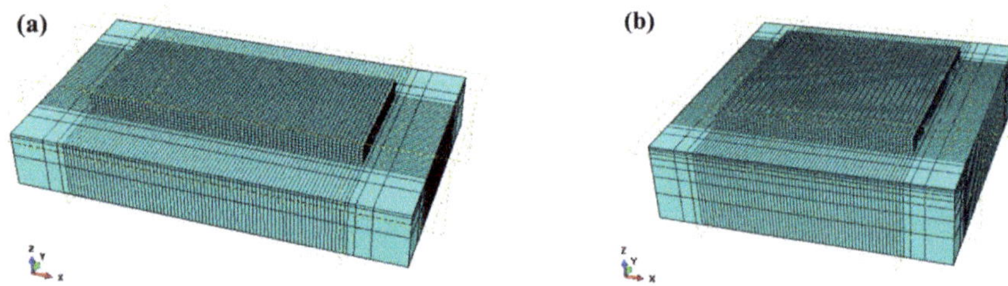

Figure 2. Finite element (FE) model: (**a**) model 1; (**b**) model 2.

2.5. Material Properties

In this paper, the substrate material is 45 steel whose specific material properties are given in Reference [17]. The metal powder is 24CrNiMo alloy steel whose chemical composition is listed in Table 1.

Table 1. Chemical composition of 24CrNiMo alloy steel powder.

Element	Fe	Cr	Ni	Mo	C	Mn	Al	P	S	Si	Nb
Content	Bal.	0.86	1.11	0.50	0.23	1.12	0.04	0.015	0.08	0.55	0.03

The physical properties of 24CrNiMo alloy steel vary significantly with temperature, which should be considered in the simulation. In this paper, the temperature-dependent physical properties of 24CrNiMo alloy steel in solid state are obtained by the commercial software JMatPro (version 9.0, Sente Software Ltd., Surrey, UK), which is a powerful and integrated software for material properties simulation [18–20], as shown in Figure 3.

It is known that the temperature-dependent physical properties of 24CrNiMo alloy steel in powder state are quite different from those in solid state, including density, thermal conductivity, elastic modulus, and thermal expansion coefficient.

The effective thermal conductivity of the powder bed can be calculated by the Yagi-Kunii model [21]. It is described as follows:

$$\frac{k_e}{k_g} = \frac{\beta(1-\varphi)}{\gamma\left(\frac{k_g}{k_s}\right) + \frac{1}{\frac{1}{\varphi} + \frac{D_p h_{rs}}{k_g}}} + \varepsilon\beta\frac{D_p h_{rv}}{k_g}, \tag{6}$$

where φ is the porosity of the powder, D_p is the average diameter of the powder particles, γ is the ratio of the solid effective length to the solid average diameter related to heat conduction, β is the ratio of the average length to average stacking diameter between two adjacent particles in the direction of heat flux, k_e, k_g, and k_s are the effective thermal conductivity of the powder bed, the gas in the pores, and the solid material, h_{rs} is the thermal emissivity between adjacent particle surfaces, and h_{rv} is the thermal emissivity between adjacent gases.

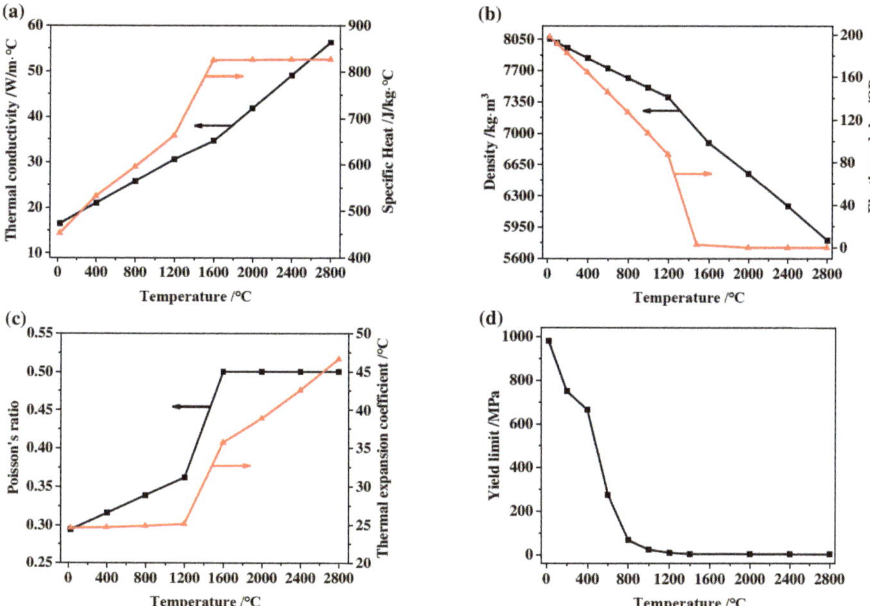

Figure 3. (a) Thermal conductivity and specific heat, (b) density and elastic modulus, (c) Poisson's ratio and thermal expansion coefficient, and (d) yield limit of 24CrNiMo alloy steel.

The density of the powder bed can be calculated by Equation (7). It is described as follows:

$$\rho_p = (1 - \varphi)\rho_s, \tag{7}$$

where ρ_p is the density of the powder material, ρ_s is the density of the solid material.

In the actual SLM process, Marangoni convection exists in the molten pool, resulting in the intensification of energy transfer, which has a significant effect on temperature and stress distribution. Therefore, the thermal conductivity above the melting point is modified by an enhancement factor to approximately consider the effects of Marangoni convection on heat transfer in the molten pool [22,23]. The modified thermal conductivity can be expressed as

$$k^*(T) = \begin{cases} k(T) & T \le T_m \\ ak(T) & T \ge T_m \end{cases}, \tag{8}$$

where $k(T)$ is the actual thermal conductivity, $k^*(T)$ is the modified thermal conductivity, and T_m is the melting temperature.

2.6. Latent Heat of Phase Transformation

In order to improve the calculation accuracy, the phase transformation latent heat of 24CrNiMo alloy steel during the SLM process is considered by applying the equivalent specific heat method.

In the temperature range of phase transformation, the specific heat of the material can be modified by using the phase transformation latent heat [24]:

$$C_P^* = C_P + \frac{L}{T_L - T_s},$$

(9)

where L is the phase transformation latent heat, T_L is the liquidus temperature, T_S is the solidus temperature, C_P is the actual specific heat, and C_P^* is the modified specific heat.

3. Results and Discussion

Firstly, it should be noted that the results of Sections 3.1 and 3.2 are based on FE model 1 (Figure 2a) and the results of Section 3.3 is based on FE model 2 (Figure 2b).

3.1. Distribution and Variation Rules of Temperature Field and Stress Field

Figure 4 shows the temperature distribution contour of the whole model and the top surface and cross section of the molten pool under P = 250 W and v = 200 mm/s when the laser beam reaches the center of the first layer. It demonstrates that the peak temperature of the molten pool is 2621 °C and the isotherms on the front of the heat source are denser than those in the back, showing that the temperature gradient on the front of the heat source is larger. The main reason is that the front material of the heat source is still in powder state, and the heat cannot be transferred quickly because of its lower thermal conductivity, which results in a larger temperature gradient and a denser isotherm. The powder in the back of the heat source was transformed into solid, which increases the thermal conductivity and makes a faster heat transfer. The dotted line is the area where the temperature is higher than the melting point of the material (1435 °C), which can be used to characterize the structure of the molten pool. The length of the molten pool is about 296 µm, the width is about 177 µm, the depth is about 66 µm, the ratio of length to width is about 1.7, and the shape of the molten pool is similar to that of water droplets.

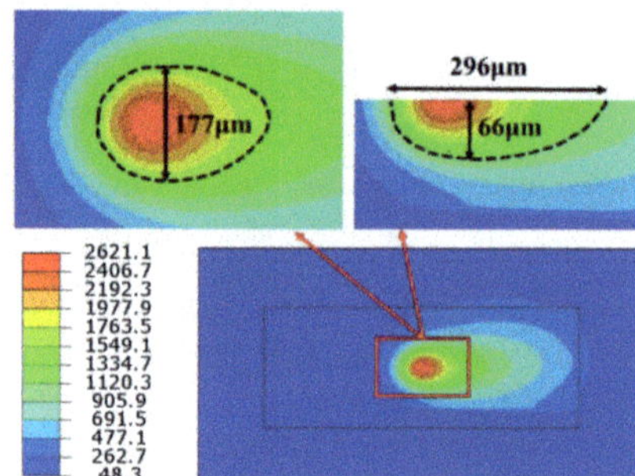

Figure 4. Temperature field distribution contour when the laser beam reaches the center of the first layer.

Figure 5a shows the schematic diagram of the midpoint (P1–P6) on the top surface of each scanning tracks. The peak temperatures at the midpoint of the three tracks in the first layer are 2466 °C, 2621 °C, and 2722 °C, respectively, and those of the three tracks in the second layer are 2605 °C, 2725 °C, and 2818 °C, respectively, as shown in Figure 5b. As the laser scanning progresses the peak temperature increases with the increase of scanning tracks and scanning layers, but the increase amplitude decreases gradually.

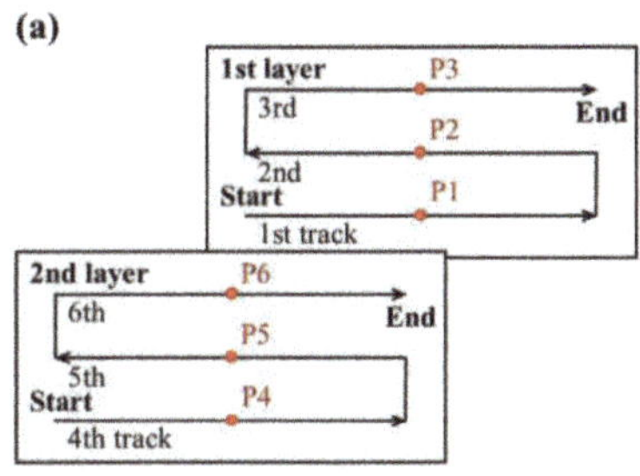

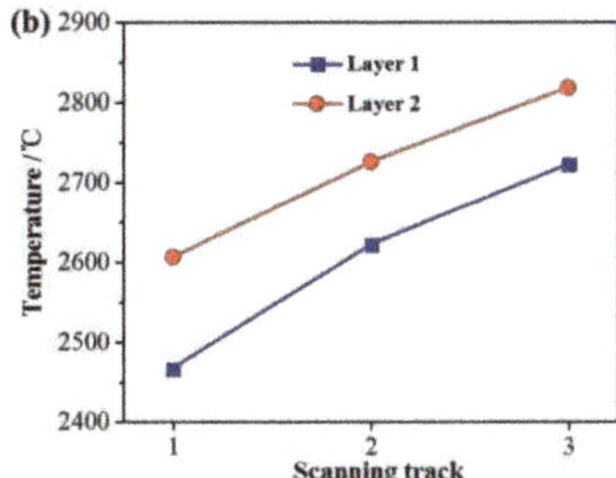

Figure 5. (**a**) Schematic diagram of the midpoint (P1–P6) on the top surface of each scanning tracks; (**b**) peak temperature at the midpoint of each scanning tracks.

Figure 6 shows the three dimensions of the molten pool at the midpoint of each scanning track. The solid lines represent the length of the molten pool at the midpoint of each scanning track, the dashed lines represent the width of the molten pool at the midpoint of each scanning track, and the dotted lines represent the depth of the molten pool at the midpoint of each scanning track. As demonstrated, with the increase of scanning tracks, the length, width, and depth of the molten pool increase gradually. Furthermore, the three dimensions of the second layer molten pool are larger than those of the first layer to a certain extent. Taking the first layer as an example, the length, width, and depth of the molten pool at the midpoint of the first scanning track are 236 μm, 162 μm, and 59 μm, respectively. The length, width, and depth of the molten pool at the midpoint of the second scanning track are 25.4%, 9.3%, and 11.9% larger than those at the midpoint of the first scanning track, respectively. Moreover, the length, width, and depth of the molten pool at the midpoint of the third track are 14.9%, 7.3%, and 10.6% larger than those at the midpoint of the first scanning track, respectively. However, the increase amplitude of the three dimensions of the molten pool decreases gradually with the increase of the scanning tracks, as shown in Figure 7. Taking the second track of the two layers as an example, the length, width, and depth of the molten pool at the midpoint of the second track in the first layer are 23.6%, 13.0%, and 13.6% larger than those at the midpoint of the second track in the second layer, respectively. Combining with the temperature field, it is found that the three dimensions of the molten pool with different scanning tracks and layers increase in accordance with the variation of the peak temperature of the molten pool (see Figure 5), which indicates that the three dimensions of the molten pool are greatly affected by the temperature distribution of the molten pool. The reason for the decrease of the increase is that the heat transfer is slowed down with the increase of scanning tracks and layers.

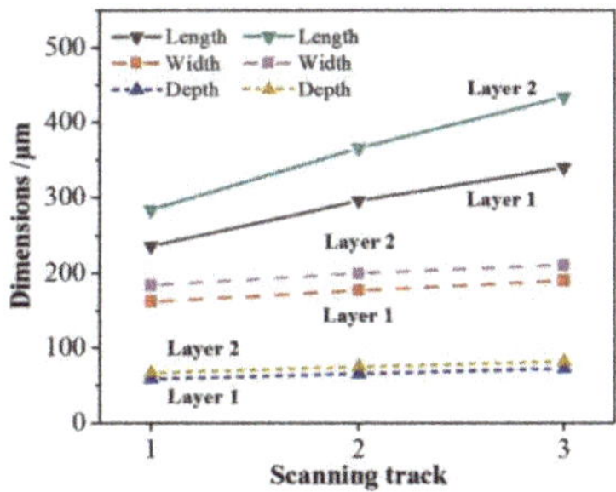

Figure 6. Three dimensions of the molten pool at the midpoint of each scanning tracks.

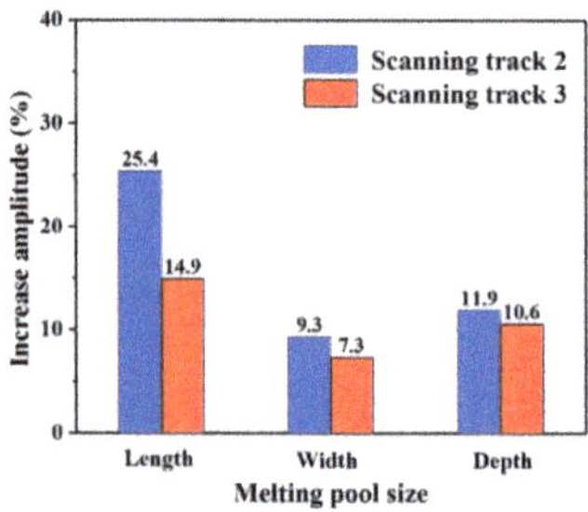

Figure 7. Increase amplitude of the molten pool size.

When the laser beam scans the powder, the irradiated area of the spot absorbs a large amount of heat so that the temperature increases rapidly. After reaching the melting point, the powder state changes to a liquid state and then a molten pool is formed. At this time, the liquid phase is purely plastic, and the Mises stress is small and is approximately considered to be zero, as shown in Figure 8. When the laser beam is far away from this area, the temperature of the position decreases rapidly and the metal of the molten pool begins to solidify gradually. The material shrinks due to "thermal expansion and contraction", and is bound by the solidified metal around it. The large temperature gradient results in the thermal stress increasing gradually. When the thermal stress is higher than the yield limit of the material, the solidified area is plastically deformed. As the molten pool moves continuously, the scanning area cools gradually. When the temperature tends to be stable, the thermal stress redistributes and remains in the SLMed parts.

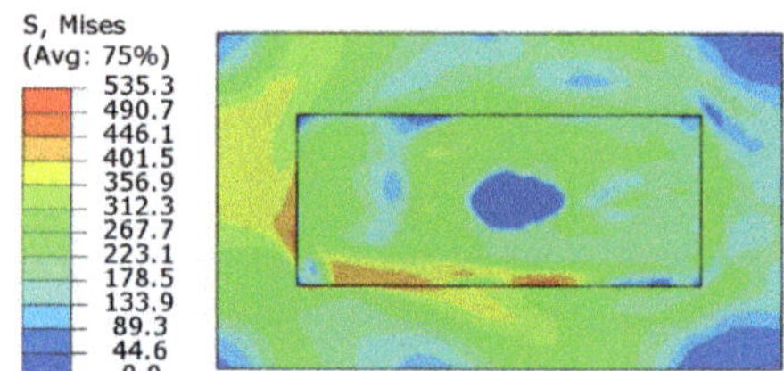

Figure 8. Stress field distribution contour when the laser beam reaches the center of the first layer.

Figure 9 shows the evolution of stress at the center of the first layer. When the laser beam scans the center of the first layer, the high temperature gradient results in a larger thermal stress before and after the molten pool, reaching more than 800 MPa. Moreover, the stress stabilizes at about 400 MPa with the cooling process at the end of the first layer. When the second layer is scanned, the position undergoes a similar temperature evolution with the first layer, but the temperature gradient decreases so that the peak stress decreases. Due to material shrinkage, the stress slightly increases during the cooling processing. The residual stress is released by a post-thermal effect and finally stabilizes at about 350 MPa.

Figure 10 shows the residual stress distribution contour after cooling. The maximum Mises stress is about 355 MPa, which appears at the joint edge of the substrate and the formed layer because of the large temperature gradient between them and the difference in thermal expansion coefficients of the materials. The high stress area is mainly concentrated in the formed layer, and the substrate stress is relatively low. It can be seen from Figure 11b,c that the residual tensile stress mainly occurs in the formed layer, and the X-direction residual stress is slightly higher than the Y-direction residual stress because the X-direction is the scanning direction, and the large cooling rate causes the stress to be parallel to the scanning direction. As can be seen from Figure 10d, the Z-direction residual stress is mainly concentrated between the substrate and the formed layer, which tends to cause warping and cracking.

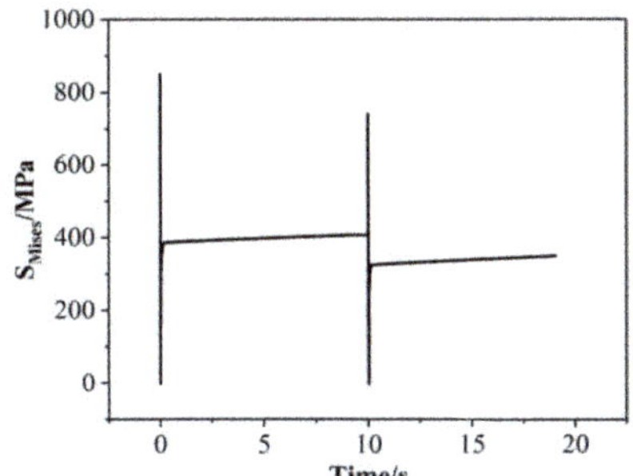

Figure 9. Evolution of stress at the center of the first layer.

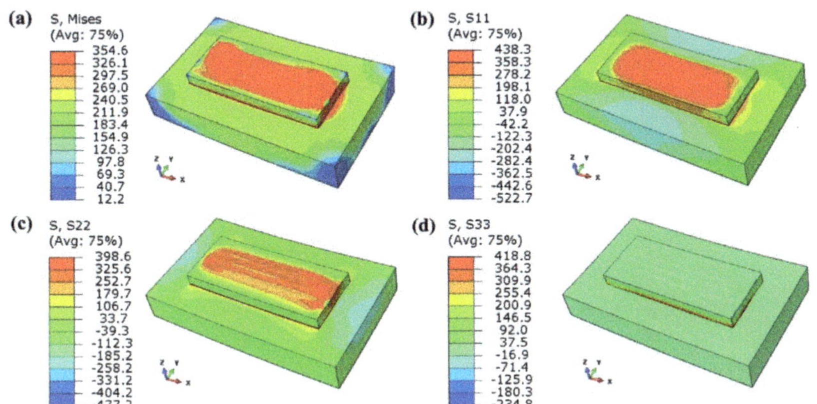

Figure 10. Residual stress distribution contour: (**a**) Von Mises stress, (**b**) X-direction stress, (**c**) Y-direction stress, and (**d**) Z-direction stress.

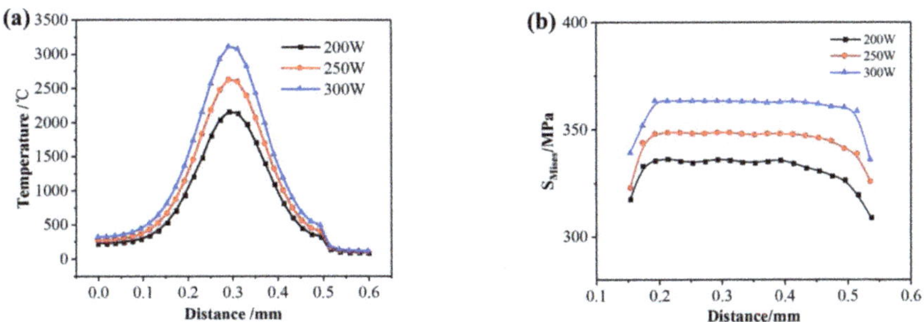

Figure 11. (**a**) Temperature distribution; (**b**) stress distribution under different laser powers.

3.2. Effects of Process Parameters on Temperature Field and Stress Field

Figure 11a shows the temperature distribution along the second scanning track under different laser powers when the laser beam scans the center of the first layer. The temperature distribution tends to follow the same rules, showing characteristics of high in the middle and low on both sides. The temperature and temperature gradient near the molten pool are relatively high in the middle area. When the laser power increases from 200 W to 300 W, the peak temperature of the molten pool increases accordingly, and the central temperature increases from 2153 °C to 3105 °C because the laser energy density and the heat input increases. Figure 11b shows the residual stress distribution of the second scanning track under different laser powers when scanning the center of the first layer. It can be seen that the Mises stress distribution rules are similar, and the stress in the middle of the scanning track is close to each other. As the laser power increases from 200 W to 300 W, the maximum of Mises stress increases from 335 MPa to 364 MPa. The main reason is that the increase of laser power has a great influence on the temperature gradient of the molten pool so that the growth rate increases from 6.9×10^3 °C/mm to 1.2×10^4 °C/mm during the cooling process.

Figure 12a shows the temperature distribution along the second scanning track under different scanning speeds when the laser beam scans the center of the first layer. When the scanning speed increases from 150 mm/s to 200 mm/s, the peak temperature of the molten pool decreases accordingly, and the central temperature decreases from 2905 °C to 2405 °C because the laser energy density and the heat input per unit time decreases. Figure 12b shows the residual stress distribution of the second scanning track under different laser speeds when the laser beam scans the center of the first layer. As the scanning speed increases from 150 mm/s to 200 mm/s, the maximum of Mises stress increases from 327 MPa to 363 MPa. The main reason is that the increase of scanning speed has a great influence on the temperature gradient of the molten pool so that the growth rate increases from 7.7×10^3 °C/mm to 1.1×10^4 °C/mm during the cooling process.

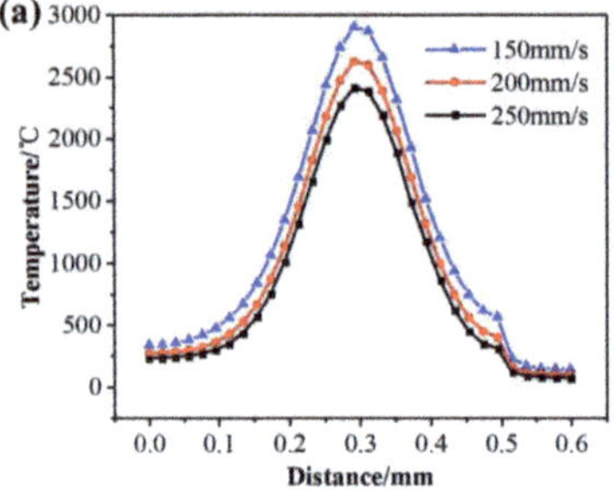

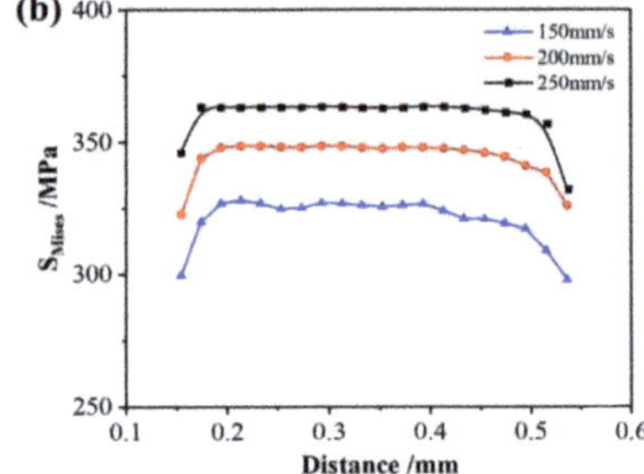

Figure 12. (**a**) Temperature distribution; (**b**) stress distribution under different scanning speeds.

Figure 13a shows the temperature distribution along the second scanning track under different preheating temperatures when the laser beam scans the center of the first layer. When the preheating temperature increases from 25 °C to 400 °C, the peak temperature of the molten pool increases accordingly, and the central temperature increases from 2621 °C to 2914 °C because of the increase of heat accumulation. Figure 13b shows the residual stress distribution of the second scanning track under different preheating temperatures when the laser beam scans the center of the first layer. As the preheating temperature increases from 25 °C to 400 °C, the maximum of Mises stress decreases from 354 MPa to 300 MPa. The main reason is that preheating of the substrate helps to reduce temperature gradient. It also can be seen that the increase of the preheating temperature in a certain range makes the overall distribution of residual stress show a significant downward trend, and the high stress range of the formed area is gradually reduced, indicating that the preheating of the substrate helps to reduce the residual stress and the cracking tendency of SLMed parts.

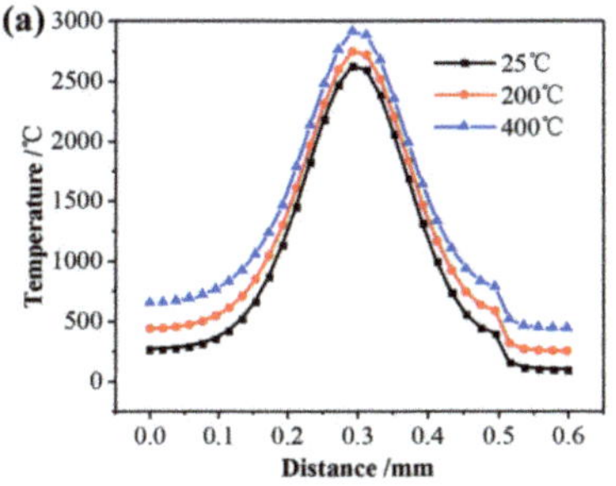
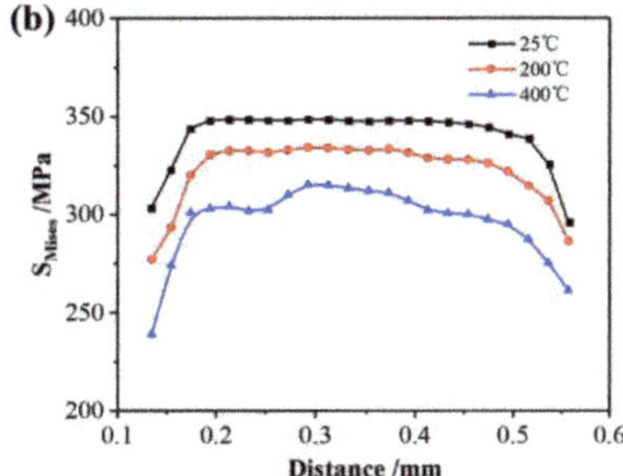

Figure 13. (**a**) Temperature distribution and (**b**) stress distribution under different preheating temperatures.

3.3. Effects of Scanning Strategy on Temperature Field and Stress Field

Some differences exist in the temperature field under different subarea scanning strategies in the SLM process. The specific subarea scanning strategies are shown in Figure 14, and the scanning area is divided into four areas in which 1, 2, 3, and 4 represent the scanning sequence, "start" represents the starting point of scanning, and points A, B, C, and D are the area centers.

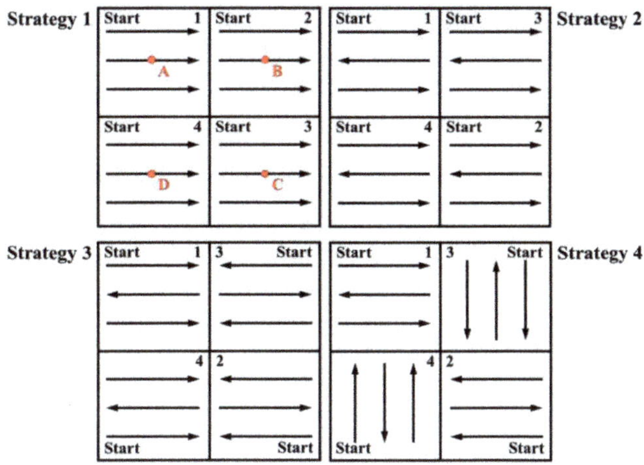

Figure 14. Schematic diagram of subarea scanning strategies.

Figure 15a shows the peak temperatures of points A, B, C, and D under different subarea scanning strategies. The temperature of strategies 1, 2, 3, and 4 gradually increases due to the heat accumulation effect. Comparing Strategy 1 with Strategy 2, Strategy 2 has a lower temperature in Area 2 and 4, which indicates that the diagonal scanning can make the scanning area as far as possible to reduce the temperature increase caused by the thermal influence of the scanned area. Comparing Strategy 2 and Strategy 3, the temperature in Area 3 and 4 of Strategy 3 slightly decreases, indicating that the method of scanning from the edge to the center can also reduce the preheating effect. Comparing Strategy 3 and Strategy 4, Strategy 4 has a lower temperature in Area 3 and 4, indicating vertical scanning direction of adjacent areas that can reduce the influence of thermal accumulation from adjacent areas. In summary, far away from the scanning area, far away from the scanning starting point and adjacent areas with vertical scanning direction are the scanning strategies, which could result in a more uniform temperature distribution and a lower average temperature (as shown in Figure 15b), and help to reduce the residual stress caused by temperature difference.

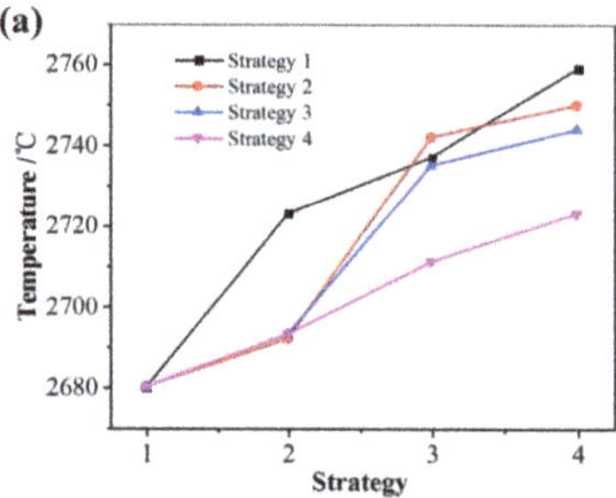
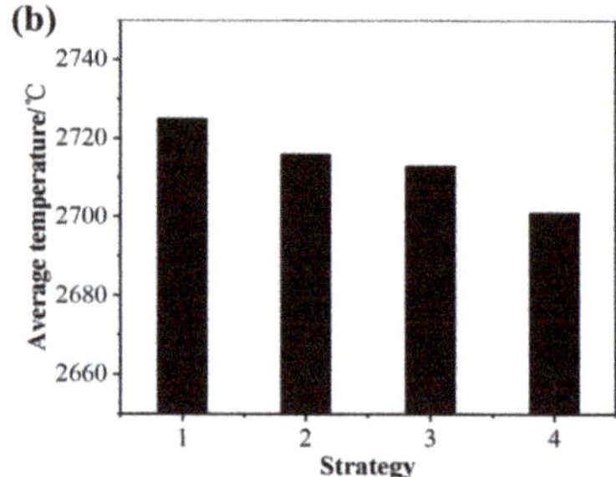

Figure 15. (**a**) Peak temperature of the area center and (**b**) average temperature under different subarea scanning strategies.

Figure 16 shows the schematic diagram of the interlayer scanning angle changing. They have the same scanning starting point, the same scanning length of each tracks, and the same number of tracks. Figure 17 shows the residual stress distribution contour for scanning only the first layer after cooling. It demonstrates that the maximum Mises stress is about 366 MPa, and the stress distribution of the formed layer is relatively uniform, basically above 300 MPa. The substrate stress is lower, less than 200 MPa. The maximum tensile stresses in the X-direction, Y-direction, and Z-direction are 435 MPa, 415 MPa, and 371 MPa, respectively. The maximum compressive stresses in the X-direction, Y-direction, and Z-direction are 337 MPa, 444 MPa, and 190 MPa, respectively. Moreover, a large stress concentration occurs in the place where the formed layer is in contact with the substrate, which may cause cracking and warping.

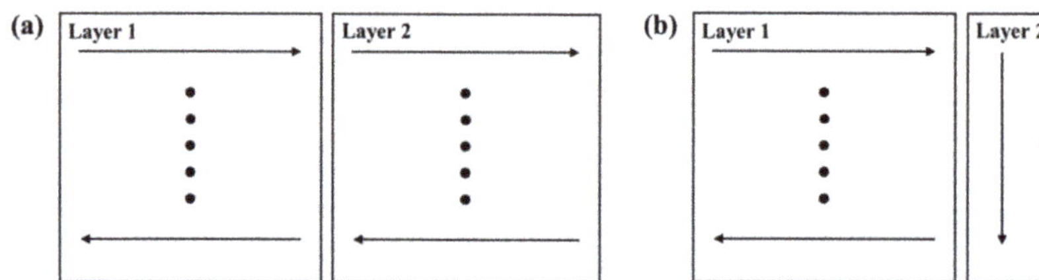

Figure 16. Schematic diagram of (**a**) no interlayer scanning angle and (**b**) 90° interlayer scanning angle.

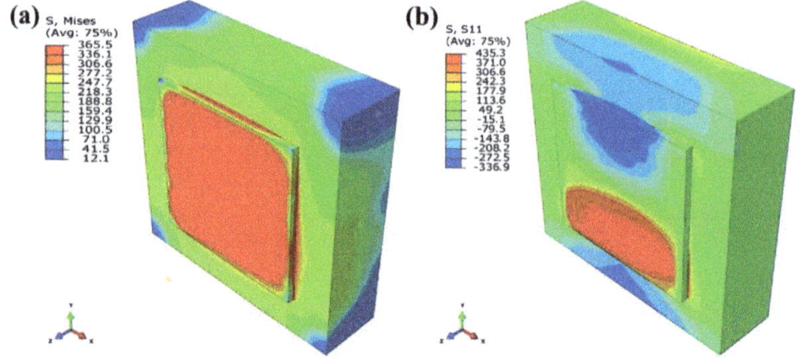

Figure 17. *Cont.*

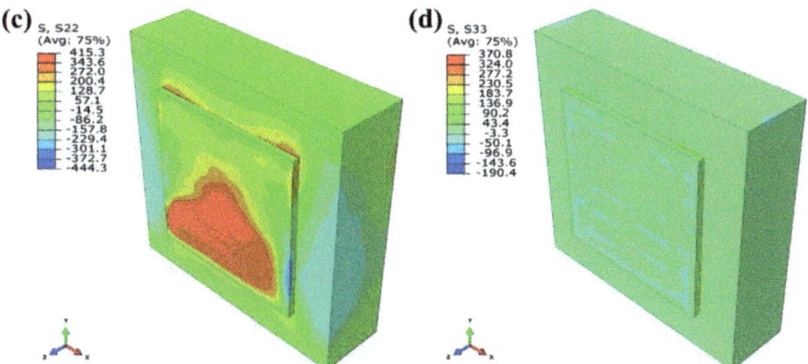

Figure 17. Residual stress distribution contour for scanning only the first layer: (**a**) Von Mises stress, (**b**) X-direction stress, (**c**) Y-direction stress, and (**d**) Z-direction stress.

Figure 18 shows the residual stress distribution contour for scanning two layers after cooling with no interlayer scanning angle and 90° interlayer scanning angle. It can be seen that the maximum Mises stress with no interlayer scanning angle is 332 MPa, and that of 90° interlayer scanning angle is 328 MPa. The maximum tensile stresses in the X-direction, Y-direction, and Z-direction with no interlayer scanning angle are 392 MPa, 376 MPa, and 392 MPa, respectively; the maximum compressive stresses in the X-direction, Y-direction, and Z-direction with no interlayer scanning angle are 329 MPa, 446 MPa, and 201 MPa, respectively; the maximum tensile stresses in the X-direction, Y-direction, and Z-direction with 90° interlayer scanning angle are 374 MPa, 398 MPa, and 398 MPa, respectively; the maximum compressive stresses in the X-direction, Y-direction, and Z-direction with 90° interlayer scanning angle are 459 MPa, 317 MPa, and 207 MPa, respectively. Meanwhile, by analyzing the residual stress values of the first layer, it is found that the stress with no interlayer scanning angle is lower than that with 90° interlayer scanning angle. The interlayer remelting under the 90° interlayer scanning angle causes the residual stress of the first layer to be released multiple times so that the stress of the first layer is reduced after redistribution. Therefore, the interlayer scanning angle changing helps to reduce the tendency of cracking by reducing the residual stress of the formed layer.

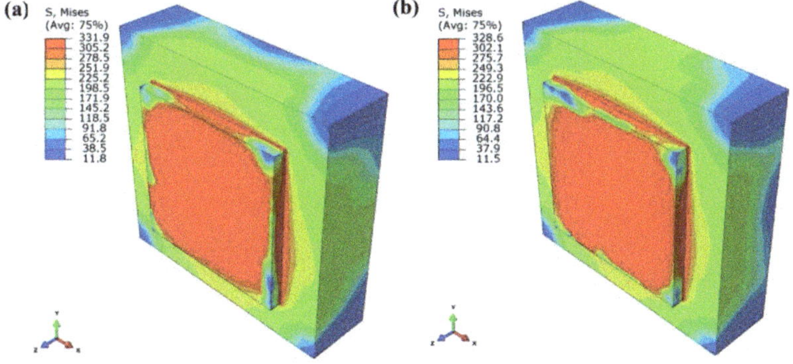

Figure 18. *Cont.*

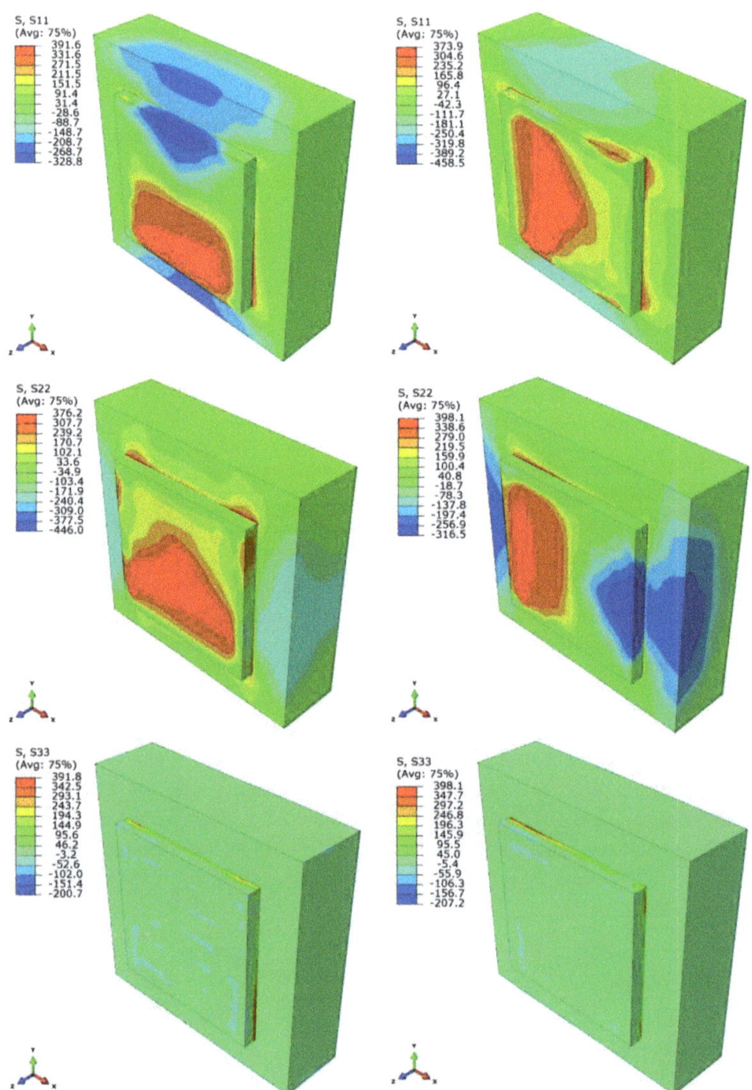

Figure 18. Residual stress distribution contour for scanning two layers with (**a**) no interlayer scanning angle and (**b**) 90° interlayer scanning angle.

3.4. Experimental Verification

Figure 19 shows the experimental and numerical molten pool morphology on the cross section of SLM processed specimens under P = 250 W and v = 200 mm/s. The average molten pool depth and half of the average molten pool width obtained by experiments are 76.5 μm and 111.5 μm, respectively, and those by numerical simulations are 79.5 μm and 109.5 μm, with errors of 5.8% and 2.9%, respectively (as shown in Table 2). It can be seen that the results of numerical simulation and experimental measurements are consistent with each other, which proves that the reliability and accuracy of the numerical analysis is good.

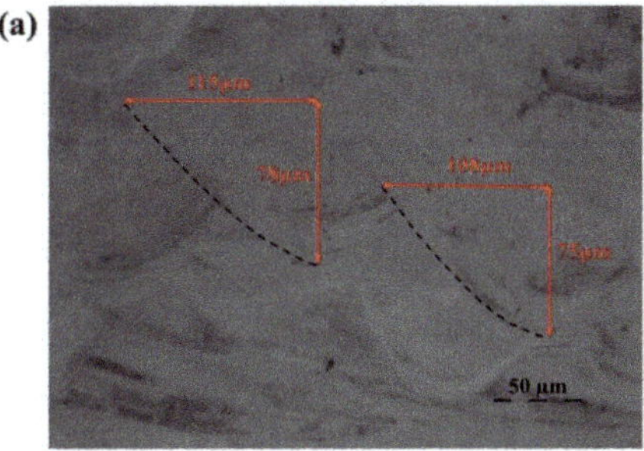
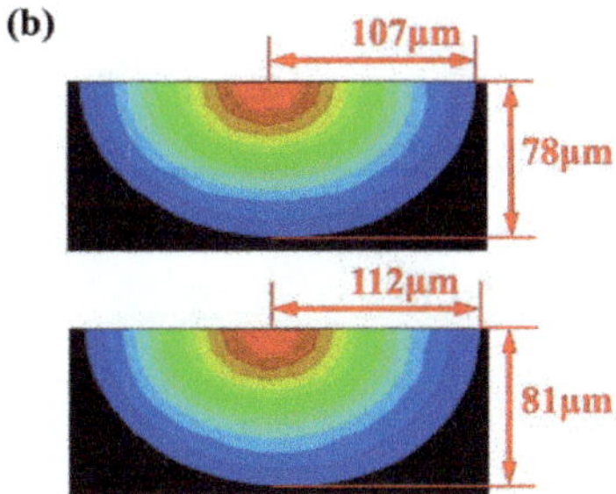

Figure 19. Comparisons of experimental (**a**) and numerical (**b**) morphology on the cross section of the molten pool.

Table 2. Comparison of experimental and numerical molten pool dimensions.

Molten Pool Dimensions	Experimental Results (µm)	Numerical Results (µm)	Error
The average molten pool depth	76.5	79.5	3.9%
Half of the average molten pool width	111.5	109.5	1.8%

4. Conclusions

In this paper, the distribution and evolution of temperature field and stress field of 24CrNiMo alloy steel by SLM are studied by numerical simulation. The influence of process parameters (laser power, scanning speed, preheating temperature, and scanning strategy) on temperature field distribution, temperature change rate, thermal stress evolution, and residual stress distribution is analyzed, and the relevant experimental verification is carried out. The following conclusions are drawn:

(1) The peak temperature at the center of the first layer is 2621 °C, the length of the molten pool is about 296 µm, the width is about 177 µm, the depth is about 66 µm, the aspect ratio is about 1.7, and the shape is similar to that of water droplets. The increase of thermal accumulation leads to the increase of the peak temperature with the increase of scanning tracks and scanning layers, but the increase of that decreases gradually. The temperature field of the molten pool has great influence on the molten pool size. Temperature gradient results in large thermal stress around the molten pool, which reaches more than 800 MPa at the center of the first layer and stabilizes at about 400 MPa after the first layer is scanned. When the second layer is scanned, the temperature gradient decreases, and the residual stress is released post-thermally and finally stabilizes at about 350 MPa.

(2) When the laser power increases from 200 W to 300 W, the peak temperature of the molten pool increases from 2153 °C to 3105 °C, the maximum of temperature gradient increases from 6.9×10^3 °C/mm to 1.2×10^4 °C/mm during the cooling process, and the Mises stress increases from 335 MPa to 364 MPa. When the scanning speed increases from 150 mm/s to 200 mm/s, the peak temperature of the molten pool decreases from 2905 °C to 2405 °C, the maximum of temperature gradient increases from 7.7×10^3 °C/mm to 1.1×10^4 °C/mm during the cooling process, and the maximum of Mises stress increases from 327 MPa to 363 MPa. When the preheating temperature increases from 25 °C to 400 °C, the peak temperature of the molten pool increases from 2621 °C to 2914 °C, and the maximum of Mises stress decreases from 354 MPa to 300 MPa. Therefore, using appropriate process parameters help to reduce the residual stress.

(3) Far away from scanning area, far away from the scanning starting point and adjacent areas with vertical scanning direction are the scanning strategies, which could result in a more uniform temperature distribution, and help to reduce the residual stress caused by temperature difference. The interlayer remelting under the 90° interlayer scanning angle causes the residual stress of the first layer to be released multiple times so that the stress of the first layer is reduced after redistribution.

Therefore, the interlayer scanning angle changing helps to reduce the tendency of cracking by reducing the residual stress.

Author Contributions: Methodology, X.L. and M.Z.; software, M.Z. and J.L.; validation, M.Z.; investigation, M.Z. and X.L.; resources, C.D.; data curation, M.Z. and J.L.; writing—original draft preparation, X.L. and M.Z.; writing—review and editing, X.L. and C.D.; visualization, M.Z. and J.L.; supervision, X.L. and C.D.; project administration, C.D.; funding acquisition, C.D. All authors have read and agreed to the published version of the manuscript.

Funding: This research was funded by the National Key Research and Development Program of China, grant number 2016YFB1100202.

Acknowledgments: The authors gratefully acknowledge the financial support from the National Key Research and Development Program of China (No.2016YFB1100202).

References

1. Buchbinder, D.; Meiners, W.; Pirch, N.; Wissenbach, K. Investigation on reducing distortion by preheating during manufacture of aluminum components using selective laser melting. *J. Laser Appl.* **2014**, *26*, 012004. [CrossRef]

2. Xia, M.J.; Gu, D.D.; Ma, C.L.; Chen, H.Y.; Zhang, H.M. Microstructure evolution, mechanical response and underlying thermodynamic mechanism of multi-phase strengthening WC/Inconel 718 composites using selective laser melting. *J. Alloy. Compd.* **2018**, *747*, 684–695. [CrossRef]

3. Zhang, X.; Yocom, C.J.; Mao, B.; Liao, Y.L. Microstructure evolution during selective laser melting of metallic materials: A review. *J. Laser Appl.* **2019**, *31*, 031201. [CrossRef]

4. Suryawanshi, J.; Prashanth, K.G.; Ramamurty, U. Mechanical behavior of selective laser melted 316L stainless steel. *Mater. Sci. Eng. A* **2017**, *696*, 113–121. [CrossRef]

5. Gu, D.D.; Dai, D.H. Role of melt behavior in modifying oxidation distribution using an interface incorporated model in selective laser melting of aluminum-based material. *J. Appl. Phys.* **2016**, *120*, 083104. [CrossRef]

6. Guan, T.T.; Chen, S.Y.; Chen, X.T.; Liang, J.; Liu, C.S.; Wang, M. Effect of laser incident energy on microstructures and mechanical properties of 12CrNi2Y alloy steel by direct laser deposition. *J. Mater. Sci. Technol.* **2019**, *35*, 395–402. [CrossRef]

7. Wang, L.Z.; Wei, W.H. Selective laser melting of 30CrMnSiA steel: Laser energy density dependence of microstructural and mechanical properties. *Acta Metall. Sin. Engl. Lett.* **2018**, *31*, 807–814. [CrossRef]

8. Mueller, M.; Riede, M.; Eberle, S.; Reutlinger, A.; Brandão, A.D.; Pambaguian, L.; Seidel, A.; López, E.; Brueckner, F.; Beyer, E.; et al. Microstructural, mechanical, and thermo-physical characterization of hypereutectic AlSi40 fabricated by selective laser melting. *J. Laser Appl.* **2019**, *31*, 022321. [CrossRef]

9. Wei, M.W.; Chen, S.Y.; Xi, L.Y.; Liang, J.; Liu, C.S. Selective laser melting of 24CrNiMo steel for brake disc: Fabrication efficiency, microstructure evolution, and properties. *Opt. Laser Technol.* **2018**, *107*, 99–109. [CrossRef]

10. Chen, C.Y.; Gu, D.D.; Dai, D.H.; Du, L.; Wang, R.; Ma, C.L.; Xia, M.J. Laser additive manufacturing of layered TiB2/Ti6Al4V multi-material parts: Understanding thermal behavior evolution. *Opt. Laser Technol.* **2019**, *119*, 105666. [CrossRef]

11. Woo, Y.; Hwang, T.; Oh, I.; Seo, D.; Moon, Y. Analysis on selective laser melting of WC-reinforced H13 steel composite powder by finite element method. *Adv. Mech. Eng.* **2019**, *11*, 1–11. [CrossRef]

12. Ali, H.; Ghadbeigi, H.; Mumtaz, K. Residual stress development in selective laser-melted Ti6Al4V: A parametric thermal modelling approach. *Int. J. Adv. Manuf. Technol.* **2018**, *97*, 2621–2633. [CrossRef]

13. Foroozmehr, A.; Badrossamay, M.; Foroozmehr, E.; Golabi, S. Finite element simulation of selective laser melting process considering optical penetration depth of laser in powder bed. *Mater. Des.* **2016**, *89*, 255–263. [CrossRef]

14. Wang, L.F.; Jiang, X.H.; Zhu, Y.H.; Zhu, X.G.; Sun, J.; Yan, B. An approach to predict the residual stress and distortion during the selective laser melting of AlSi10Mg parts. *Int. J. Adv. Manuf. Technol.* **2018**, *97*, 3535–3546. [CrossRef]

15. Wen, S.; Dong, A.P.; Lu, H.Y.; Zhu, G.L.; Shu, D.; Sun, B.D. Finite element simulation of the temperature field and residual stress in GH536 superalloy treated by selective laser melting. *Acta Metall. Sin.* **2018**, *54*, 393–403.

16. Goldak, J.; Chakravarti, A.; Bibby, M. A new finite element model for welding heat sources. *Metall. Trans. B* **1984**, *15*, 299–305. [CrossRef]

17. Duan, C.H.; Hao, X.J.; Luo, X.P. Study on temperature field of selective laser melting 316L. *Appl. Laser* **2018**, *38*, 48–53.

18. Alhafadhi, M.H.; Krallics, G. Numerical simulation prediction and validation two dimensional model weld pipe. *Mach. Technol. Mater.* **2019**, *13*, 447–450.

19. Behrens, B.; Schröder, J.; Brands, D.; Scheunemann, L.; Niekamp, R.; Chugreev, A.; Sarhil, M.; Uebing, S.; Kock, C. Experimental and numerical investigations of the development of residual stresses in thermo-mechanically processed Cr-alloyed steel 1.3505. *Metals* **2019**, *9*, 480. [CrossRef]

20. Zhao, G.X.; Du, J.; Wei, Z.Y.; Geng, R.W.; Xu, S.Y. Numerical analysis of arc driving forces and temperature distribution in pulsed TIG welding. *J. Braz. Soc. Mech. Sci. Eng.* **2019**, *41*, 60. [CrossRef]

21. Yagi, S.; Kunii, D.; Wakao, N. Studies on axial effective thermal conductivities in packed beds. *AIChE J.* **1960**, *6*, 373–381. [CrossRef]

22. Du, L.; Gu, D.D.; Dai, D.H.; Shi, Q.M.; Ma, C.L.; Xia, M.J. Relation of thermal behavior and microstructure evolution during multi-track laser melting deposition of Ni-based material. *Opt. Laser Technol.* **2018**, *108*, 207–217. [CrossRef]

23. Kumar, S.; Roy, S.; Paul, C.P.; Nath, A.K. Three-dimensional conduction heat transfer model for laser cladding process. *Numer. Heat Transf. Part B Fundam.* **2008**, *53*, 271–287. [CrossRef]

24. Toyserkani, E.; Khajepour, A.; Corbin, S. 3-D finite element modeling of laser cladding by powder injection: Effects of laser pulse shaping on the process. *Opt. Laser Eng.* **2004**, *41*, 849–867. [CrossRef]

Article

Tri-Planar Geometric Dimensioning and Tolerancing Characteristics of SS 316L Laser Powder Bed Fusion Process Test Artifacts and Effect of Base Plate Removal

Baltej Singh Rupal [1,2], Tegbir Singh [3], Tonya Wolfe [3], Marc Secanell [2] and Ahmed Jawad Qureshi [1,*]

1   Additive Design and Manufacturing Systems (ADaMS) Lab, Department of Mechanical Engineering, University of Alberta, Edmonton, AB T6G 2G8, Canada; baltej@ualberta.ca
2   Energy Systems Design Lab (ESDL), Department of Mechanical Engineering, University of Alberta, Edmonton, AB T6G 2G8, Canada; secanell@ualberta.ca
3   InnoTech Alberta, Surface Engineering, Edmonton, AB T6N 1E4, Canada; Tegbir.singh@innotechalberta.ca (T.S.); tonya.wolfe@innotechalberta.ca (T.W.)
*   Correspondence: ajquresh@ualberta.ca; Tel.: +(780)-492-3609

Abstract: The precision of LPBF manufactured parts is quantified by characterizing the geometric tolerances based on the ISO 1101 standard. However, there are research gaps in the characterization of geometric tolerance of LPBF parts. A literature survey reveals three significant research gaps: (1) systematic design of benchmarks for geometric tolerance characterization with minimum experimentation; (2) holistic geometric tolerance characterization in different orientations and with varying feature sizes; and (3) a comparison of results, with and without the base plate. This research article focuses on addressing these issues by systematically designing a benchmark that can characterize geometric tolerances in three principal planar directions. The designed benchmark was simulated using the finite element method, manufactured using a commercial LPBF process using stainless steel (SS 316L) powder, and the geometric tolerances were characterized. The effect of base plate removal on the geometric tolerances was quantified. Simulation and experimental results were compared to understand tolerance variations using process variations such as base plate removal, orientation, and size. The tolerance zone variations not only validate the need for systematically designed benchmarks, but also for tri-planar characterization. Simulation and experimental result comparisons provide quantitative information about the applicability of numerical simulation for geometric tolerance prediction for the LPBF process.

Keywords: laser powder bed fusion (LPBF); benchmark test artifact; numerical simulations; dimensional metrology; geometric dimensioning and tolerancing (GD&T)

Citation: Rupal, B.S.; Singh, T.; Wolfe, T.; Secanell, M.; Qureshi, A.J. Tri-Planar Geometric Dimensioning and Tolerancing Characteristics of SS 316L Laser Powder Bed Fusion Process Test Artifacts and Effect of Base Plate Removal. *Materials* **2021**, *14*, 3575. https://doi.org/10.3390/ma14133575

Academic Editor: Mika Salmi

Received: 29 April 2021
Accepted: 17 June 2021
Published: 26 June 2021

Publisher's Note: MDPI stays neutral with regard to jurisdictional claims in published maps and institutional affiliations.

1. Introduction and Background

According to ISO/ASTM 52900:2015 [1], additive manufacturing (AM) is defined as the "process of joining materials to make parts from 3D model data, usually layer upon layer, as opposed to subtractive manufacturing and formative manufacturing methodologies." AM is also commonly known as "3D printing". AM was developed in the early 1980s and has shown great promise and the ability to produce complex and custom geometrical shapes from a wide range of materials [2]. Common AM processes are material extrusion, vat photopolymerization, directed energy deposition, and laser powder bed fusion (LPBF) [3]. The LPBF process for metals uses a laser as a source of energy to fuse layers of metal powder. The laser melts a defined path on each layer of the deposited metal powder. The bed then moves down, and another layer of metal powder with a thickness of 20–100 µm is deposited. This process is repeated until the part is completely manufactured. No binders are used in this process. The process can produce fully dense metal parts ready for industrial use and has limited post-processing needs. It is a widely used metal AM process with various applications in the automotive, aerospace, and biomedical industries [4]. Although AM has made the transition from research labs to industry in the last decade, various challenges, such as geometric tolerance control of manufactured parts, low repeatability, and controlling deviations due to residual stresses still need to be addressed [2]. To define the challenges, the output characteristics of AM parts must be understood and can be categorized as follows:

- Mechanical characteristics, such as tensile, compressive, flexural strength;
- Material characteristics, such as microstructure and porosity;
- Surface characteristics, such as surface roughness and wear properties;
- Geometric characteristics, such as dimensional accuracy and geometric tolerances.

These output characteristics are dependent on various process parameters, such as layer thickness, scan strategy, and part orientation. Standardized procedures have already been established for mechanical, material, and surface characteristic testing and quantification [5–7]; however, standardized methods for quantification and assessment of geometric characteristics are still in their infancy for metal AM [8,9]. Like any other manufacturing process, additively manufactured parts also have deviations and errors compared to the input CAD data. Understanding these deviations or variations is critical in engineering design and manufacturing. For example, a higher frequency of part rejections is typical in precision applications, as a high degree of geometric control is challenging. Variations in a final part's geometry occur due to process physics, various process parameters, and manufacturing conditions. Resulting "geometric deviations" or "geometric errors" need to be quantified and assessed in advance to ensure the required dimensions and geometric tolerances of the AM part are met.

To date, methods for geometric tolerance assessment of AM processes are not yet completely defined or standardized. Experimental methods based on geometric benchmark test artifacts, or GBTAs [10,11], aim at providing quantitative information about various geometric tolerance characteristics. A brief comparison of common GBTAs available in the literature is presented from a geometric tolerance perspective in Table 1. The features of the manufactured GBTAs are used to characterize different dimensional and geometric tolerances [12–15]. The GBTA experiments are followed by ranking and optimizing manufacturing parameters that influence the geometric tolerances, and this information is used to optimize these parameters for the actual manufacturing. For example, Shahrain et al. [16] varied 13 process parameters of a fused deposition modeling printer and, based on the experimental results for flatness and cylindricity, the parameters were ranked and optimized to meet the required geometric tolerance of subsequent prints. References [12,13,17–22] provide similar studies, where the geometric tolerances of manufactured GBTAs are characterized.

Even though these studies provided a great starting point for further study, their application in predicting geometric tolerances for other parts manufactured on the same machine is questionable. This is due to the lack of features in the GBTA to quantify certain geometric dimensioning and tolerancing (GD&T) characteristics as per ISO 1101 [23]. Such as step cylinders for concentricity, parallel flat features for parallelism, and so on. The inability of available GBTAs to provide accurate and reliable information about the geometric tolerance of actual parts makes them unsatisfactory for geometric quality characterization. A comparative study of various GBTAs tested on LPBF and the corresponding geometric tolerances is shown in Table 1. The comparison shows that each GBTA has characterized only some of the geometric tolerances, and those characteristics are not quantified in all three principal planar directions. The removal of support structures and the base plate is also not considered during GBTA design or measurements. Moreover, numerical simulation tools were not used to conduct geometric tolerance predictions for the GBTAs to be manufactured.

Table 1. Comparison of selected GBTAs according to geometric tolerance characteristics.

Ref.	Geometric Tolerance Characteristics	Tolerance Directionality	Effect of Base Plate Removal Studied	Prior Numerical Simulation Conducted	Comments on GBTA Design
[12]	Straightness, flatness, roundness, cylindricity, perpendicularity, parallelism, profile, concentricity, position	Positive Z direction, XY plane, and few negative features in the Y direction	No	No	• Widely accepted NIST standard test artifact (generic) • Process-specific geometric evaluation requirement ignored • Very thick base plate
[24]	Straightness, flatness, circularity	Positive Z direction and XY plane	No	No	• Eight different samples regarded as one GBTA • Manufactured in different batches (a potential source of deviation)
[25]	Flatness, squareness, parallelism, circularity	Positive Z direction and XY plane	No	No	• The positional deviation is observed in all features • Lack of process-specific pre-processing analysis
[26]	Straightness, flatness, roundness, concentricity, total runout	For flat features: XY plane For cylindrical features: positive Z-axis	No	No	• Surface roughness and topographies measured • Geometry linked to scan strategy • Focussed on flat features only
[27]	Flatness, cylindricity, true position	Features in bi-planar directions	No	No	• Parts stacked, not ideal for repeatable results • Geometric tolerances not studied even when the features are present such as cylindricity
[28]	Straightness, flatness, roundness, cylindricity, perpendicularity, parallelism, angularity, profile, concentricity, position	Positive features in the bi-planar direction	No	No	• Directional true position not studied • Scalability issue • Feature accessibility issue

The comparative study highlights state-of-the-art GBTA design limitations to characterize critical GD&T, as well as other crucial geometric quality parameters, such as the minimum feature size and fit for assembly. Most of these experimental methods are not conducted on systematically designed benchmarks, lack proper design for the experiments, are based directly on application geometry, and are trial-and-error based. As such, multiple repetitions are required with various parametric changes, some of which do not actually add to any geometric information and eventually add up to the increasing costs. Thus, for very costly metals and large build volumes, repeated benchmark experimentation is not feasible. Significant limitations of the various benchmark artifacts present in the literature are:

- Inability to provide the geometric evaluation of major GD&T characteristics;
- Lack of features aligned with x, y, z directions for a robust single setup, tri-planar characterization;
- Lack of variation in feature size, specifically for medium to larger mechanical parts;
- No consideration of the effect of removal of the base plate towards the design and measurement phase or consideration of a very thick base feature not capable of capturing base surface warping effects;
- Inclusion of repetitive and/or redundant features that result in material and measurement time wastage;
- Artifact feature size incompatible with available measurement equipment;
- Lack of numerical simulation tools for predicting the geometric tolerances before actual manufacturing.

The solution is to systematically design GBTAs and conduct only minimal required experimentation for geometric tolerance characterization purposes to obtain useful data for quality characterization of the metal AM process and for prediction purposes. A "feature-based methodology" [29], which links primitive geometric features to GD&T characteristics, and that enables straightforward and systematic design should be used. The AM standard ISO 17296-3 [30] outlines the geometric characteristics that need to be measured, quantified, and controlled to define the geometric properties of an AM part:

- Size, length, angle dimensions, and dimensional tolerances;
- Geometrical tolerancing (deviations in form, orientation, and position).

The above-mentioned geometric characteristics are further standardized by geometric dimensioning and tolerancing (GD&T). The ISO 1101: Geometrical Product Specifications [23] standard defines GD&T in detail. GD&T characteristics enable the systematic quantification of form, location, and orientation tolerances of the features of a part. The 14 GD&T characteristics in ISO 1101 are explained briefly along with their drawing symbols in Table A1 of Appendix A. The generic guidelines outlined in ISO/ASTM 52902:2019 [31] for test artifact design should be followed carefully. Any new GBTA should quantify all the major GD&T characteristics in the tri-planar directions. Further, to minimize material and manufacturing time, GD&T estimation should be conducted using numerical simulations instead of experimental GBTA fabrication and characterization. Numerical simulation tools are rarely used to estimate GD&T and, as a result, the reliability of these tools to predict GD&T has not yet been assessed. Most research work is focused on deviation simulations, residual stress simulations, microstructure simulations, and topology optimization [32–34]. There is little to no research on the use of macro-scale finite element simulations to predict the geometric tolerances of LPBF parts according to the ISO 1101 standards. Thus, along with experimental datasets, a framework for simulation-based geometric tolerance prediction and analysis is also required.

This article aims to present a new GBTA design with a "feature-based methodology" [29], linking primitive geometric features to the GD&T characteristics and quantifying all the major GD&T characteristics in the tri-planar directions. GBTA will be used, not only to quantify GD&T, but also to assess the reliability of the proposed methodology to quantify GD&T based on numerical simulation tools. The aim is that, in the future, GD&T can be assessed using numerical simulations with the proposed GBTA being used for model calibration and verification. The detailed design process is explained in Section 2. It enables a generic quality characterization of the LPBF process under consideration from a GD&T perspective. The corresponding experimental plan and the measurement procedures are explained in Section 3. Section 4 outlines the numerical simulation process, which is used to extract deviated geometries, with and without a base plate, for a virtual GD&T characterization for comparison with the experimental results. Finally, the experimental and simulation results are presented in Section 5, followed by conclusions in Section 6.

2. Geometric Benchmark Test Artifact Design

The geometric benchmark test artifact (GBTA) was designed to fulfill two major objectives:

- To perform the tri-planar GD&T quantification of the metal AM process;
- To study the effect of base plate removal on GD&T characteristics.

To achieve the first objective, all major GD&T characteristics were quantified on a range of features and sizes on the GBTA. Also, the feature orientation was selected to enable the three principal planar directions for the tri-planar normative quantification of the AM process in consideration. Given the orthotropic, layer-by-layer fabrication of the AM process, and the coupled process physics, it is essential to consider tri-planar repeatability as the mere assumptions of tolerance band applicability across different axes is not valid. The features on GBTA are selected according to the feature-based GBTA design methodology [29], which gives systematic guidelines for selecting features according to the geometric tolerance characterization requirement. The ability to manufacture minimum feature sizes and overhangs was also considered. To fulfill the second objective, i.e., to characterize the effect of removing the base plate, GD&T quantifiers were measured before and after removing the base plate.

Features of the GBTA were selected based on the GD&T requirements for the normative quantification of the geometric quality characteristics of the metal AM process. The bounding box size of the GBTA was decided based on the build volume of the LPBF system on which it had to be manufactured. Since the build volume was $250 \times 250 \times 300$ mm, the size of the GBTA base was kept at 200×200 mm and then features were spread on that surface. In the case of LPBF, the raster direction changes with a change in orientation of the part in the AM machine. Therefore, the orientation of the part has a significant effect on the geometric properties of the features. There are two different ways that this variation can be studied and characterized:

- Printing the features in different orientations on the same GBTA;
- Printing the complete GBTA in different orientations.

The former method is used in this study because, for the latter method, the support structure design for the features in different orientations can vary significantly and, as a result, it would not allow for a robust measurement and comparison of the benchmark artifacts samples. Therefore, the GBTA used in this study was built by printing features in different orientations on a common base. The features to be present on the GBTA base are shown in Figure 1 and are explained below:

- Cuboids: Hollow and solid cuboids selected to quantify straightness, flatness, perpendicularity, and parallelism. Since these are large volume features, the effect of the process on the deviations will be different compared to small volume features, such as thin walls.
- Thin walls: Thin walls of varying dimensions to quantify straightness, flatness, and the effect of residual stresses on the GD&T quantifiers for thin features.
- Cylindrical features: Solid and hollow cylindrical features placed with axes in all three principal directions (X, Y, and Z) to quantify circularity and cylindricity.
- Axially stacked cylinders: Stacked cylinders for measuring concentricity and runout in the X, Y, and Z directions.
- Conical features: For measuring the circularity of features other than cylindrical forms.
- Positional features: Features positioned at specified distances from each other to find positional tolerance zones of features.

- Profile: Profile features with curves to quantify the profiles of a surface and of a line tolerance.
- Minimum size features with diameters and external sizes ranging from 0.5 to 2 mm. This will help to check the ability of the metal AM process in consideration of printing minimum-sized features as well as the effect of the size on the tolerance.
- Thin overhangs: Thin overhang features to understand the effect of support and base removal on thin and fragile features.

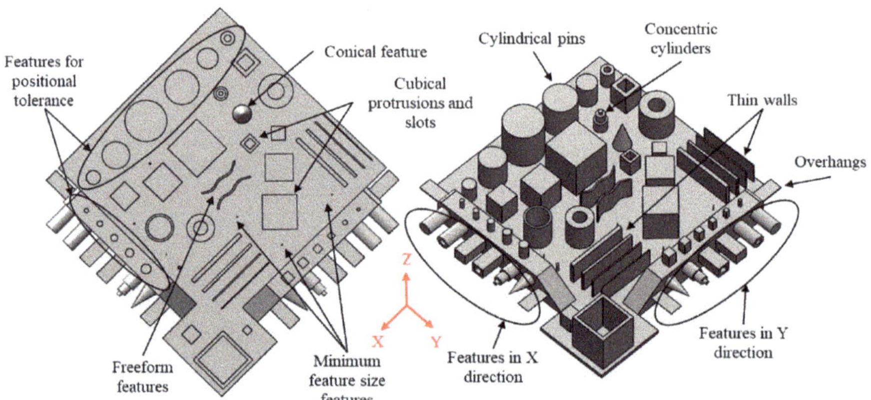

Figure 1. Nominal GBTA design and corresponding features. (**Left**) Top view and (**Right**) isometric view.

In addition to the GD&T specifications and the other geometric quantifiers, the design needed to take into account the metal AM process constraints. The GBTA features were placed such that the maximum area on the base plate was covered to understand the spatial variations. The effect of variations in terms of feature size on the geometric tolerances is also an important aspect to consider, as the size tolerances significantly influence the design of metal AM in precision applications. The features are shown and explained in greater detail in Figure A1 and Table A2 of Appendix B, showing the numbering of the features on the GBTA, and the relevant geometric tolerances of the features. Complete drawings with size dimensions, location dimensions, and geometric tolerances are also provided in Supplementary Data.

3. Experimental Procedure

3.1. GBTA Manufacturing

Three newly designed GBTAs were manufactured using an AM250 LPBF machine (Renishaw PLC, Wotton-under-Edge, Gloucestershire, UK) [35]. The STL file of the GBTA was converted to a sliced file by specifying the process parameters, such as layer height, hatch style, and support structure, using the manufacturer-provided software (Renishaw QuantAM 2018). The process parameters were decided as per previous experience and manufactured using best practices. The process parameters for the manufactured parts were kept constant to investigate the variations in the process and are listed in Table 2.

Table 2. Machine parameters and manufacturing parameters for the experimentation.

Machine Settings	Value	Manufacturing Parameters	Value
Layer thickness	50 μm	Powder material	Stainless steel (SS 316L)
Laser power	200 W	Powder particle size	15–45 μm
Base plate size	250 × 250 × 15 mm	Hatching style	Chessboard style
Scan speed	600 mm/s	Hatch spacing	150 μm
Laser beam diameter	70 μm	Interlayer angle	67°

The build chamber was sealed, and argon gas was introduced to maintain an inert environment for laser-based manufacturing. Throughout the build process, the oxygen level was kept at less than 0.1%. The constant argon gas flow acted to keep the laser lens clean, reduce any oxygen or nitrogen entrapment, and moderate the vapor flow of the processed metal powder. Stainless steel (SS 316L) grade powder from Renishaw™ (Wotton-under-Edge, Gloucestershire, UK) was used as a raw material for manufacturing. Upon completion of the build, the part was left in the chamber to cool down to room temperature for approximately 2–3 h and then removed. One of the manufactured samples, as manufactured and attached to the build plate, is shown in Figure 2. To understand the as-built geometric properties of the GBTA, the manufactured samples were assessed without any kind of post-processing.

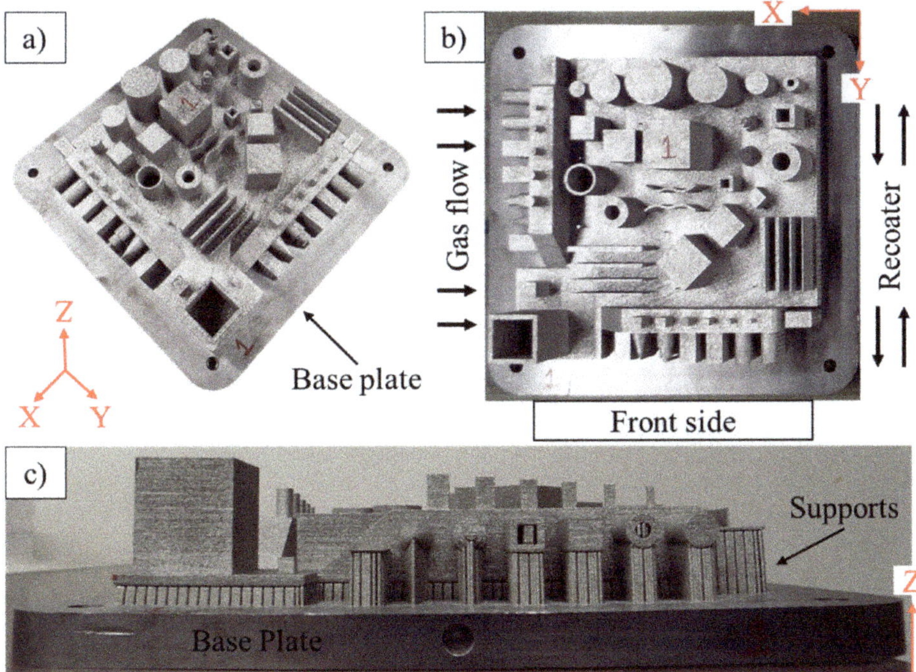

Figure 2. Example of manufactured sample: (**a**) Isometric view of the GBTA, (**b**) top view of the manufactured GBTA showing the "front side" of the LPBF system, gas flow direction, and recoater direction, and (**c**) side view of the manufactured GBTA showing the depicting and the base plate.

3.2. Measurement Procedure

The features on the manufactured samples were measured for various GD&T characteristics using a Mitutoyo Crysta-Plus M443 (Mitutoyo, Kawasaki, Japan) coordinate measuring machine (CMM). The CMM has a resolution of 0.5 µm. For measurements with the base plate, the GBTA was stationed on the CMM table with fixtures to restrict its movement. The measurements were conducted keeping in mind the tight tolerance zones of the LPBF process under consideration, so reducing measurement variability was a priority. For each characteristic measurement, at least ten data points were taken using the CMM and each measurement was repeated thrice to achieve repeatable readings. For one flatness tolerance zone of a feature, ten data points were collected from the surface to calculate it. The measurement process for this study was divided into four major geometric categories, i.e., circle, cylinder, line, and surface. The procedure adopted for the measurement for each geometry is listed below [36]:

- Circle: Measured using 10 equally spaced points around the circular cross-section.
- Cylinder: Measured using a total of 15 points with 5 points, equally spaced, distributed at three cross-sections of the cylinder at heights h1, h2 and h3 (see Figure 3).
- Line: Measured using 10 points, equally spaced, along the length of the line to cover at least 90% of the length of the line.
- Surface: Measured using 10 points with their random distribution over the surface to cover the maximum area of the surface.

The schematics of the measurement strategies are shown in Figure 3. The red dots represent the locations where the CMM probe recorded the measurement data.

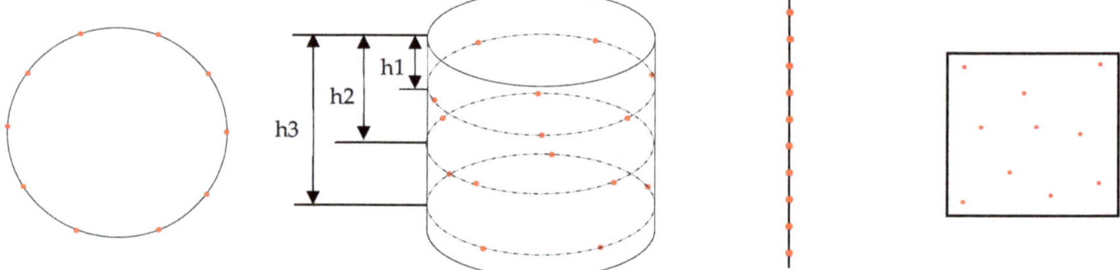

Figure 3. Measurement strategies for different features. The red dots represent the locations where the CMM probe recorded the measurement data.

The method of least squares was used to define planes and cylinders in the CMM for computing the GD&T tolerance zones. Two coordinate systems were defined to perform the measurements of the GBTA in the CMM. One is the coordinate system of the CMM itself, which defines the motion of the CMM probe in three directions, referred to as the X, Y, and Z directions. The second is the part of the coordinate system that is defined using the GBTA. It is the orthogonal frame of reference that defines the orientation of the features with respect to each other and is used for computing the orientation- and location-based geometric tolerances. The part coordinate system is shown in Figure 2. To eliminate ambiguity, both frames were aligned with each other on the same axes (X–X, Y–Y, and Z–Z) with each other. In the CMM, the surface–line–line method was used to set the part coordinate system [37]. The part coordinate system is depicted in Figure 2. The top surface of the base of the GBTA, on which most of the features are built, is defined as the XY plane. The features categorization for measurements, as per the GD&T, is shown in Table 3. The results are discussed in Section 5 according to the GD&T characteristic.

Table 3. Feature categorization for measurements based on GD&T characterization.

#	GD&T Characteristics	Features Categorization for Measurement
1	Straightness	Features with straight-line–flat surfaces of cuboids and thin sheets.
2	Flatness	Features with flat surfaces-cuboids, the top surface of cylinders.
3	Perpendicularity	Flat surfaces of cuboids and thin sheets with respect to a planar surface as a datum feature.
4	Parallelism	Features with flat surfaces parallel to a planar surface as a datum feature
5	Angularity	Angular features with respect to a planar surface as a datum feature.
6	True Position	Cylindrical and planar features with respect to a datum coordinate system.
7	Circularity	Cylindrical features–circular periphery on the outer surface of cylinders and hollow cylinders.
8	Cylindricity	Cylindrical features.
9	Concentricity	Stacked cylindrical features.
10	Circular run-out	Stacked cylindrical features.
11	Total run-out	Stacked cylindrical features.

For the features with support structures, such as features 1–8, the area covered by the support structures was excluded from the measurements. The supports were kept intact to make sure that the features were hinged on the base plate when the measurements were performed. For the second round of measurements without the base plate, first, the base plate was removed from the GBTA using a bandsaw. The support structures were also removed manually using hand tools. The GBTA, after the removal of the base plate and support structures, is shown in Figure 4. In addition, for fixing the GBTA after the removal of the base plate on the CMM table, special fixtures were used. These fixtures were designed according to the GBTA dimensions and manufactured using polymer 3D printing.

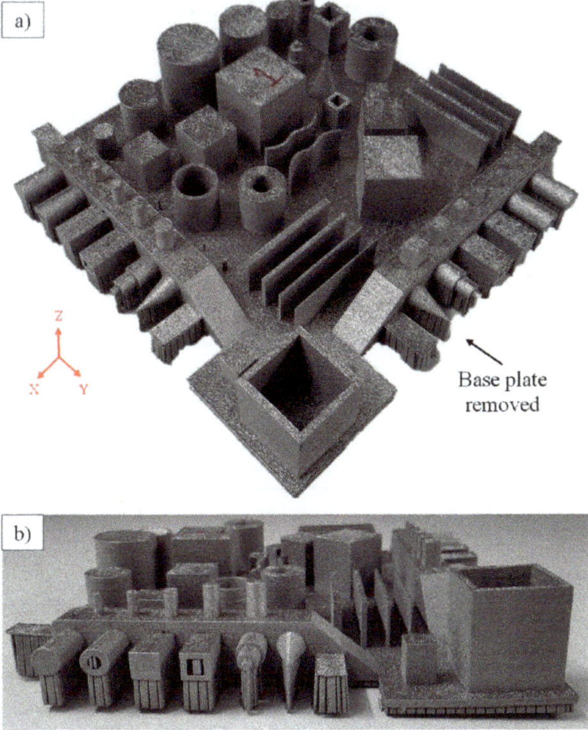

Figure 4. GBTA after the removal of the base plate: (**a**) Isometric view and (**b**) side view showing warped features.

4. Numerical Simulation

A numerical simulation of the GBTA was conducted to understand the effects of temperature and stress fields on the specific planar GD&T characteristics, as well as to assess the reliability of the thermo-elastic finite element model implemented in Netfabb Local Simulation (Autodesk Inc., San Rafael, CA, USA) for GD&T quantification [38]. First, the part and the support structures were imported into Netfabb Local Simulation. Then, a process parameter file (PRM) was generated by conducting a micro-scale simulation using the experimental parameters for the SS 316L material. The process parameters used were the same as those used in the experimental study and are described in Table 2. The simulation parameters are shown in Table 4. The temperature-dependent material properties of SS 316L were based on data from References [39,40]. The PRM file was used to conduct a macroscale simulation of the GBTA part [41,42]. The boundary conditions for the macroscale model were as follows:

- Mechanical boundary conditions: The bottom nodes of the base plate were fixed, i.e., zero displacements was imposed in all directions. At the other boundaries, a zero normal stress boundary condition was used.
- Thermal boundary conditions: A convective heat transfer boundary condition was considered between the powder and the environment. A heat loss coefficient value of 20 W/m^2/K was used [43].

The input STL, along with supports, is shown in Figure 5a and the meshing strategy is shown in Figure 5b. After a solution was obtained, the predicted displacement field was used to construct a deviated GBTA part. The deviated STL files were imported to GOM Inspect (GOM GmBH, Braunschweig, Germany) [44] for extracting GD&T quantifiers for a one-to-one comparison with the experimental results. Figure 6 shows the process of constructing fitted features on the deviated STLs from simulations and extracting the GD&T characteristics. The deviated STL file was first converted into a point cloud. Standard geometric shapes were then fitted into the individual features of the deviated point cloud, such as cylinder or plane. GD&T definitions were then used to find the tolerance zones on these fitted features. The GD&T results from the simulations (both with and without base plate) are discussed alongside experimental results in the next section for analytical comparison and physical interpretation.

Table 4. Simulation parameters.

Simulation Parameter	Value
Heat source absorption efficiency (%)	35
Analysis type	Thermal and mechanical
Structural plasticity calculation	After the part is cooled down and before removing the base plate
Mesh approach	Layer based
Maximum mesh adaptively levels	5
Coarsening generations	1
Layers per mesh element	20

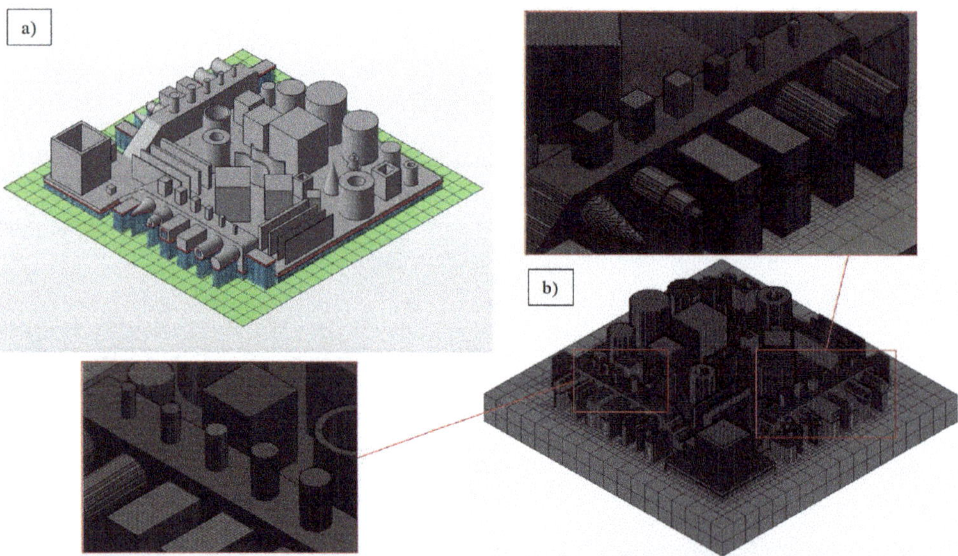

Figure 5. (**a**) GBTA with supports (in blue) and (**b**) layer-based adaptive coarsening mesh.

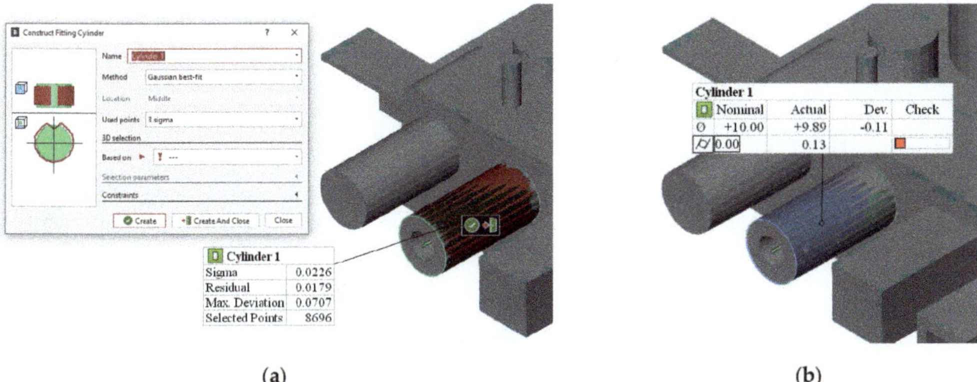

(a)   (b)

Figure 6. Extraction of GD&T: (**a**) construction of fitting feature on deviated STL and (**b**) extraction of GD&T characteristics based on the constructed feature.

5. Results and Discussion

The measurement data from the experiments and the predicted deformed geometry from simulations are used to extract the GD&T characteristics and are presented in the following sections. The data from the samples after removing the base plate are also presented to have a comparative analysis and discussion.

5.1. Form Tolerances

5.1.1. Straightness

The straightness tolerance zones were computed for the features and categorized according to the feature's orientation to the reference planes—XY, YZ, and ZX. Results for features parallel to reference plane XY are shown with bars labeled "XY" in Figure 7. A similar approach is used for YZ and ZX. For the experimental measurements, the sample space contained 13 features with a feature axis aligning with the XY plane and 8 for each of the YZ and ZX planes. For example, the first XY bar in Figure 7 shows the experimental results for the average tolerance zone for the 13 features aligned to the XY plane with the base plate intact on the GBTA.

Experimental results showed that the straightness tolerance was anisotropic with respect to the reference planes since the results were different in all three principal planar directions, along with variations in standard deviation. The straightness tolerance was higher in the XY planar direction because the measurements were performed on the as-built parts without any post-processing; therefore, the hatching effect cannot be ignored on this plane. The effect is increased by chessboard (or checkerboard) style hatching, in which the laser movement is multi-directional and creates a crisscross waviness on the surface. The hatching effect was maximum on surfaces parallel to the XY plane as it is the plane on which the laser moved while the part was built. In general, the straightness tolerance results increased by 15 to 20 μm in all three planar directions after the removal of the base plate, which was anticipated as the residual stresses were released after the base plate removal, which, in turn, increased the deviations. Even with a larger curvature on the base feature of the GBTA after the removal of the base plate, the form tolerances, such as straightness, were less affected, as they were independent of datums. The largest straightness tolerance zone was observed on a thin sheet (feature 9) on the XY plane. It was 158 μm with the base plate and 183 μm without the base plate. The large tolerance zone resulted due to the low thickness value, i.e., 500 μm, of the sheet, which led to larger deformations during the rapid cooling of the part throughout the process.

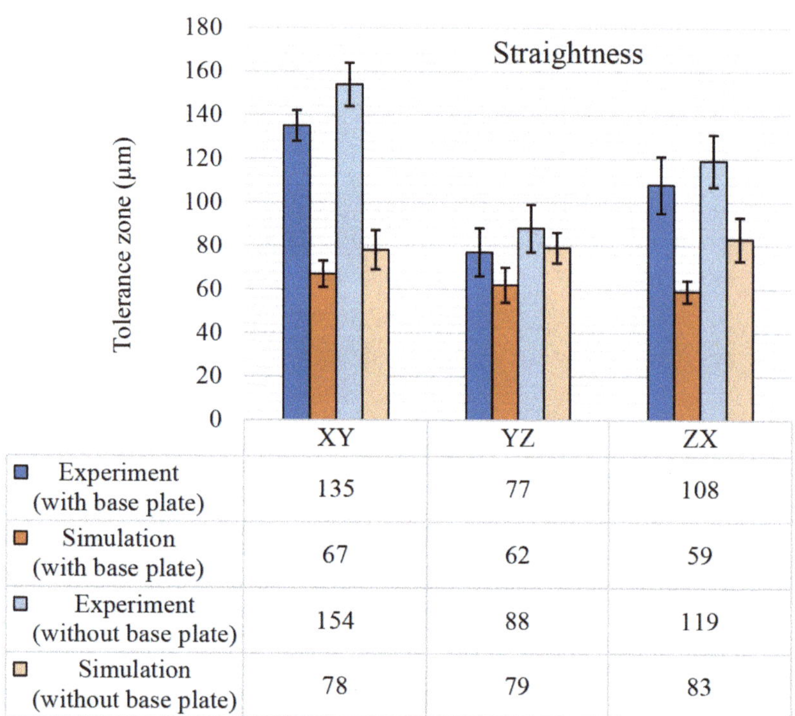

	XY	YZ	ZX
▣ Experiment (with base plate)	135	77	108
▣ Simulation (with base plate)	67	62	59
▣ Experiment (without base plate)	154	88	119
▣ Simulation (without base plate)	78	79	83

Figure 7. Straightness tolerance results (µm), with and without the base plate.

Figure 7 shows that the predicted straightness tolerances by numerical simulations. The simulation results closely followed the experimental results in the YZ plane; however, there is a dip in simulation results in the XY plane, which could be because the hatching phenomena was not properly captured in simulations. In all three directions, the overall tolerance band is 40 µm lower for both (with and without base plate) results and follows the same pattern. It is hypothesized that the reason for the major variation in the straightness tolerance of the XY plane was that the surface waviness due to hatching was not captured well by the simulations. The difference in results is in a similar range with the generic surface waviness values for LPBF-manufactured SS 316L parts [5]. The difference in the experimental results in the YZ and ZX planes was also not completely captured by the simulations. Since straightness is a form tolerance and the boundary conditions have a maximum effect on the form tolerances, the "uniform heat loss condition" could be one reason for this similarity in simulation results for the YZ and ZX planes. However, the standard deviations for the XY plane are almost the same for the experimental and numerical results, with and without base plate, i.e., only ±1 µm.

5.1.2. Flatness

The sample space for flatness tolerance contained 20 features for the XY plane, 25 features for the YZ plane, and 24 features for the ZX plane. The flatness results for both experiments and simulations are shown in Figure 8, with a comparison of with and without base plate removal. Much like the straightness results, the anisotropy of the experimental results was observed for the flatness tolerance as well. The largest tolerance bands with and without the base plate were observed for the XY plane and the smallest for the YZ plane. Maximum tolerance bands are observed for the XY plane for both with and without base plate conditions. This is because the XY plane was the building plane, which also bore the effect of in-plane shrinkage during the building of the part. Further, the release of the residual stress with the removal of the base plate also affected the XY plane to the highest degree as the base feature of the GBTA was parallel to the XY plane, on which all the features were mounted. Since flatness tolerance was spread over the complete region of the feature, even minor curvatures and warpage due to the removal of the base plate affected the tolerance zone. A large standard deviation in the results for the ZX plane without the base plate was observed, i.e., a standard deviation of 56 μm. This could be due to the recoater direction, which was perpendicular to the ZX plane and affected the layer edges during the building, hence reducing the surface quality of the ZX walls.

Looking at the simulation results in Figure 8, an overall lower range of flatness tolerance is observed as compared to experimental results, except for the YZ plane. For both experiments and simulations, the standard deviation of the results increased with the removal of the base plate. This shows that with the removal of the base plate, some of the surfaces were warped to a greater extent, leading to a rise in standard deviation. For instance, features 4, 5, 18, and 19 show an approximate increase of 20 μm in flatness for the XY plane after the removal of the base plate as they were closer to the edge and were most affected by the warpage of the GBTA base feature. An interesting result to notice is the standard deviation of the ZX plane with the base plate, which reduces from 26 μm to 16 μm. The possible reasoning behind this could be that the features that had larger flatness tolerance bands experienced warpage in such a way after the base plate removal that the flatness values dropped.

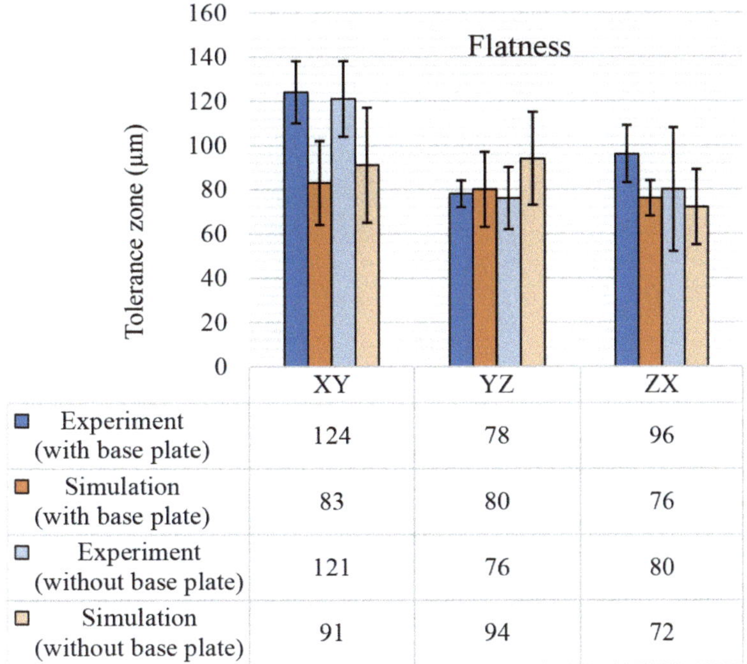

	XY	YZ	ZX
▣ Experiment (with base plate)	124	78	96
▣ Simulation (with base plate)	83	80	76
▢ Experiment (without base plate)	121	76	80
▢ Simulation (without base plate)	91	94	72

Figure 8. Flatness tolerance results (μm) with and without the base plate.

5.1.3. Circularity

Measurements were done for different cylindrical and conical features with their axes aligned to their respective reference axial directions—X, Y, and Z. The tolerance zones were computed for these categories of features, and the same are represented in the bar chart in Figure 9. The sample space contained six features each for the X and Y axes directions and 26 features for the Z-axis direction. The larger number of features in the Z direction varied in terms of size and volume to get a detailed account of the circularity variation in the principal build axis, i.e., the Z-axis.

For the experimental results, the circularity tolerance zone with the base plate intact was around 104 μm for all three axial directions. However, the standard deviation was highest for the Z-axis, i.e., 40 μm, and without the base plate, the standard deviation increased to 52 μm. The large standard deviation observed in Z-axis was due to the removal of the base plate and because a large number of features were characterized in the Z-axis direction. However, since circularity is measured on features of revolutions, the change in the diametric value of the feature was also correlated to circularity tolerance. Trends for circularity vs. diameter in the Z-axis direction are shown in Figure 10. Similar results were observed for experimental circularity tolerance after the removal of the base plate.

For circularity with the base plate, a clear relationship was observed, i.e., the circularity tolerance zone increased with the increase in diameter. It was also depicted by a linear trend line with an R-squared coefficient value of 0.8272. However, for circularity results after removing the base plate, no particular trend was observed. The most reasonable explanation is the uneven distribution of the residual stress led to variations in the deviation. Since with the change in the diameter, the surface area of the feature changed, which led to a change in residual stress and the resulting deviations or the tolerance zone.

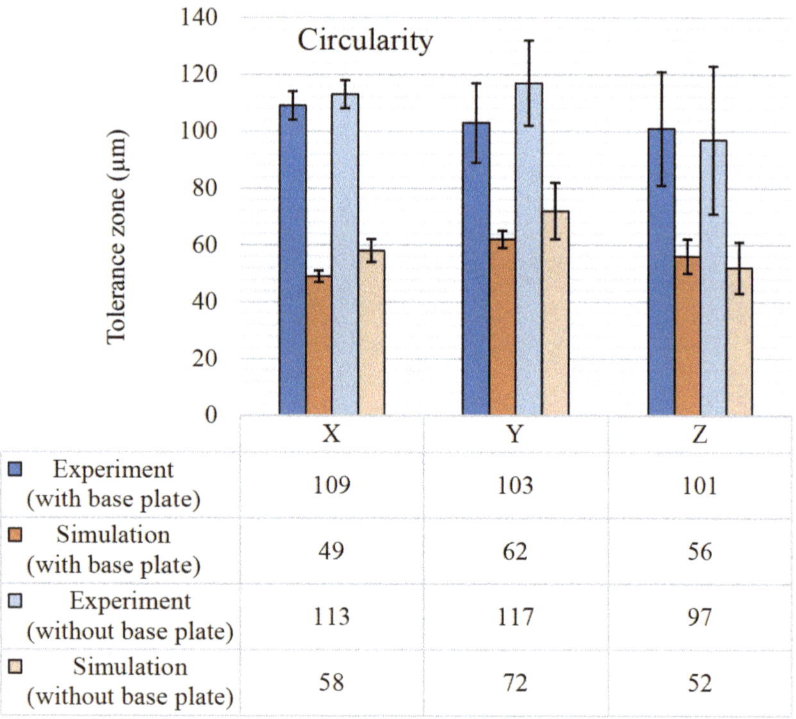

	X	Y	Z
▣ Experiment (with base plate)	109	103	101
▣ Simulation (with base plate)	49	62	56
▢ Experiment (without base plate)	113	117	97
▢ Simulation (without base plate)	58	72	52

Figure 9. Circularity (μm) tolerance results, with and without the base plate.

In the simulation results, an overall mean shift of around 40 μm was observed in all directions for results with and without the base plate. The major reason for this mean shift could be the generic under-prediction observed in the deviation simulation and the inability of the simulation to capture meso-level surface variations. Apart from this mean shift and low standard deviation, the simulation results followed a very similar trend to the experimental results, especially for results after removal of the base plate.

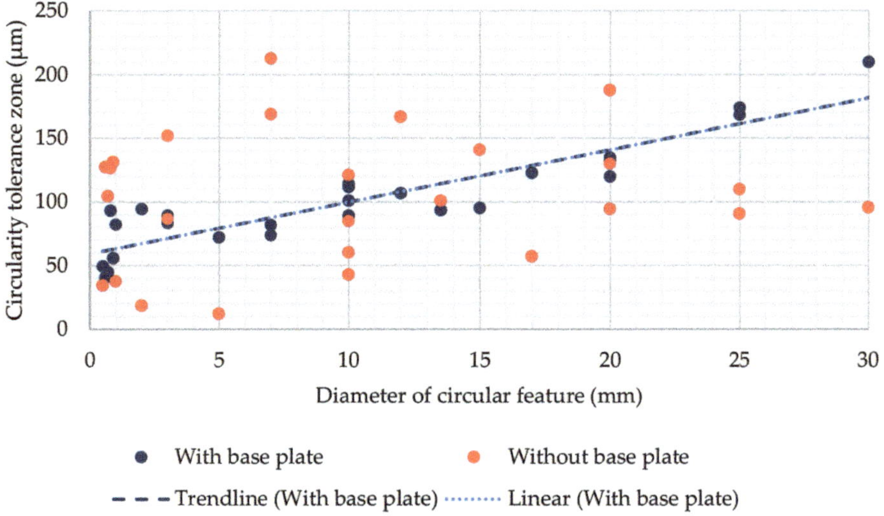

Figure 10. Diameter vs. circularity (experimental).

5.1.4. Cylindricity

Cylindricity tolerance results are shown in Figure 11. The sample space contained six features each for the X and Y direction and 25 features for the Z direction. The diameter of the cylindrical features varied from 0.5 mm to 30 mm. Unlike circularity, the directional results after the removal of the base plate were not in a similar range for the three principal axial directions. The results in axial direction X showed the highest peak in the experimental cylindricity tolerance zone, as well as a standard deviation hike. One reason for this huge variation is the large support structures, which decreased the dimensional fidelity of the features, and along with the residual stress release, another reason could be the cantilever effect due to the nature of the geometric forms of the feature [45]. The effect of the diameter was also studied for the cylindricity tolerance zones, and the results are shown in Figure 12. A linear trend is observed with the cylindricity tolerance zone, which gets wider with an increase in the diameter of the cylindrical feature. Even with a few outliers with small diameter features, both sets of results (with and without the base plate) showed a similar trend. The trendline (with base plate) has an R-squared coefficient value of 0.7734, and for the trendline (without the base plate) the R-squared coefficient value is 0.5607. The results without the base plate were offset with an additional 40–50 μm.

Simulation results for cylindricity with the base plate intact are close to the experimental results. Unlike the large rise of the cylindricity observed in the X-axis direction for the experimental results without the base plate, the simulation does not show any such anomaly. This could be because the surface defects induced by the removal of the support structures are not captured by simulations. Moreover, a rise in cylindricity for the features in the X direction (such as features 2, 3, and 6) is still observed in the simulation results, similar to experimental results, due to the removal of the base plate and the support structures. However, the rise in the cylindricity tolerance with the removal of the base plate is still on the lower side compared to the tolerance increase seen for other form tolerances. This is because cylindricity tolerance is based on the form of the cylindrical feature, but not on the orientation, and with the removal of the base plate, the maximum effect is seen on the orientation as compared to the form of the cylindrical features.

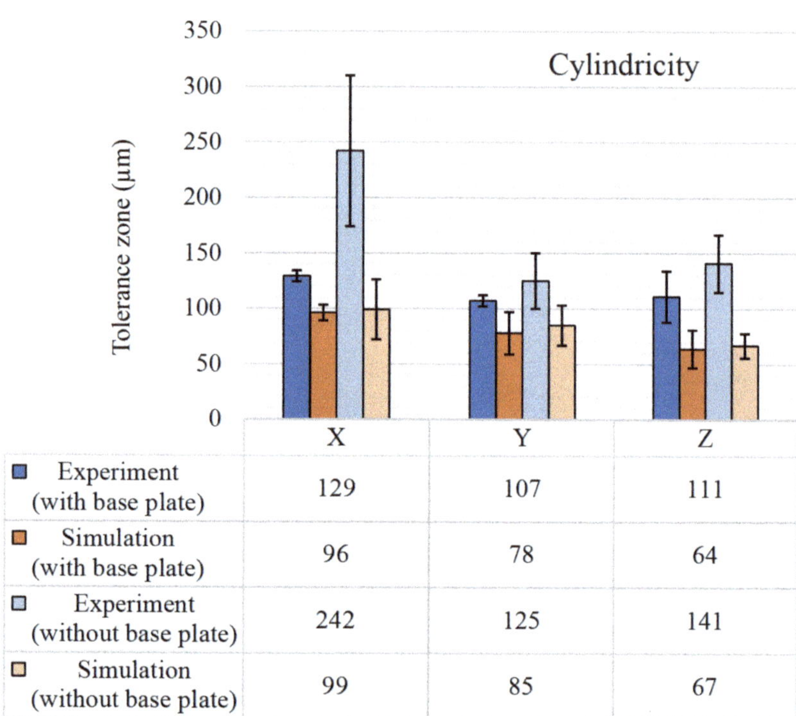

	X	Y	Z
■ Experiment (with base plate)	129	107	111
■ Simulation (with base plate)	96	78	64
▫ Experiment (without base plate)	242	125	141
▫ Simulation (without base plate)	99	85	67

Figure 11. Cylindricity (µm) tolerance results, with and without the base plate.

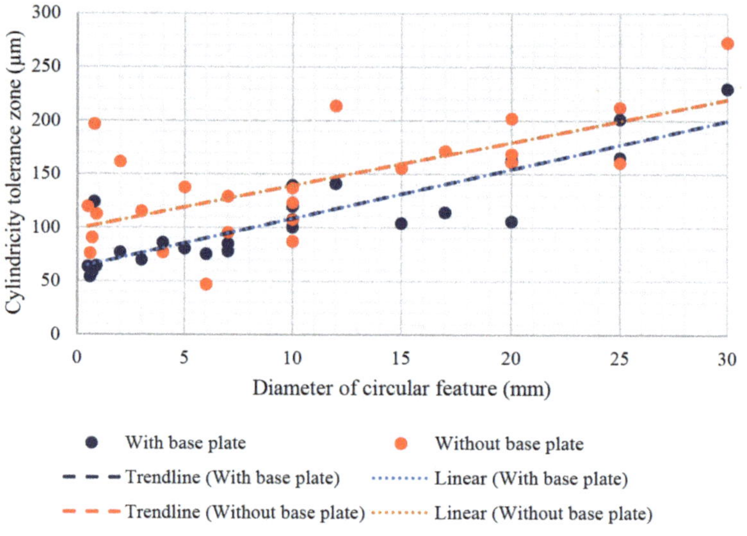

Figure 12. Diameter vs. cylindricity (experimental).

Overall, analysis of the results for the form tolerances shows a general increase in tolerance zones after the GBTA is removed from the base plate. It can directly be linked to the release of the residual stress within the GBTA. Additionally, the increase in standard deviation suggests a similar reason. Apart from this, the process parameters and laser-material interaction have a significant effect, which will become clearer when the results are analyzed for a set of process parametric variations. For now, it can be stated that for the selected process parameters and SS 316L material, the form tolerances are generally in the range of 50 to 130 μm with a few exceptions, such as the spike in cylindricity in the X-axis direction after removal of the base plate.

5.2. Orientation Tolerances

5.2.1. Perpendicularity

The GBTA contained 16 features for the XY plane and 18 features for both the YZ and ZX planes on which perpendicularity is measured. The perpendicularity tolerance zone results are shown in Figure 13. For experimental perpendicularity, the standard deviation increased by a factor of two when the base plate was removed. It increased to 54 μm from 24 μm. The maximum experimental tolerance bands were observed in the XY plane. The simulation results for perpendicularity were very close to those of the experimental results with few exceptions. The major variation consistently observed in the perpendicularity results was that the perpendicularity was minimum in the YZ plane for the experimental results, but for simulation results it was for the ZX plane. In addition, the simulation results in the YZ plane were higher than the experimental results, which was not the usual case in all other tolerance characterizations and simulation results. Since perpendicularity is an orientation tolerance, the flatness of the datum plane affected the tolerance zone, which was seen in both the experiments and the simulations. The variation in the flatness of the datum planes also propagated the tolerance zone to the perpendicularity results. This could be the major reason for the inconsistencies and variations in the results.

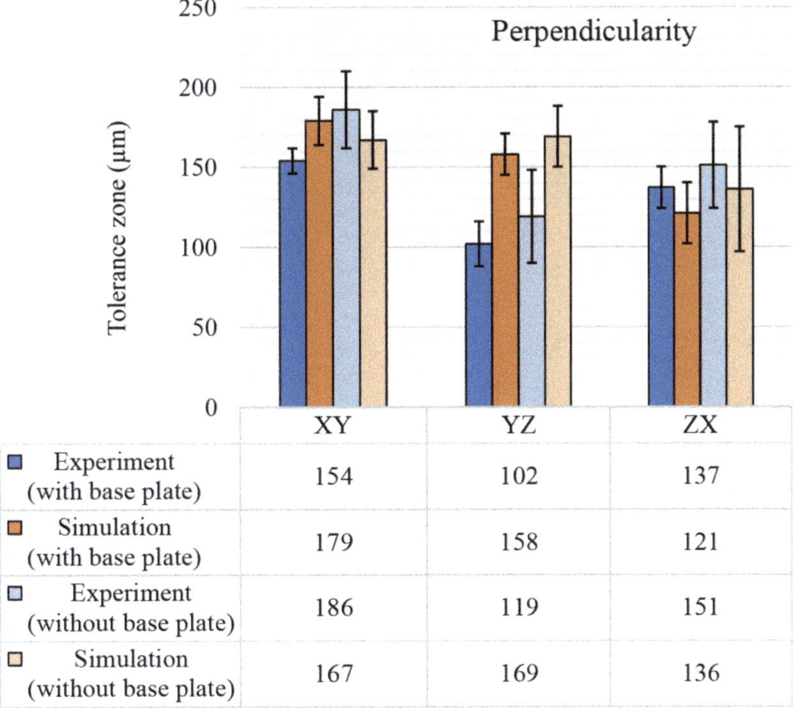

	XY	YZ	ZX
Experiment (with base plate)	154	102	137
Simulation (with base plate)	179	158	121
Experiment (without base plate)	186	119	151
Simulation (without base plate)	167	169	136

Figure 13. Perpendicularity (μm) tolerance results, with and without the base plate.

5.2.2. Parallelism

The sample space for parallelism tolerance contained 15 features for the XY plane and 13 features each for the YZ and ZX planes. Parallelism results from the experiments and simulations are shown in Figure 14. The experimental results show the anticipated increase of the tolerance zone after the removal of the base plate. The standard deviation values are different for each planar direction, but the average standard deviation for the three planes remained similar for both with and without base plate results. The results with the base plate for XY and YZ are comparable; however, there is a rise of around 20 μm in the ZX plane. The results without the base plate show the highest average tolerance zone of 215 μm for the YZ plane. In general, the small-sized features showed a larger parallelism tolerance zone with the base plate intact. However, after removing the base plate, the tolerance zone variations in the small features were less than the tolerance zone variations in the large features. The increase in the average tolerance zone after removing the base plate was due mainly to the larger size features as they are susceptible to more warpage and deformations with the release of stress. The simulation results showed a very similar trend for variations between the planes, especially for results with the base plate. The planar variation follows the same trend as the highest ZX plane tolerance zone with a similar standard deviation. However, the zone is 36 μm smaller than the experimental results. The same variations were not observed in the results without the base plate. In addition, the standard deviation for the results without the base plate increase by 14 μm from the experimental to simulation results.

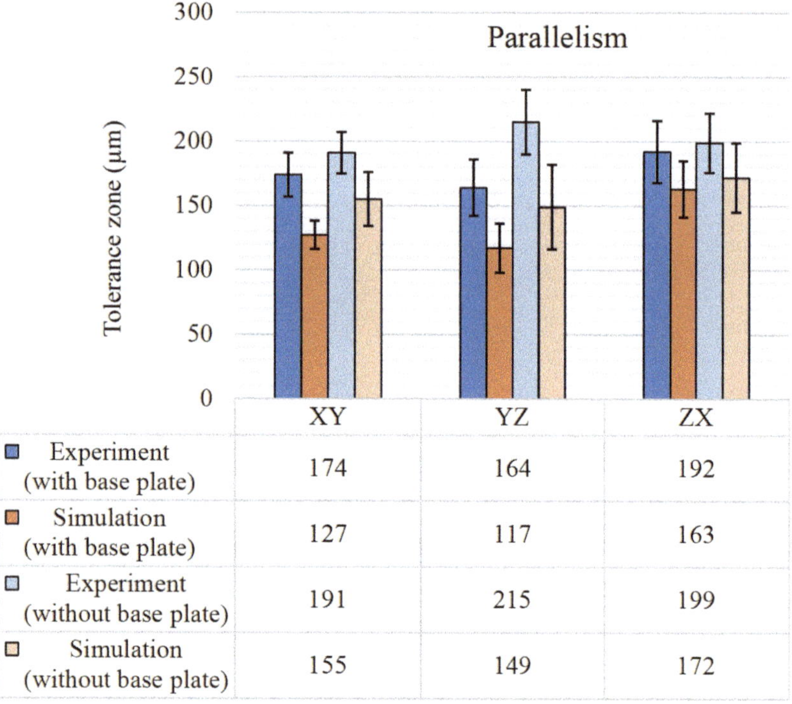

	XY	YZ	ZX
▣ Experiment (with base plate)	174	164	192
▣ Simulation (with base plate)	127	117	163
▣ Experiment (without base plate)	191	215	199
▣ Simulation (without base plate)	155	149	172

Figure 14. Parallelism (μm) tolerance results, with and without the base plate.

5.2.3. Angularity

Angularity measurements were done for two planar features at 40-degree angles to the XY plane and are shown in Figure 15. The experimental angularity increased after removing the base plate. For simulations, the rise in the tolerance zone after removal of the base plate was similar at 3 μm. However, the average tolerance zone was higher by 8 μm, i.e., at 91 μm with the base plate compared to 83 μm for the experimental value.

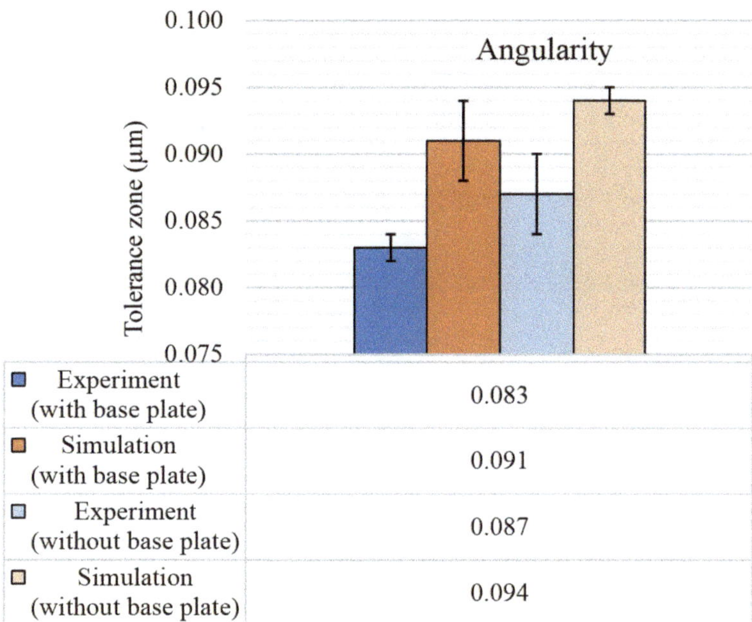

▣ Experiment (with base plate)	0.083
▣ Simulation (with base plate)	0.091
▢ Experiment (without base plate)	0.087
▢ Simulation (without base plate)	0.094

Figure 15. Angularity (μm) tolerance results, with and without the base plate.

5.3. Location Tolerances

5.3.1. Concentricity

Concentricity tolerance is an important location tolerance for deciding the location of one cylindrical feature with respect to another and plays a major role during assembly operations. The concentricity results are shown in Figure 16. Experimental concentricity results for the base plate intact and without the base plate follow similar trends with a maximum tolerance zone in the Y-axis and a minimum in the Z-axis. It is worth noting that among all the GD&T characteristics discussed above, concentricity had the largest tolerance zones. This is due to the composite nature of concentricity tolerance, as it also depends on the form of the cylinders and the axial shift of the cylindrical features with respect to each other. Even with two features per axis and three measurements per feature, the standard deviation is large. The simulation results also follow suit with average concentricity tolerances being highest along the Y-axis and minimum along the Z-axis with an increase in the tolerance zone after the removal of the base plate. The standard deviations for the simulation results closely followed the experimental results.

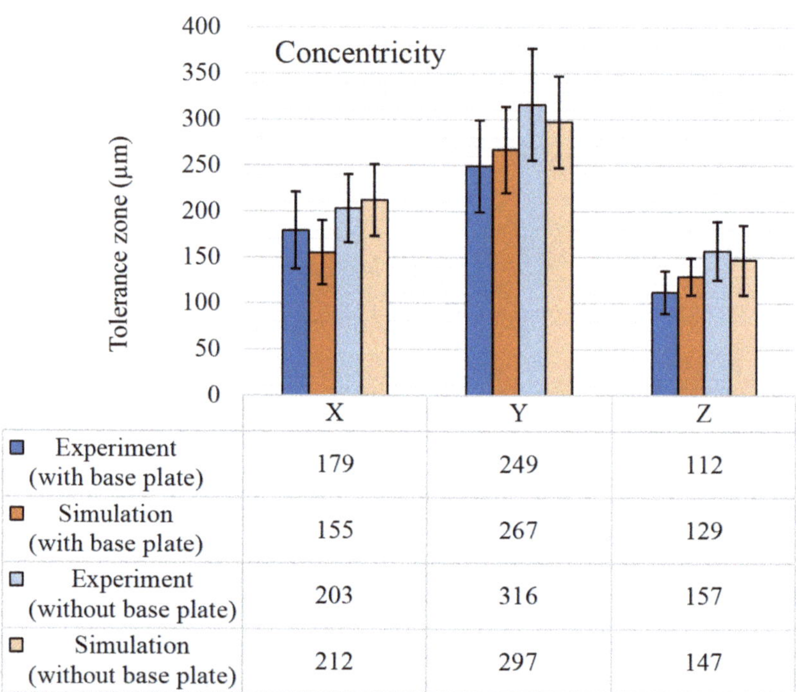

Figure 16. Concentricity (μm) tolerance results, with and without the base plate.

	X	Y	Z
Experiment (with base plate)	179	249	112
Simulation (with base plate)	155	267	129
Experiment (without base plate)	203	316	157
Simulation (without base plate)	212	297	147

5.3.2. Position

The GBTA features were positioned evenly on the base to study the position tolerance both for the planar and axial positions and the results are shown in Figure 17. The planar position variation was highest in the YZ plane for all results. The YZ plane is perpendicular to the recoater direction and the recoater movement and powder spread could be the most significant reason for these large tolerance zones. A similar trend was observed in the simulations as the simulation solver also considered the recoater direction and the recoater interference. The standard deviations also got wider for the YZ planar direction results along with a rise in the tolerance zone. For the axial position tolerance, cylindrical features were used across the entire GBTA and the results are shown in Figure 18. The experimental and simulation results are consistent in terms of variations across the axes. In the axial position, the Y-axis tolerance zone was highest for both experimental and simulation results. Simulation results have underpredicted the position tolerance for both planar and axial features. The standard deviations were much higher than any standard deviations observed for other GD&T characteristics, which is because the tolerance values are quite a bit higher than other tolerance characteristics for some of the positional features. The exact reasoning behind this significant shift is still unknown, but the most logical reason is the recoater direction effect and the micro-mirror orientations governing the laser beam.

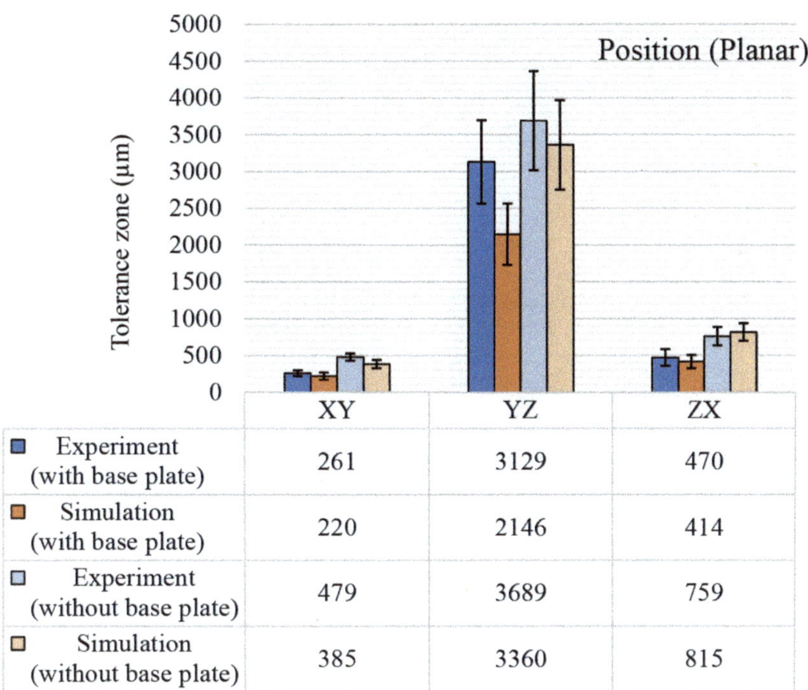

	XY	YZ	ZX
☐ Experiment (with base plate)	261	3129	470
☐ Simulation (with base plate)	220	2146	414
☐ Experiment (without base plate)	479	3689	759
☐ Simulation (without base plate)	385	3360	815

Figure 17. Position tolerance (planar) results (μm), with and without the base plate.

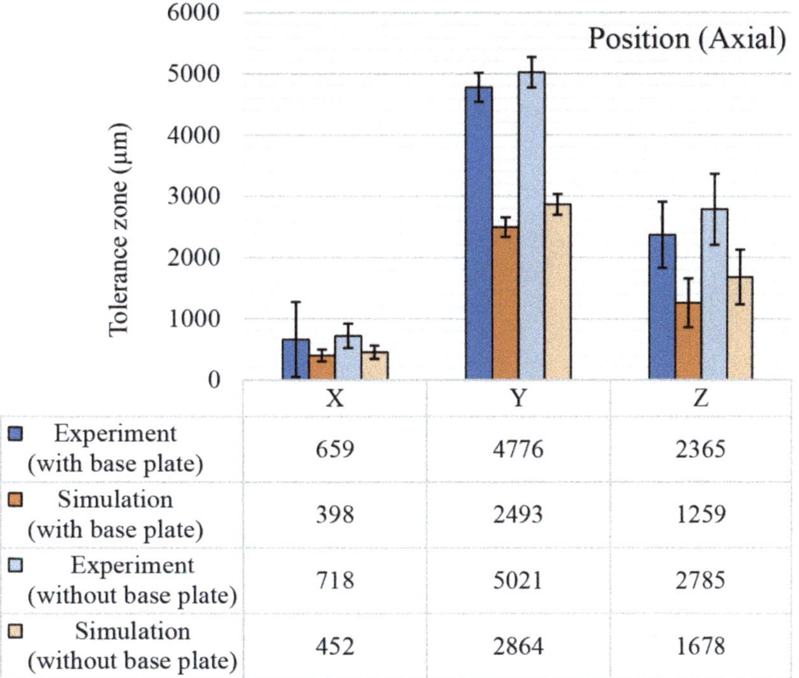

	X	Y	Z
☐ Experiment (with base plate)	659	4776	2365
☐ Simulation (with base plate)	398	2493	1259
☐ Experiment (without base plate)	718	5021	2785
☐ Simulation (without base plate)	452	2864	1678

Figure 18. Position tolerance (axial) results (μm), with and without the base plate.

5.4. Runout Tolerances

Circular runout and total runout tolerance results from both experiments and simulations are shown in Figures 19 and 20, respectively. Circular runout, in theory, is a more precise way of measuring circularity as it considers all the points on the periphery of the circular feature that is measured. This is why almost all the circular runout tolerance zones shown in Figure 19 are 50–100 μm more than the respective circularity tolerance results (Figure 9). The average tolerance band for experimental circular runout is 197 μm with the base plate intact. After the removal of the base plate, the average value reached 217 μm. The standard deviation also increased from 99 μm to 109 μm. The maximum runout was observed for the Y-axis features, and the same trend was observed in the simulation results. However, the simulation results for the circular runout showed a downward mean shift of around 10 μm in average tolerance values and the standard deviation was also reduced by at least 30 μm. The average change was mostly due to the drop in the values in the directions of the Y and X axes. For experimental results, due to the removal of the base plate, surface defects come into play, which is not the case in simulations. This effect is not significant in total runout results, as it considers the complete feature, and the effect diminishes compared to the tolerance zone of the complete feature. As shown in Figure 20, the results for total runout are quite a bit higher than the circular runout and even higher than concentricity results except for the Y-axis features.

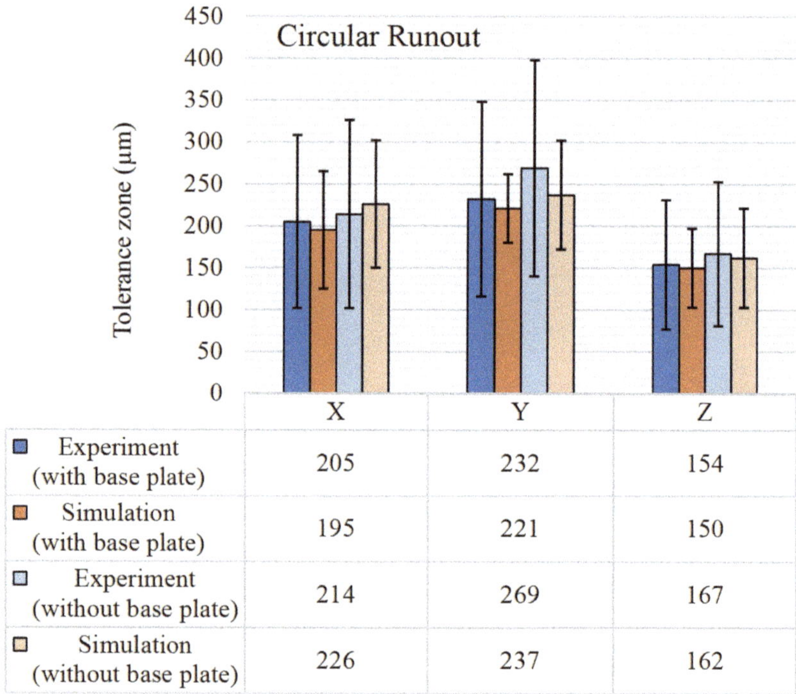

	X	Y	Z
◼ Experiment (with base plate)	205	232	154
◼ Simulation (with base plate)	195	221	150
◻ Experiment (without base plate)	214	269	167
◻ Simulation (without base plate)	226	237	162

Figure 19. Circular runout (μm) tolerance results, with and without the base plate.

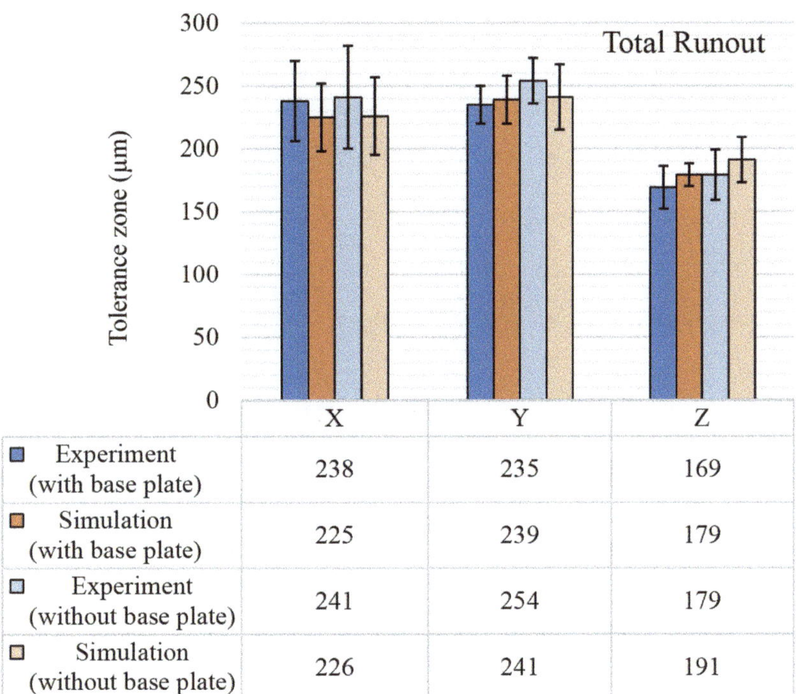

	X	Y	Z
Experiment (with base plate)	238	235	169
Simulation (with base plate)	225	239	179
Experiment (without base plate)	241	254	179
Simulation (without base plate)	226	241	191

Figure 20. Total runout (µm) tolerance results, with and without the base plate.

For total runout, the experimental and simulation results were quite similar for the with and without base plate results, i.e., 214 µm and 224 µm, respectively. The standard deviation increased from 21 µm to 26 µm. Similarly, for the simulation results, the average tolerance values increased from 214 µm to 219 µm and standard deviation values rose from 18 µm to 25 µm. The trend was similar to the trend observed for circular runout and concentricity and the major reasons for this were the support removal and warpage due to residual stresses.

6. Conclusions and Future Scope

Geometric tolerance characterization is necessary for AM processes due to the inherent residual stresses during the build process that leads to variations in geometric tolerances. A new feature-based design of a geometric benchmark test artifact (GBTA) for LPBF is presented, which can characterize major geometric tolerance characteristics in three principal planar directions. The GBTA was manufactured using stainless steel (SS 316L) using a LPBF process and subjected to geometric tolerance measurements. The measurements were conducted on the GBTA in two stages: first with the base plate intact and then with the base plate removed. The geometric tolerance zones for features aligned with principal planar and axial directions were presented. The variations in the geometric tolerance zones with orientations and sizes were discussed.

Major conclusions from this research work are summarized below:

- The results not only justify the new GBTA design and its features, but also gives a quantitative outlook on the variation of the geometric tolerances according to orientation, sizes, and base plate condition;
- The results show that minimizing the residual stress and overall deviations do not lead to minimum tolerance zones for various geometric characteristics that dictate the functionality of the part;

- The circularity and cylindricity tolerance zones show a direct proportionality linkage to the diameter of the feature;
- The orientations of the features lead to a wide range of variation in geometric tolerance results, the average form tolerance increases after the removal of the base plate;
- The orientation and position tolerances also show an increase but, in some cases, the combined effect of the stress relief from removal of the base plate and the variation in the tolerance zone of the datum features minimize the overall tolerance variation;
- The results prove the need for directionality-based analysis of geometric tolerances and the need to consider the removal of the base plate;
- The simulation results are utilized to get an understanding of the tolerance variations and the effect of the residual stresses, with and without the base plate.

However, for specific applications, custom features and analysis are required to ascertain if given tolerance specifications are met. Moreover, since the simulation results are not in 100% agreement with the experimental results, this leads to the conclusion that more research is required to make sure that simulations can precisely predict the geometric tolerances for the LPBF process and can take into account the uncertainty of the process. It is hypothesized that the geometric tolerances will be affected by the selection of process parameters. Therefore, a process parameter optimization for specific geometric tolerances is should be performed after a normative benchmark analysis. The authors' future work will be to create a training database for a machine learning algorithm that can predict variations in geometric tolerances with process parameter changes.

Supplementary Materials: The following are available online at https://www.mdpi.com/article/10.3390/ma14133575/s1, A PDF file containing a set of drawings depicting the size dimensions and geometric tolerances of the geometric benchmark test artifact (GBTA).

Author Contributions: B.S.R.: conceptualization, methodology, formal analysis, investigation, data curation, validation, writing—original draft; T.S.: formal analysis, investigation, data curation; T.W.: conceptualization, writing—review and editing, funding acquisition; M.S.: validation, writing—review and editing, supervision; A.J.Q.: conceptualization, resources, writing—review and editing, analysis, supervision, project administration, funding acquisition. All authors have read and agreed to the published version of the manuscript.

Funding: This research was funded by the Natural Sciences and Engineering Research Council of Canada (NSERC), Government of Canada, under strategic partnership grant number 494158: Holistic Innovation in Additive Manufacturing (HI-AM).

Institutional Review Board Statement: Not applicable.

Informed Consent Statement: Not applicable.

Data Availability Statement: The raw/processed data required to reproduce these findings cannot be shared at this time as the data also form part of an ongoing study.

Acknowledgments: Authors acknowledge Innotech Alberta for providing the samples for the experimental work.

Conflicts of Interest: The authors declare no conflict of interest.

Appendix A. GD&T Characteristics

Table A1. GD&T characteristics, symbols and corresponding descriptions.

GD&T Characteristic Symbol	Control Type	Name	Summary Description
—	Form	Straightness	Controls the straightness of a feature in a relation to its perfect form
▱	Form	Flatness	Controls the flatness of a surface in relation to its own perfect form
○	Form	Circularity (Roundness)	Controls the form of a revolved surface in relation to its perfect form by independent cross-sections
⌀	Form	Cylindricity	Circularity' applied to the entire revolved surface
⊥	Orientation	Perpendicularity	Controls the orientation of a feature that is nominally perpendicular to the primary datum of its datum reference frame
//	Orientation	Parallelism	Controls orientation of a feature that is nominally parallel to the primary datum of its datum reference frame
∠	Orientation	Angularity	Controls orientation of a feature at a specific angle in relation to the primary datum of its datum reference frame
◎	Location	Concentricity	Controls concentricity of a surface of revolution to a central datum
⌖	Location	Position	Controls the location and orientation of a feature in relation to its datum reference frame
≡	Location	Symmetry	Controls the symmetry of two surfaces about a central datum
⌒	Profile	Profile of a line	Controls the size and form of a freeform feature. Additionally controls the location and orientation when a datum reference frame is used
⌓	Profile	Profile of a surface	Profile of a line' applied to the complete feature surface
↗	Runout	Circular Runout	Controls circularity and coaxiality of each circular segment of a surface independently about a coaxial datum
↗↗	Runout	Total Runout	Controls circularity, straightness, coaxiality, and taper of a cylindrical surface about a coaxial datum

Appendix B. GBTA Features

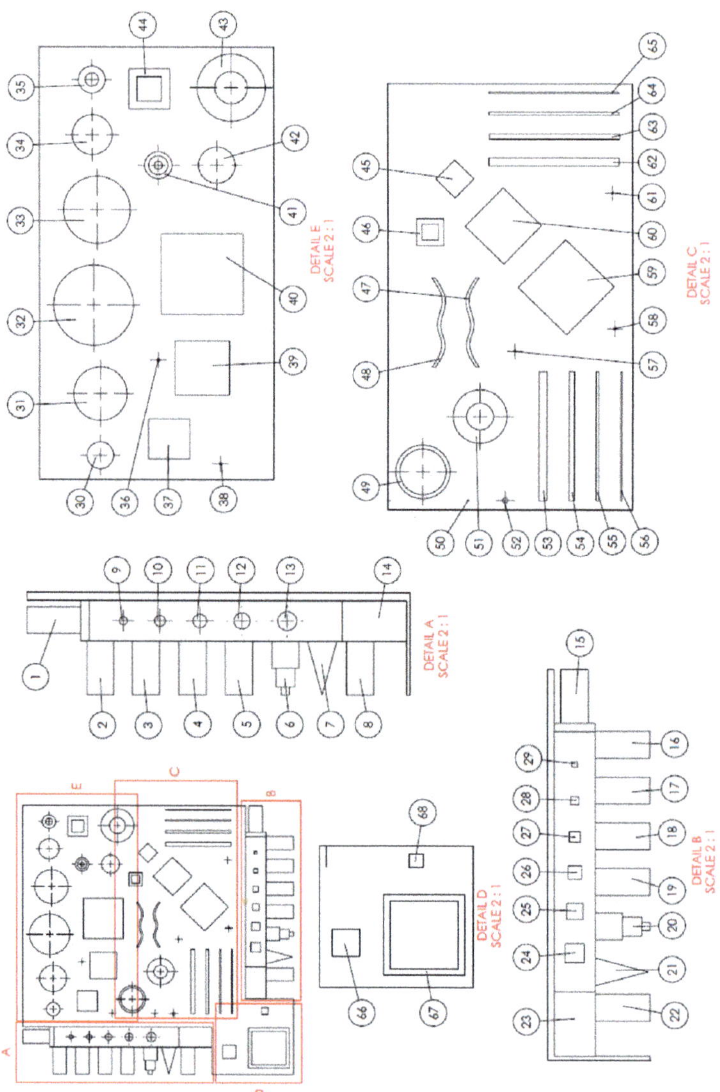

Figure A1. GBTA drawing depicting all the features along with their numbering.

Table A2. GBTA features description as per the numbering.

GD&T Characteristics	Plane/Axis	Features
Flatness	XY	1, 4, 5, 8, 13, 15, 18, 19, 22, 24, 30, 31, 32, 33, 37, 39, 40, 45, 60, 66, 68
	YZ	2, 4, 18, 19, 24, 37, 39, 40, 44, 62, 63, 64, 65, 66
	ZX	4, 5, 16, 18, 24, 37, 39, 40, 44, 53, 54, 55, 56, 66
Straightness	XY	1, 4, 5, 8, 15, 18, 19, 22, 37, 39, 40, 53, 62
	YZ	24, 37, 39, 40, 62, 65, 66, 67
	ZX	24, 37, 39, 40, 53, 56, 66, 67
Parallelism	XY	4, 5, 8, 13, 18, 22, 24, 30, 31, 32, 37, 39, 40, 60, 66
	YZ	2, 4, 18, 19, 24, 37, 39, 40, 62, 63, 64, 65, 66
	ZX	4, 5, 16, 18, 24, 37, 39, 40, 53, 54, 55, 56, 66
Perpendicularity	XY	4, 5, 8, 18, 19, 22, 24, 37, 39, 40, 66
	YZ	24, 37, 39, 40, 62, 63, 64, 65, 66
	ZX	24, 37, 39, 40, 53, 54, 55, 56, 66
Angularity		14, 23
Circularity	X	2, 3, 6B, 6M, 6T, 7
	Y	16, 17, 20B, 20M, 20T, 21
	Z	9, 11, 13, 30, 31, 32, 33, 34, 35, 36, 38, 41B, 41M, 41T, 42, 43O, 43I, 49O, 49I, 50, 51O, 51I, 52, 57, 58, 61
Cylindricity	X	2, 3, 6B, 6M
	Y	16, 17, 20B, 20M
	Z	9, 10, 11, 12, 13, 30, 31, 32, 33, 34, 35, 38, 41B, 41M, 43O, 43I, 49O, 49I, 50, 51O, 51I, 52, 57, 58, 61
Concentricity	X	6M, 6T

References

1. *ISO/ASTM 52900 (ASTM F2792). Additive Manufacturing-General Principles-Terminology*; ASTM International: West Conshohocken, PA, USA, 2015.
2. Thompson, M.K.; Moroni, G.; Vaneker, T.; Fadel, G.; Campbell, R.I.; Gibson, I.; Bernard, A.; Schulz, J.; Graf, P.; Ahuja, B.; et al. Design for Additive Manufacturing: Trends, opportunities, considerations, and constraints. *CIRP Ann. Manuf. Technol.* **2016**, *65*, 737–760. [CrossRef]
3. ISO. *ISO 17296-2. Additive Manufacturing-General Principles-Part 2: Overview of Process Categories and Feedstock*; ISO: Geneva, Switzerland, 2015.
4. Gibson, I.; Rosen, D.; Stucker, B. *Additive Manufacturing Technologies: 3D Printing, Rapid Prototyping, and Direct Digital Manufacturing*; Springer: New York, NY, USA, 2016.
5. Barari, A.; Kishawy, H.A.; Kaji, F.; Elbestawi, M.A. On the surface quality of additive manufactured parts. *Int. J. Adv. Manuf. Technol.* **2017**, *89*, 1969–1974. [CrossRef]
6. Lou, S.; Jiang, X.; Sun, W.; Zeng, W.; Pagani, L.; Scott, P.J. Characterisation methods for powder bed fusion processed surface topography. *Precis. Eng.* **2019**, *57*, 1–15. [CrossRef]
7. Choo, H.; Sham, K.L.; Bohling, J.; Ngo, A.; Xiao, X.; Ren, Y.; Depond, P.J.; Matthews, M.J.; Garlea, E. Effect of laser power on defect, texture, and microstructure of a laser powder bed fusion processed 316L stainless steel. *Mater. Des.* **2019**, *164*, 107534. [CrossRef]
8. Leach, R.K.; Bourell, D.; Carmignato, S.; Donmez, A.; Senin, N.; Dewulf, W. Geometrical metrology for metal additive manufacturing. *CIRP Ann.* **2019**, *68*, 677–700. [CrossRef]
9. Rupal, B.S.; Qureshi, A.J. Geometric Deviation Modeling and Tolerancing in Additive manufacturing: A GDT Perspective. In Proceedings of the 1st Conference of NSERC Network for Holistic Innovation in Additive Manufacturing (HI-AM), Waterloo, ON, Canada, 22–23 May 2018; pp. 1–6.
10. Toguem, S.C.T.; Rupal, B.S.; Mehdi-Souzani, C.; Qureshi, A.J.; Anwer, N. A review of AM artifact design methods. In Proceedings of the 2018 ASPE and Euspen Summer Topical Meeting: Advancing Precision in Additive Manufacturing, Berkeley, CA, USA, 22–25 July 2018.
11. Toguem, S.C.T.; Mehdi-Souzani, C.; Nouira, H.; Anwer, N. Axiomatic design of customised additive manufacturing artefacts. *Procedia CIRP* **2020**, *91*, 899–904. [CrossRef]
12. Moylan, S.; Slotwinski, J.; Cooke, A.; Jurrens, K.; Donmez, M.A. Proposal for a standardized test artifact for additive manufacturing machines and processes. In Proceedings of the 2012 Annual International Solid Freeform Fabrication Symposium, Austin, TX, USA, 6–8 August 2012.

13. Mahesh, M.; Wong, Y.S.; Fuh, J.Y.H.; Loh, H.T. Benchmarking for comparative evaluation of RP systems and processes. *Rapid Prototyp. J.* **2004**, *10*, 123–135. [CrossRef]
14. Shahrain, M.; Didier, T.; Lim, G.K.; Qureshi, A.J. Fast Deviation Simulation for 'Fused Deposition Modeling' Process. *Procedia CIRP* **2016**, *43*, 327–332. [CrossRef]
15. Rebaioli, L.; Fassi, I. A review on benchmark artifacts for evaluating the geometrical performance of additive manufacturing processes. *Int. J. Adv. Manuf. Technol.* **2017**, *93*, 1–28. [CrossRef]
16. Mahmood, S.; Qureshi, A.J.; Goh, K.L.; Talamona, D. Tensile strength of partially filled FFF printed parts: Experimental results. *Rapid Prototyp. J.* **2017**, *23*, 122–128. [CrossRef]
17. Saqib, S.; Urbanic, J. An Experimental Study to Determine Geometric and Dimensional Accuracy Impact Factors for Fused Deposition Modelled Parts. In *Enabling Manufacturing Competitiveness and Economic Sustainability*; Springer: Berlin/Heidelberg, Germany, 2012; pp. 293–298.
18. Mahmood, S.; Qureshi, A.J.; Talamona, D. Taguchi based process optimization for dimension and tolerance control for fused deposition modelling. *Addit. Manuf.* **2018**, *21*, 183–190. [CrossRef]
19. Kruth, J.-P.; Vandenbroucke, B.; van Vaerenbergh, J.; Mercelis, P. Benchmarking of different sls/slm processes as rapid manufacturing techniques. In Proceedings of the International Conference Polymers & Moulds Innovations PMI 2005, Gent, Belgium, 20–24 April 2005.
20. Galati, M.; Minetola, P. Analysis of Density, Roughness, and Accuracy of the Atomic Diffusion Additive Manufacturing (ADAM) Process for Metal Parts. *Materials* **2019**, *12*, 4122. [CrossRef]
21. Teeter, M.G.; Kopacz, A.J.; Nikolov, H.N.; Holdsworth, D.W. Metrology test object for dimensional verification in additive manufacturing of metals for biomedical applications. *Proc. Inst. Mech. Eng. Part H J. Eng. Med.* **2015**, *229*, 20–27. [CrossRef]
22. Delgado, J.; Ciurana, J.; Reguant, C.; Cavallini, B. Studying the repeatability in DMLS technology using a complete geometry test part. Innovative Developments in Design and Manufacturing: Advanced Research in Virtual and Rapid Prototyping. In Proceedings of the VRP4, Leiria, Portugal, 22 September 2009; pp. 367–372.
23. ISO. *ISO 1101. Geometrical product specifications (GPS)—Geometrical Tolerancing—Tolerances of Form, Orientation, Location and Run-Out*; ISO: Geneva, Switzerland, 2017.
24. Hanumaiah, N.; Ravi, B. Rapid tooling form accuracy estimation using region elimination adaptive search based sampling technique. *Rapid Prototyp. J.* **2007**, *13*, 182–190. [CrossRef]
25. Cooke, A.L.; Soons, J.A. Variability in the Geometric Accuracy of Additively Manufactured Test Parts. In Proceedings of the 21st Annual Solid Freeform Fabrication Symposium: An Additive Manufacturing Conference, Austin, TX, USA, 9–11 August 2010; pp. 1–12.
26. Han, J.; Wu, M.; Ge, Y.; Wu, J. Optimizing the structure accuracy by changing the scanning strategy using selective laser melting. *Int. J. Adv. Manuf. Technol.* **2018**, *95*, 4439–4447. [CrossRef]
27. Fahad, M.; Hopkinson, N. Evaluation and comparison of geometrical accuracy of parts produced by sintering-based additive manufacturing processes. *Int. J. Adv. Manuf. Technol.* **2017**, *88*, 3389–3394. [CrossRef]
28. Toguem, S.-C.T.; Mehdi-Souzani, C.; Anwer, N.; Nouira, H. Customized design of artefacts for additive manufacturing. In Proceedings of the Euspen's 19th International Conference & Exhibition, Bilbao, Spain, 3–7 June 2019.
29. Rupal, B.S.; Ahmad, R.; Qureshi, A.J. Feature-Based Methodology for Design of Geometric Benchmark Test Artifacts for Additive Manufacturing. *Procedia Cirp* **2018**, *70*, 84–89. [CrossRef]
30. ISO. *ISO 17296-3. Additive Manufacturing-General Principles-Part 3: Main Characteristics and Corresponding Test Methods*; ISO: Geneva, Switzerland, 2014.
31. *ISO/ASTM 52902:2019 [ASTM F42]. Additive manufacturing—Test artifacts—Geometric Capability Assessment of Additive Manufacturing Systems*; ASTM International: West Conshohocken, PA, USA, 2019.
32. Bertini, L.; Bucchi, F.; Frendo, F.; Moda, M.; Monelli, B.D. Residual stress prediction in selective laser melting. *Int. J. Adv. Manuf. Technol.* **2019**, *105*, 609–636. [CrossRef]
33. Tan, J.H.K.; Sing, S.L.; Yeong, W.Y. Microstructure modelling for metallic additive manufacturing: A review. *Virtual Phys. Prototyp.* **2020**, *15*, 87–105. [CrossRef]
34. Plocher, J.; Panesar, A. Review on design and structural optimisation in additive manufacturing: Towards next-generation lightweight structures. *Mater. Des.* **2019**, *183*, 108164. [CrossRef]
35. Renishaw PLC. "AM250". Available online: https://www.renishaw.com/en/am250--15253 (accessed on 19 November 2020).
36. *BS 7172: Guide to Assessment of Position, Size and Departure from Nominal Form of Geometric Features*; BSI: London, UK, 1989.
37. Hocken, R.J.; Pereira, P.H. (Eds.) *Coordinate Measuring Machines and Systems*; CRC Press: Boca Raton, FL, USA, 2016.
38. Autodesk Inc. "Netfabb". Available online: https://www.autodesk.com/products/netfabb/ (accessed on 10 January 2021).
39. Hodge, N.E.; Ferencz, R.M.; Solberg, J.M. Implementation of a thermomechanical model for the simulation of selective laser melting. *Comput. Mech.* **2014**, *54*, 33–51. [CrossRef]
40. Peng, H.; Ghasri-Khouzani, M.; Gong, S.; Attardo, R.; Ostiguy, P.; Rogge, R.B.; Gatrell, B.A.; Budzinski, J.; Tomonto, C.; Neidig, J.; et al. Fast prediction of thermal distortion in metal powder bed fusion additive manufacturing: Part 2, a quasi-static thermo-mechanical model. *Addit. Manuf.* **2018**, *22*, 869–882. [CrossRef]
41. Rupal, B.S.; Anwer, N.; Secanell, M.; Qureshi, A.J. Geometric Tolerance and Manufacturing Assemblability Estimation of Metal Additive Manufacturing (AM) Processes. *Mater. Des.* **2020**, *194*, 108842. [CrossRef]

42. Gouge, M.; Denlinger, E.; Irwin, J.; Li, C.; Michaleris, P. Experimental validation of thermo-mechanical part-scale modeling for laser powder bed fusion processes. *Addit. Manuf.* **2019**, *29*, 100771. [CrossRef]
43. Zhang, Y.; Zhang, J. Finite element simulation and experimental validation of distortion and cracking failure phenomena in direct metal laser sintering fabricated component. *Addit. Manuf.* **2017**, *16*, 49–57. [CrossRef]
44. GOM GmBH. GOM Inspect. Available online: https://www.gom.com/3d-software/gom-inspect.html (accessed on 26 February 2021).
45. Li, C.; Liu, J.F.; Fang, X.Y.; Guo, Y.B. Efficient predictive model of part distortion and residual stress in selective laser melting. *Addit. Manuf.* **2017**, *17*, 157–168. [CrossRef]

MDPI

Article

Methodology for the Quality Control Process of Additive Manufacturing Products Made of Polymer Materials

Grzegorz Budzik [1], Joanna Woźniak [2], Andrzej Paszkiewicz [3,*], Łukasz Przeszłowski [1], Tomasz Dziubek [1] and Mariusz Dębski [1]

[1] Department of Machine Design, The Faculty of Mechanical Engineering and Aeronautics, Rzeszow University of Technology, al. Powstańców Warszawy 12, 35-959 Rzeszów, Poland; gbudzik@prz.edu.pl (G.B.); lprzeszl@prz.edu.pl (Ł.P.); tdziubek@prz.edu.pl (T.D.); m.debski@prz.edu.pl (M.D.)

[2] Department of Management Systems and Logistics, The Faculty of Management, Rzeszow University of Technology, al. Powstańców Warszawy 12, 35-959 Rzeszów, Poland; j.wozniak@prz.edu.pl

[3] Department of Complex Systems, The Faculty of Electrical and Computer Engineering, Rzeszow University of Technology, al. Powstańców Warszawy 12, 35-959 Rzeszów, Poland

* Correspondence: andrzejp@prz.edu.pl

Abstract: The objective of this publication is to present a quality control methodology for additive manufacturing products made of polymer materials, where the methodology varies depending on the intended use. The models presented in this paper are divided into those that are manufactured for the purpose of visual presentation and those that directly serve the needs of the manufacturing process. The authors also a propose a comprehensive control system for the additive manufacturing process to meet the needs of Industry 4.0. Depending on the intended use of the models, the quality control process is divided into three stages: data control, manufacturing control, and post-processing control. Research models were made from the following materials: RGD 720 photopolymer resin (PolyJet method), ABS M30 thermoplastic (FDM method), E-Partial photopolymer resin (DLP method), PLA thermoplastic (FFF method), and ABS thermoplastic (MEM method). The applied measuring tools had an accuracy of at least an order of magnitude higher than that of the manufacturing technologies used. The results show that the PolyJet method is the most accurate, and the MEM method is the least accurate. The findings also confirm that the selection of materials, 3D printing methods, and measurement methods should always account not only for the specificity and purpose of the model but also for economic aspects, as not all products require high accuracy and durability.

Keywords: additive manufacturing; quality control management; process optimization; computing techniques; ICT systems

check for **updates**

Citation: Budzik, G.; Woźniak, J.; Paszkiewicz, A.; Przeszłowski, Ł.; Dziubek, T.; Dębski, M. Methodology for the Quality Control Process of Additive Manufacturing Products Made of Polymer Materials. *Materials* **2021**, *14*, 2202. https://doi.org/10.3390/ma14092202

Academic Editor: Mika Salmi

Received: 31 March 2021
Accepted: 23 April 2021
Published: 25 April 2021

Publisher's Note: MDPI stays neutral with regard to jurisdictional claims in published maps and institutional affiliations.

1. Introduction

Quality control is one of the key elements of the manufacturing process, regardless of the number of manufactured products. Depending on the stage of the technological process, methods of quality control include different activities with varying scopes, such as ensuring the correctness of designed 3D CAD models; verifying prepared process data; performing visual product control; and controlling dimensional and shape accuracy, surface geometric structure, and material internal structure, particularly in the case of safety-critical elements of vehicles and aircrafts [1–4]. Quality control processes in manufacturing systems have been the subject of numerous publications. Many of them are related to computing techniques, such as contact coordinate computing [5,6], geometric measurement and analysis using optical computing systems [7–10], and geometric accuracy analyses using volume-based methods, including computed tomography [11–13]. Product research that applies CT can produce comprehensive results that reveal the geometric accuracy of both the external and internal surfaces of the product, as well as its material structure. However, it is costly and time-consuming and requires specialized equipment. Therefore, CT is reserved for particular product groups, such as aircraft engine blades [14–16]. Quality control in the manufacturing process can be accelerated by applying gauges, but these instruments provide information relevant to a particular product dimension, for instance, the measurement necessary for the mounting or functioning of an element in a machine set [17]. One of the methods for accelerating computing is the application of automated 3D scanning systems, including hardware and software automation, based on robotized measuring sockets that apply rapid scanning procedures. A perfect solution in this case is the application of blue light 3D scanners [18], which enable the quality control of both small-dimension elements and large-dimension artifacts directly on the production line: for instance, these instruments are useful for ensuring the accuracy of vehicle bodywork.

The applied additive manufacturing models and 3D printing method affect several aspects of product quality control and should be considered when developing quality assessment processes. For the quality control of an additive manufacturing model, it is important to consider its place in the technological process. A model that has the properties of an end product will be subjected to quality control before it can be mounted or sold. Visual models are often indispensable at the stages of product conception or the determination of ergonomic characteristics [19,20], but quality control is not usually particularly strict in such cases. For the additive manufacturing of products made of polymer materials, additional mechanical or thermal-chemical processing is not usually applied. Taking this into account, after finishing 3D printing and post-processing, an element has to be subjected to thorough quality control, including both visual control and the analysis of dimensional and shape accuracy [21]. Additive manufacturing technologies that are currently applied to the elements of metal powders (SLS/SLM/DMLS/PBF) can produce semi-finished products that require further mechanical processing and, frequently, thermal-chemical processing. This results in the need for quality control of both the semi-finished product and end product of the additive manufacturing process [22]. Therefore, the material of a product, as well as the additive manufacturing method, can be an important basis for determining the appropriate quality control methodology, which is also defined in particular standards [23–31].

Visual prototype assessment plays a significant role in the quality control of additive manufacturing products made of polymer materials. These assessments include the analysis of model structure continuity, correctness in reference to a 3D CAD model, and the colors of elements made of numerous materials that come in various colors or external color textures [32]. Dimensional and shape accuracy must be analyzed with the appropriate tools, devices, and computing systems [18]. In the majority of cases, the dimensional accuracy of products made of polymer materials does not exceed $\pm$ 0.05 mm. For this reason, these products may be effectively evaluated with workshop measuring instruments, such as a caliper or micrometer [33]. For products with complex shapes, blue light 3D scanning systems, such as 3D scanners from the GOM Company (Braunschweig, Germany), can be applied to the computing process [34]. These systems enable the relatively rapid quality controls of products in real-time or offline with the application of, for instance, computing results saved in a data cloud. With this approach, 3D scanners can be ideal measuring instruments that simultaneously perform detailed quality control and geometric analyses, which can significantly accelerate the quality control process, particularly in the case of artifacts with complex shapes. The methodology of such an approach to quality control can be developed based on the principles of Industry 4.0, in which available data are used for multi-level and parallel control computing. The application of 3D scanning as a control tool for additive manufacturing products has yet another aspect relevant to data processing: an additive manufacturing product is frequently manufactured on the basis of a 3D CAD/3D STL model, and the STL format is primarily used in the analysis of product dimensional accuracy. In this case, data in STL format from the computing point cloud are compared with data saved in the form of a nominal STL model based on a 3D CAD model [35]. This is a significant element of the quality control methodology, a major part of which is analyzing and processing numerical data.

In the literature on this subject, the quality control of products fabricated via incremental manufacturing is most often dominated by technical aspects. However, there is still a lack of research and analysis on the management and systematic ordering of processes [12,36].

In line with the above factors, the principal objective of this publication is to present a quality control methodology for additive manufacturing products made of polymer materials, where the methodology varies depending on the intended use. In this paper, the 3D models used for quality control procedures are divided into those that are manufactured for the purpose of visual presentation and those that directly serve the needs of the manufacturing process. For each group, an algorithm of the developed quality control methodology is presented graphically. The authors also propose a comprehensive control system for the additive manufacturing process to meet the needs of Industry 4.0, which applies to both direct control by a qualified specialist and complete automation, as well as remote supervision.

The findings presented in this paper can be used in practice by academia and manufacturing companies that use additive manufacturing technologies in their production processes. In the context of Industry 4.0, the developed quality process for additive manufacturing products made from polymeric materials includes access to a large production database. In the initial phase, an image is formed during production to identify geometry deviations, and on this basis, factors that influence their formation can be determined. The measurement data obtained during inspections can be used to adjust both the geometry of the 3D CAD models and the parameters of the manufacturing processes, thus increasing the accuracy of manufactured parts. Due to its rapid response to data, an Industry 4.0 system is able to immediately react to phenomena that occur during production, which has a widespread impact on the dimensional and shape accuracy of manufactured products.

2. Standards Applied in 3D Printing Quality Control

With the increasing application of additive manufacturing technologies in the fabrication of end products, it became necessary to establish standards for 3D printing quality and the organization of processes within the entire delivery chain [37]. International activity in the scope of additive manufacturing technologies commenced at the end of 2011 when the Technical Committee for Additive Manufacturing, designated ISO/TC 261, was established. Since the very first days of its activity, it has worked in close cooperation and formed joint committees with ASTM F42 Additive Manufacturing Technologies. The fruits of this work are numerous standards in the ISO/ASTM series. Publications and projects developed in the field of additive manufacturing are primarily related to general principles and terminology, process categorization, profiles and research methods, computer process description (with the application of a determined standard data saving mode and data in the form of a file), and the assessment of the geometric accuracy of additive manufacturing processes.

Major publications in the field of 3D printing are as follows:

- PN/EN ISO/ASTM 52900:2017-06 Additive Manufacturing—General Principles—Terminology. This includes terms and definitions related to additive manufacturing technologies, in which physical structures (geometries) are developed by adding additional material layers.
- PN/EN ISO/ASTM 52901:2019-01 Additive Manufacturing—General Principles—Requirements relevant to parts manufactured by means of additive manufacturing (AM) processes. This document is applicable to purchasing parts produced with additive manufacturing technologies; it sets the minimum requirements that must be met for the acceptance of products. The document recommends that additional stricter requirements be determined in the course of placing an order.
- PN/EN ISO 17296-2:2016-10 Additive Manufacturing—General Principles—Part 2: Overview of process categories and feedstock. This document presents the foundations

of the additive manufacturing process and a review of existing process categories. It also describes how to apply different kinds of materials to shape the geometry of a given product in different process categories.

- PN/EN ISO 17296-3:2016-10 Additive Manufacturing—General Principles—Part 3: principal characteristics and appropriate research methods. This standard is primarily dedicated to machine producers and users, feedstock and part suppliers, and recipients and customers. It contains basic requirements that ought to be applied to parts produced with additive manufacturing technologies.
- PN/EN ISO 17296-4:2016-10 Additive Manufacturing—General Principles—Part 4: Overview of data processing. This standard contains issues related to data exchange. It also describes terms and definitions applied to the exchange of information about the geometry of parts (products) manufactured with additive manufacturing.
- ISO/ASTM 52902:2019 Additive Manufacturing—Test Artifacts—Geometric capability assessment of additive manufacturing systems. This standard contains a general description of model specimen geometries, together with the quantitative and qualitative computing to be conducted on a test specimen(s) in order to assess the efficiency of additive manufacturing (AM) systems.
- ISO/ASTM TR 52912:2020 Additive manufacturing—Design—Functionally graded additive manufacturing. The objective of this document is to present a conceptual understanding of Functionally Graded Additive Manufacturing (FGAM).
- ISO/ASTM 52911-1:2019 Additive manufacturing—Design—Part 1: Laser-based powder bed fusion of metals. This standard defines the characteristics of the laser-based powder bed fusion of metals (PBF-LB/M) and contains detailed design recommendations.
- ISO/ASTM 52911-2:2019 Additive manufacturing—Design—Part 2: Laser-based powder bed fusion of polymers. This document defines the characteristics of the laser-based powder bed fusion of polymers (LB-PBF/P) and contains detailed design recommendations.
- ISO/ASTM 52902:2019 Additive Manufacturing—Test Artifacts—Geometric capability assessment of additive manufacturing systems. This standard contains a general description of model specimen geometries, together with the quantitative and qualitative computing to be conducted on a test specimen(s) in order to assess the efficiency of additive manufacturing (AM) systems.

Additionally, the Committee for Standardization continually works on further standards applicable to additive manufacturing.

3. Research Methodology

3.1. Division of Models According to Their Intended Use

On the basis of work conducted in the Rapid Prototyping System Laboratory of the Rzeszów University of Technology, the authors divided additive manufacturing products according to their intended use:

(1) Models that are manufactured for the purpose of visual presentation, including the following:

- Conceptual prototypes: models that present the simplified constructional and functional assumptions of a product; any technique can be applied to manufacture them in order to present a general conception relatively rapidly and economically. The basis for making it may be a product sketch made by, for instance, a fine artist.
- Numerical prototypes: models that are developed in the software environment. Numerical prototypes include a model or a set of models to be visualized for kinematic simulations, load simulations, preparation of data for manufacturing, and verification based on computer-assisted systems.
- Visual prototypes: models that present the actual dimensions or an assumed scale, geometric characteristics, and the colors and/or quality of a product surface.

- Ergonomic prototypes: models that specify the conditions of a product functioning in its intended environment in reference to its future users, taking into account ergonomic technical assumptions.

(2) Models that are directly connected to the manufacturing process, which are divided into the following categories:

- Technological prototypes: models that are used to develop and verify technological assumptions for the manufacturing process of a product. Technological prototypes may be developed for and specific to particular stages of the technological process.
- Construction prototypes: models that are intended for a comprehensive assessment of a construction solution on the basis of target functionality and the assumptions of a highly detailed technological process.
- Functional prototypes: models that enable the assessment of the main product functions in conditions similar to the actual ones, taking into account operating processes in a simplified configuration. They may be made from materials with properties that are similar to those of materials of the end product.
- Technical prototypes: models that have any or all characteristics (functional and visual) of an end product, so they can be tested in real operating conditions. These models are fabricated with the materials used to make the end product. This makes it possible to prepare the target technological process for manufacturing conditions.

3.2. Developing Algorithms of Quality Control Processes for Particular Model Groups

(1) Algorithm of the quality control process for models manufactured for the purpose of visual presentation

A good practice of enterprises in the 3D industry is to develop an algorithm for completing a work order for a 3D production model, as well as activities related to quality control. The algorithm presented in Figure 1 may be a sui generis mode of action for manufacturing models intended for visual presentation, in which dimensional and shape accuracy is less significant.

In the algorithm presented in Figure 1, the manufactured 3D model is checked to ensure the correctness of the designed 3D CAD models and confirm the prepared process data, and only visual control is conducted.

(2) Algorithm of the quality control process for models that are directly connected to the manufacturing process

The algorithm presented in Figure 2 is specific to models that are directly connected to the manufacturing process. In this case, it is very important to verify the dimensional and shape accuracy of the product.

For manufacturing models that are directly connected to the manufacturing process, visual control is the first stage of verifying quality compliance after manufacturing a physical artifact, and the result determines further courses of action. If damage to the model is visible to the 'naked eye', then it is usually treated as a non-compliant product. Then, the discovered errors are analyzed, and the printing process has to be recommenced. If no such damage is apparent, then additional quality verification activities are conducted by means of measuring instruments, gauges, automated and manual scanners, tomography, or coordinate computing techniques. Therefore, the course of 3D print quality control may be highly tailored and differ between products.

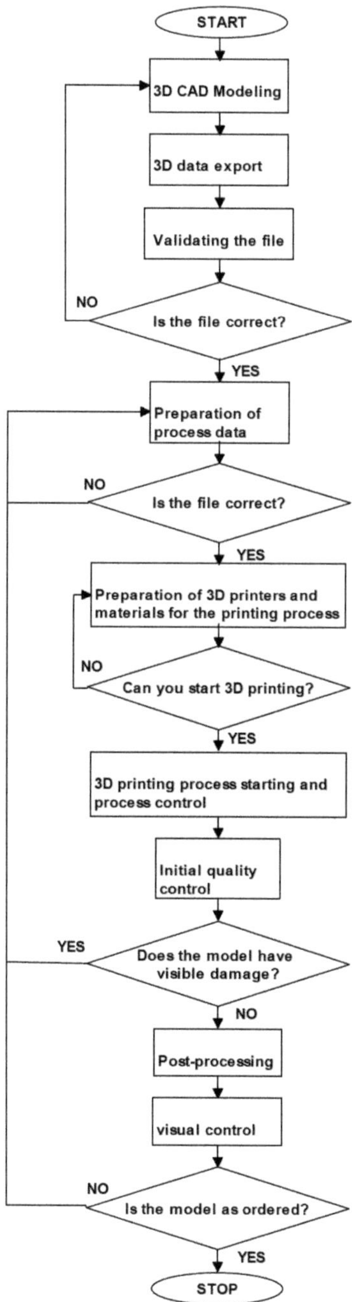

Figure 1. Algorithm of the quality control process for models manufactured for the purpose of visual presentation.

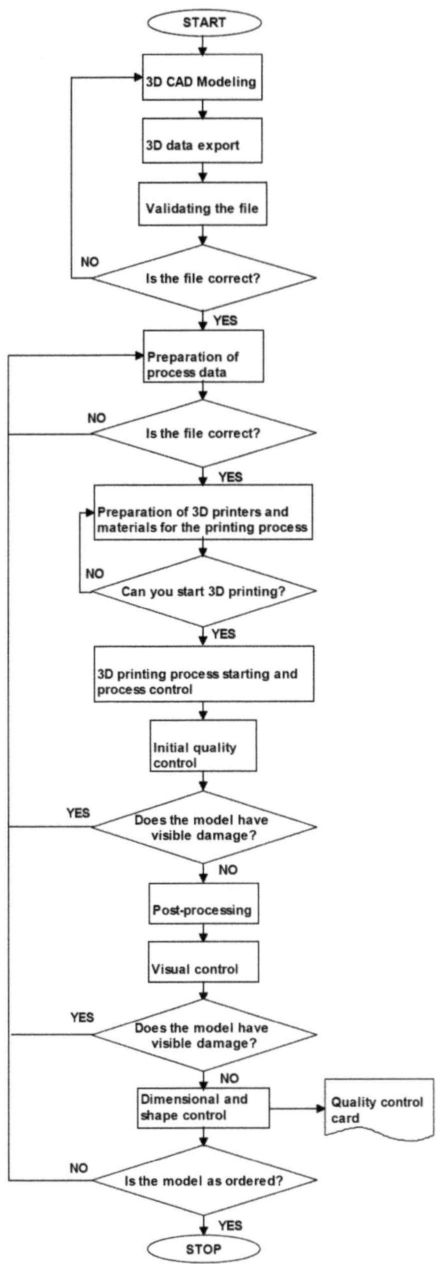

Figure 2. Algorithm of the quality control process for models that are directly connected to the manufacturing process.

3.3. Proposal of the Quality Control System

In the course of this work, the authors developed a system that can be applied to various quality control methods and is adaptable to different measuring instruments and manufacturing process supervision methods. The general concept of the proposed system is presented in Figure 3. Its key feature is its readiness to be completely integrated with ICT solutions within the framework of Industry 4.0. This system is divided into three separate control phases: data control (3D CAD models and process data), visual manufacturing control, and post-processing control. Post-processing control is further divided into two stages: control with specialized measuring instruments and control with a contactless optical system.

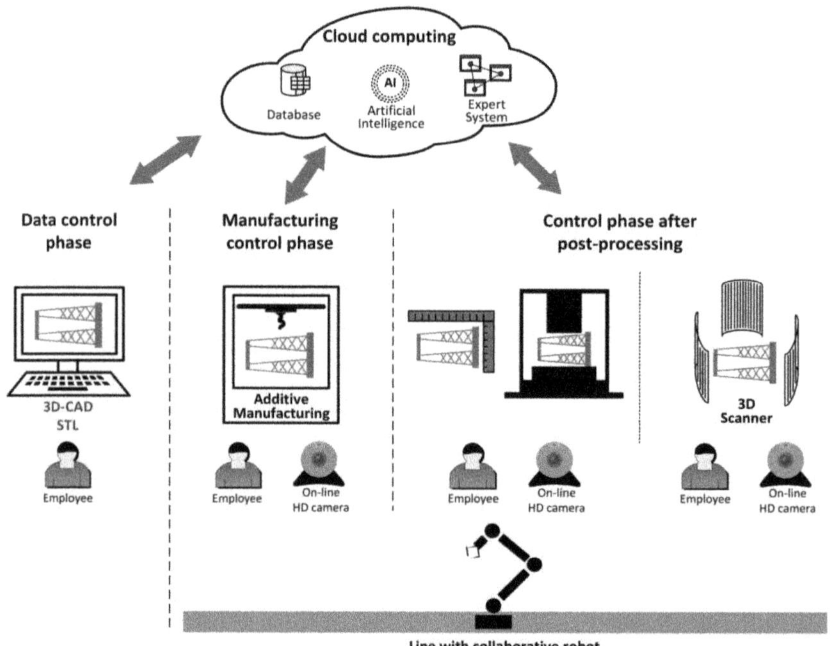

Figure 3. Manufacturing process control in additive manufacturing in the context of Industry 4.0.

The proposed architecture of the comprehensive system for additive manufacturing control accounts for both direct control by qualified specialists and complete automation, as well as remote supervision, to meet the needs of Industry 4.0. Both forms of supervision can be applied simultaneously: one to meet in-house needs and the other in the case of supervision by a customer. In particular, remote supervision has enormous potential within the scope of dispersed production lines, which are typical in the context of the global economy [32,38,39]. Furthermore, the presented system architecture involves local repositories of product and component quality records, and it collects, synchronizes, and processes data in a computing cloud. Thus, the features of the proposed system meet the needs of Industry 4.0, as it combines and integrates individual design and manufacturing processes, available knowledge, and the experience and skills of a dispersed team of specialists. The fulfillment of these needs is aimed at supporting diverse production systems and increasing the availability and efficiency of these processes with the use of currently available computer network technologies, IT systems, and the automation and control of manufacturing processes. Therefore, this system meets the requirements of Industry 4.0 in all indicated areas. It enables the integration of remote resources of both individual elements and entire lines, the use of information compiled in knowledge bases, and the use of modern systems based on artificial intelligence. Moreover, it is adapted to exploit the potential for remote operation and consultation. Additionally, the proposed architecture has features of open systems, including scalability, customization, and interoperability between different systems and resources.

This system was implemented in the Rapid Prototyping System Laboratory of the Rzeszów University of Technology under conditions of complete automation and remote supervision. The proposed IT system enables remote access to design tools and software for supervising the operation of a 3D printer. Within this system, the storage of digital quality records in a cloud was developed in order to ensure constant access to, and verification of, data in all phases of process control. Therefore, possible deviations from the initially set values of quality parameters can be rapidly and remotely assessed. The manufacturing and post-processing phases were integrated with the application of a robot (Universal Robot UR3, Universal Robots, Katowice, Poland).

The described approach ensures the 24/7 automation of the control process and, simultaneously, enables the completion of work despite restrictions connected to the SARS-CoV-2 pandemic. Additionally, applying a rotating high-resolution camera ensures the maintenance of constant visual control [40,41]. The contemporaneous use of cameras can ensure a high approximation factor, and thus, quality control can be performed in the course of printing and in the post-processing phase. To this end, a turntable, among other equipment, is used during operation to enable the accurate (visual) verification of a manufactured component from every direction. Due to the large approximation factor, inaccuracies that are invisible to the human eye can be captured with the application of specialized tools prior to subsequent stages of control. Early identification of such deviations can reduce the cost and time involved in further control processes.

Figure 4a depicts a control process with the application of a camera. In Figure 4b, a collaborative robot working with a 3D printer (Prusa i3 MK3S) is presented.

(a)

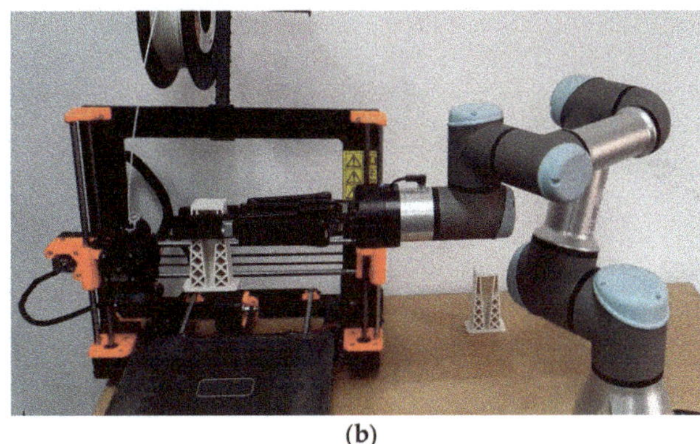

(b)

Figure 4. (**a**) Process control with the application of a camera; (**b**) Collaborative robot (Universal Robot UR3) with a 3D printer.

In this study, the presented system was applied to the manufacturing of a car mirror.

As presented in Figure 3, the architecture of the control system can include specialized IT tools such as export systems and AI mechanisms, which, in the future, will support control processes, particularly those in the specialized production of elements for the aviation and automotive industry.

4. Quality Control of a Car Mirror Holder

A 3D CAD model of a car mirror holder was applied in the analysis of print quality control. The authors' intention was to ensure that the chosen element was universal enough to be used by companies in the automotive industry. When commencing the rapid prototyping process, the dimensions of the designed artifact should be controlled (verify if the model was not scaled). Similarly, it is very important to conduct an initial verification of the model after data processing; in this study, initial verification was performed with the 3D-Tool program (Figure 5).

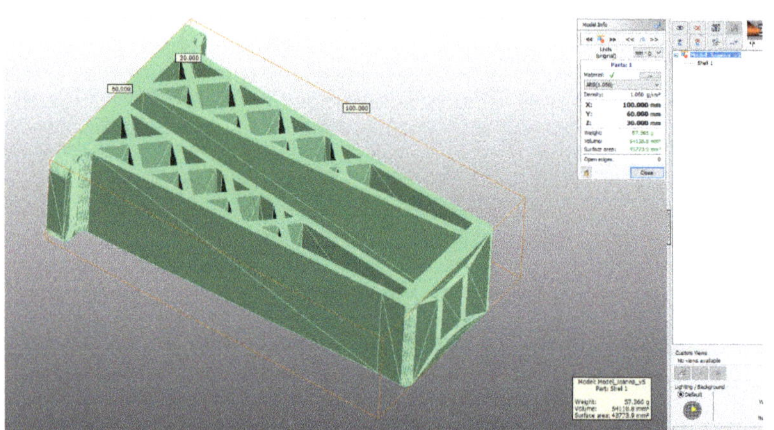

Figure 5. The 3D model (CAD) of a car mirror holder in the 3D-Tool program.

The subsequent stage was the selection of methods and materials for printing. The materials used for printing were the RGD720 photopolymer resin (applied method: PolyJet), ABS M30 thermoplastic (applied method: FDM), E-Partial photopolymer resin (applied method: DLP), PLA thermoplastic (applied method: FFF), and ABS thermoplastic (applied method: MEM). The materials used for printing, together with the model weight, thickness, and area, are presented in Table 1.

Table 1. Materials applied in the printing of particular models, presented as designed in the 3D-TOOL program.

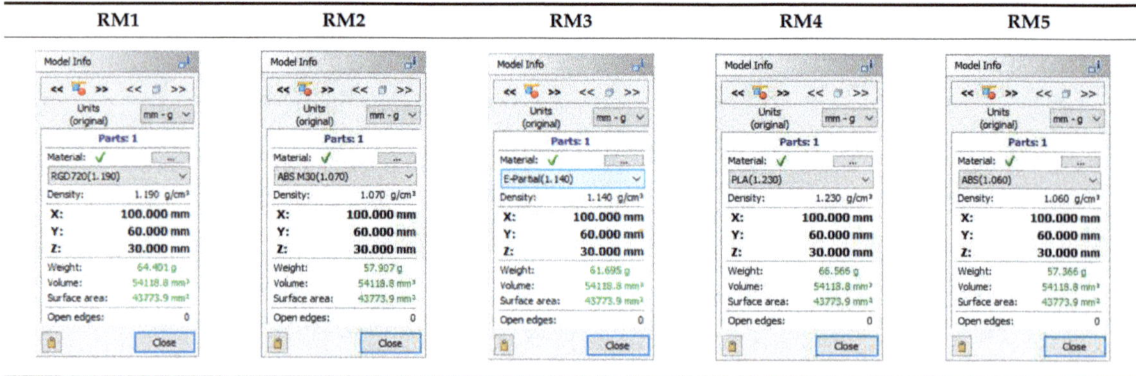

RM1	RM2	RM3	RM4	RM5
Material: RGD720(1.190)	Material: ABS M30(1.070)	Material: E-Partial(1.140)	Material: PLA(1.230)	Material: ABS(1.060)
Density: 1.190 g/cm³	Density: 1.070 g/cm³	Density: 1.140 g/cm³	Density: 1.230 g/cm³	Density: 1.060 g/cm³
X: 100.000 mm	X: 100.000 mm	X: 100.000 mm	X: 100.000 mm	X: 100.000 mm
Y: 60.000 mm	Y: 60.000 mm	Y: 60.000 mm	Y: 60.000 mm	Y: 60.000 mm
Z: 30.000 mm	Z: 30.000 mm	Z: 30.000 mm	Z: 30.000 mm	Z: 30.000 mm
Weight: 64.401 g	Weight: 57.907 g	Weight: 61.695 g	Weight: 66.566 g	Weight: 57.366 g
Volume: 54118.8 mm³	Volume: 54118.8 mm³	Volume: 54118.8 mm³	Volume: 54118.8 mm³	Volume: 54118.8 mm³
Surface area: 43773.9 mm²	Surface area: 43773.9 mm²	Surface area: 43773.9 mm²	Surface area: 43773.9 mm²	Surface area: 43773.9 mm²
Open edges: 0	Open edges: 0	Open edges: 0	Open edges: 0	Open edges: 0

In the manufacturing process, the following 3D printers were applied: Object Eden 260V, 3D STRATASYS F170, EnvisionTEC Vida, Prusa i3 MK3, and UP BOX+.

The prototype was included in the group of models that are directly connected to the manufacturing process. Then, the quality control process for models after post-processing was divided into three stages: visual control, control with a caliper, and control with a contactless optical system. The computing techniques were selected on the basis of the specific profile and the intended use of the model, as well as financial aspects.

The measuring tools used to control the dimensional and shape accuracy should be selected in such a way as to ensure that reliable results are obtained. It is necessary to maintain an appropriate level of accuracy in the measurement process, which will eliminate possible measurement errors while minimizing uncertainty. For these reasons, the applied measuring devices had an accuracy of at least an order of magnitude higher than that of the manufacturing technologies used. Therefore, the analysis of measurement errors was unnecessary, and the dimensional and shape accuracy of the research models could be reliably assessed.

4.1. Visual Control

Visual control constitutes the most cost-effective qualitative assessment method. It does not require costly equipment, nor does it result in the destruction of the assessed item. Despite its numerous assets, visual control is not flawless, principally because it is impossible to present the obtained results in SI units. It should also be noted that visual control entails some bias, as it depends on a given employee's predispositions and expertise; therefore, correctness is not guaranteed with this method [42].

Visual control is the most widely used nondestructive technique because it is simple and can be conducted quickly [13]. In principle, visual investigation is the first to be conducted among all qualitative assessments. Thus, the research models were verified by checking for imperfections that were visible to the naked eye. In this research project, the visual investigation was conducted to assess the following model characteristics: general model representation, surface deformations, surface state, 'cobweb effect', layer relocation, and broken sections.

In Table 2, photographs from the visual control process are presented. The initial visual assessment of the five 3D models did not reveal any imperfections or damage that would have made them unsuitable.

Table 2. Photographs from visual control.

No.	Description	Symbol	View
1.	A car mirror holder made with the application of the PolyJet method; material: RGD 720	RM1	
2.	A car mirror holder made with the application of FDM; material: ABS M30	RM2	

Table 2. *Cont.*

No.	Description	Symbol	View
3.	A car mirror holder made with the application of the DLP method; material E-Partial	RM3	
4.	A car mirror holder made with the application of FFF; material: PLA	RM4	
5.	A car mirror holder made with the application of MEM; material: ABS	RM5	

4.2. Control with the Application of a Caliper

The next stage of the quality control study was the dimensional control of the models, which was conducted with an electronic caliper, the accuracy of which was 0.02 mm. In order to determine the dimensional compliance of the models, four typical dimensions were measured and compared with tolerance parameters. The results are presented in Table 3.

Table 3. Results of model dimension control.

a=30±0.2 b=60±0.2 c=100±0.2 d=35±0.2

Model	Name	Measured Value	Unit	Tolerance	Actual Value (Mean of 3 Measurements)	Deviation	Compliance
RM1	Dimension a	30.00	mm	0.20	29.89	−0.11	OK
RM1	Dimension b	60.00	mm	0.20	60.00	0.00	OK
RM1	Dimension c	100.00	mm	0.20	100.13	0.13	OK
RM1	Dimension d	35.00	mm	0.20	35.03	0.03	OK
RM2	Dimension a	30.00	mm	0.20	30.28	0.28 ↑	NOK
RM2	Dimension b	60.00	mm	0.20	60.13	0.13	OK
RM2	Dimension c	100.00	mm	0.20	100,18	0.18	OK
RM2	Dimension d	35.00	mm	0.20	35.12	0.2	OK
RM3	Dimension a	30.00	mm	0.20	29.85	−0.15	OK
RM3	Dimension b	60.00	mm	0.20	59.78	−0.22 ↓	NOK
RM3	Dimension c	100.00	mm	0.20	99.73	−0.27 ↓	NOK
RM3	Dimension d	35.00	mm	0.20	34.90	−0.10	OK
RM4	Dimension a	30.00	mm	0.20	29.96	−0.04	OK
RM4	Dimension b	60.00	mm	0.20	59.80	−0.20	OK
RM4	Dimension c	100.00	mm	0.20	99.76	−0.24 ↓	NOK
RM4	Dimension d	35.00	mm	0.20	34.93	−0.07	OK
RM5	Dimension a	30.00	mm	0.20	30.15	0.15	OK
RM5	Dimension b	60.00	mm	0.20	60.04	0.04	OK
RM5	Dimension c	100.00	mm	0.20	99.72	−0.28 ↓	NOK
RM5	Dimension d	35.00	mm	0.20	34.83	−0.17	OK

Notes: sign ↑ means that the measured value is above the tolerance range, sign ↓ means that the measured value is below the tolerance range.

4.3. Control with the Application of a Contactless Optical System

The accuracy of the geometric shape of the manufactured research models was analyzed with a contactless optical system based on a coordinate optical scanner (ATOS Triple Scan II Blue Light of the GOM company). For the data analysis, the software of an ATOS Professional V7 scanner was applied. By applying blue light, this scanner enables virtual computing regardless of the intensity of daylight or artificial (white) light. Moreover, it can significantly reduce the computing time owing to the turntable, among other components, integrated with the ATOS computing system. Automation significantly accelerates the computing process because it partially eliminates the need to manually reposition the scanned artifact. The methodology of contactless measurements with the use of the Atos II Triple Scan 3D optical scanner was developed after many trials to determine the appropriate measurement process. A strategy was adopted in which two independent measurement series were carried out in two positions of the target in relation to the measuring table. The process was configured so that each of the two series was measured with the part placed in a given plane of the table and with the plane rotated by 180 degrees. However, this required reference points to be distributed so that at least three were visible in both measurement series. This enabled the program to compile measurement data obtained in both measurement series, thereby obtaining information about the entire measured geometry. This significantly reduced the duration of the digitization process, and based on the inspection of the geometry of the scanned model, the appropriate number of steps for dividing the full rotation of the measuring table for a single measurement series was determined.

By applying the GOM Inspect V8 program, 3D maps of the deviations of actual model surfaces from the designed model were plotted, and two files were output: the first was in STL format, which was exported from CAD software, and the other was a 3D grid reflecting the actual geometry of the research models and representing the models measured by an optical 3D scanner. The developed 3D CAD models are also applicable to quality control processes based on coordinate optical scanners, where an industrial robot can be used. This approach eliminates the need to manually adjust the specimen during the measurement process, which significantly shortens the inspection stage. Due to such a measurement system not being available in this study, this solution was omitted. Nevertheless, the presented measurement methodology allows for the implementation of such a control procedure.

On the basis of the presented issues and a number of previous studies, optical measurements were used in this study. Data visualization and the capabilities of the GOM Inspect software provide advantages over the use of contact measurements. Having the complete geometry of the research models obtained during the measurement process, rather than only cross-sections or a point cloud obtained by contact methods, makes it much easier to apply corrections. Due to the effective integration of CAD/CMM/RP systems, a more effective process for minimizing geometric errors can be implemented using data obtained from optical scanning.

In order to determine the accuracy of the actual geometry, analyses presented in the form of color deviation maps were conducted. The results of the analyses are presented in Table 4. The conducted analyses were focused on the geometry of the research models and were performed in order to assess the dimensional and shape accuracy. Global geometry analyses were performed. On the basis of the results, deviation values were obtained at the selected inspection points.

The presented results of the geometric analysis of car mirror holder models show imperfections that were not identified at the earlier stages of quality control.

The research results indicate that the most precise method is PolyJet. However, this technology is the most expensive of those described in this paper; therefore, every time a 3D printing method is selected, its cost-effectiveness should be evaluated. The data reported above also reveal that FFF has the optimal price–quality ratio. Due to its price and straightforward operation, the Prusa i3 MK3 printer is popular with both regular consumers and legal entities. MEM is the least precise representation, as an uneven surface and significant deviations are clearly visible.

Table 4. Maps of deviations in the accuracy of the research models.

Isometric View RM1

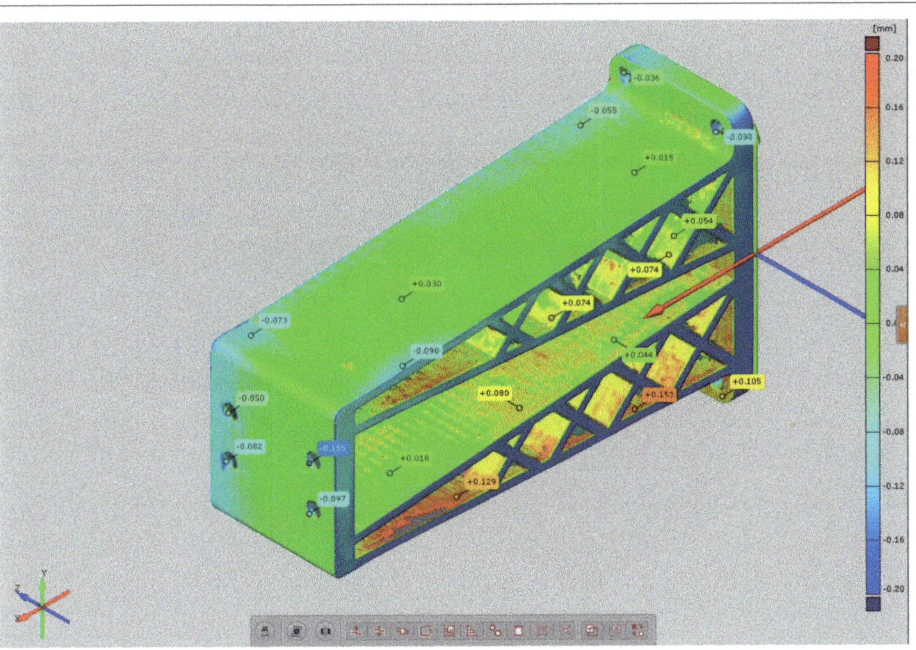

Isometric View RM2

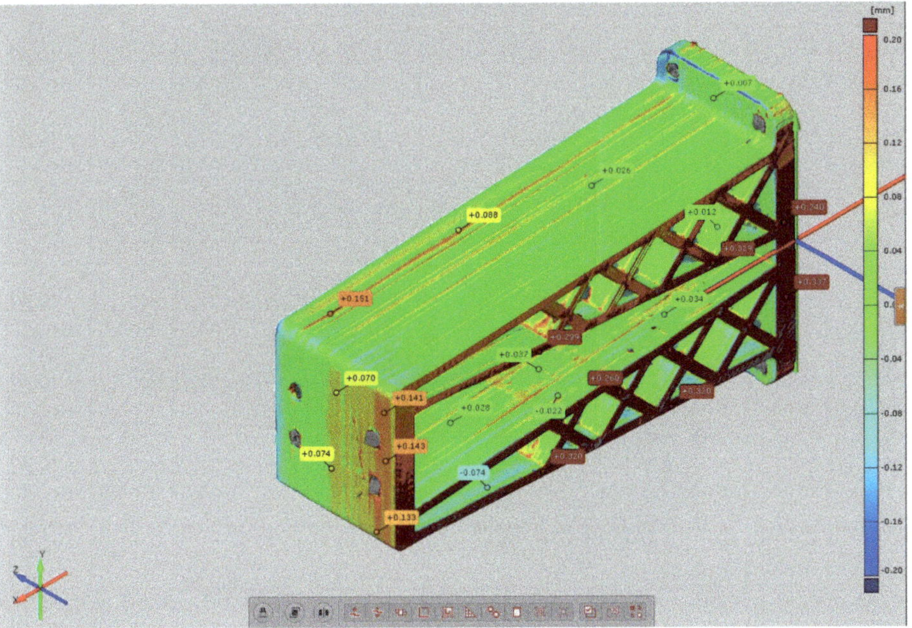

Table 4. *Cont.*

Isometric View RM3

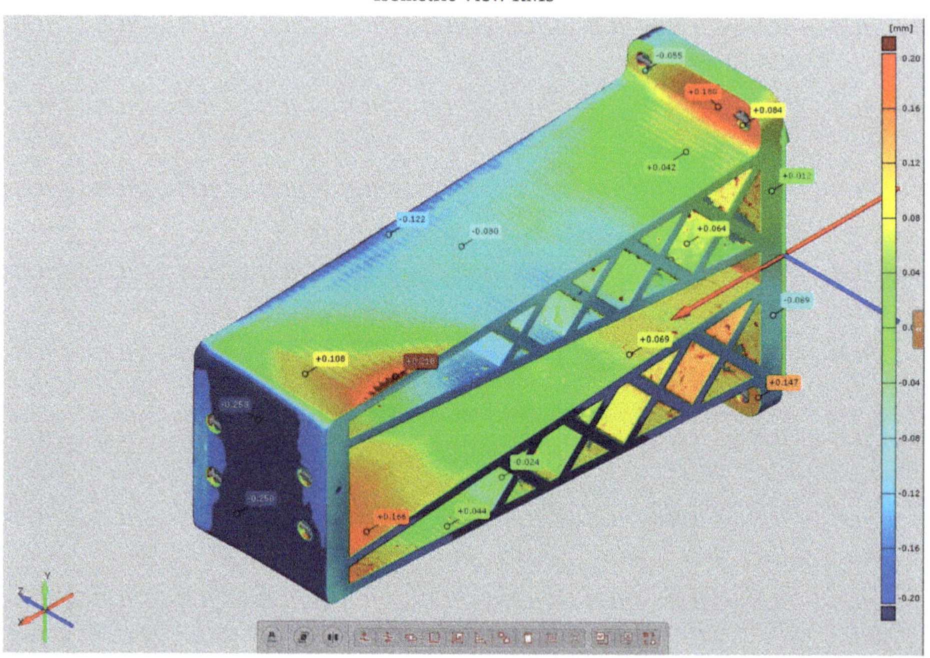

Isometric View RM4

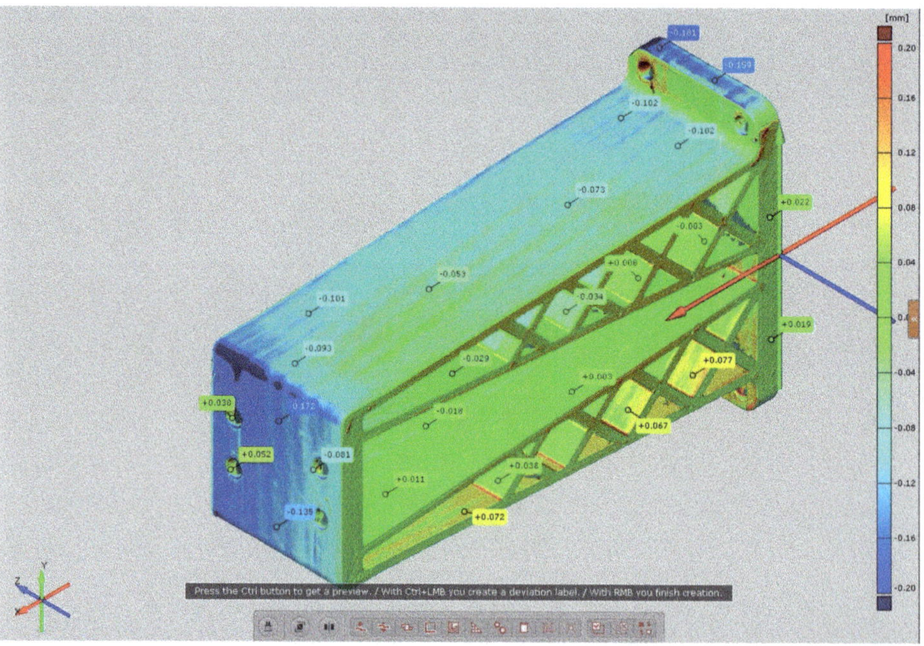

Table 4. *Cont.*

Isometric View RM5

5. Conclusions

The quality control methodology presented in this article, developed on the basis of a car mirror prototype, can be successfully applied to products manufactured additively from polymeric materials. The tests and analyses reveal that the intended use of the prototype is important, and verifications should be carried out in the following order:

- verification of the correctness of the numerical data of the 3D CAD model,
- confirmation of process data intended directly for 3D printing (analysis of support structures, layers, and transition paths in subsequent layers),
- verification of the correctness of model construction during incremental process, carried out directly or remotely,
- visual inspection of the prototype immediately after its construction,
- visual inspection after the completion of post-processing,
- dimensional and shape accuracy control with the use of measuring tools,
- dimensional and shape accuracy control with the use of computer-controlled measuring machines,
- confirmation of the accuracy and internal structure with the use of computer tomography.

The application of 3D scanners to quality control makes it possible to complete the manufacturing process using numerical data from computing. As a result, the geometric accuracy of a product can be completely documented, records can be created, and remedial actions can be prepared. This approach is compatible with manufacturing systems in the context of Industry 4.0. Additionally, certain computing systems, such as those of the GOM company, come with special-purpose applications that allow a 3D STL model to be corrected using data from the computing process, resulting in a product with improved dimensional and shape accuracy.

In summary, models that are fabricated with additive manufacturing may be subjected to different quality control processes depending on their intended use, which should thus be considered when selecting quality assessment methods. It is also important to ensure that cost aspects are always taken into account, as not all products require high precision and durability.

Author Contributions: Conceptualization, G.B. and J.W.; methodology, G.B., J.W. and A.P.; software, A.P. and T.D.; validation, G.B. and Ł.P.; formal analysis, M.D.; writing—review and editing, J.W. All authors have read and agreed to the published version of the manuscript.

Funding: This research received no external funding.

Institutional Review Board Statement: Not applicable.

Informed Consent Statement: Not applicable.

Data Availability Statement: Not applicable.

Conflicts of Interest: The authors declare no conflict of interest.

References

1. Gisario, A.; Kazarian, M.; Martina, F.; Mehrpouya, M. Metal additive manufacturing in the commercial aviation industry: A review. *J. Manuf. Syst.* **2019**, *53*, 124–149. [CrossRef]
2. Elakkad, A.S. Ain Shams University 3D Technology in the Automotive Industry. *Int. J. Eng. Res.* **2019**, *8*, 248–251. [CrossRef]
3. Kim, H.; Lin, Y.; Tseng, T.-L.B. A review on quality control in additive manufacturing. *Rapid Prototyp. J.* **2018**, *24*, 645–669. [CrossRef]
4. Turek, P. Automating the process of designing and manufacturing polymeric models of anatomical structures of mandible with Industry 4.0 convention. *Polimery* **2019**, *64*, 522–529. [CrossRef]
5. Kacmarcik, J.; Spahic, D.; Varda, K.; Porca, E.; Zaimovic-Uzunovic, N. An investigation of geometrical accuracy of desktop 3D printers using CMM. *IOP Conf. Ser. Mater. Sci. Eng.* **2018**, *393*, 012085. [CrossRef]
6. Msallem, B.; Sharma, N.; Cao, S.; Halbeisen, F.S.; Zeilhofer, H.-F.; Thieringer, F.M. Evaluation of the Dimensional Accuracy of 3D-Printed Anatomical Mandibular Models Using FFF, SLA, SLS, MJ, and BJ Printing Technology. *J. Clin. Med.* **2020**, *9*, 817. [CrossRef]
7. Yoo, S.-Y.; Kim, S.-K.; Heo, S.-J.; Koak, J.-Y.; Kim, J.-G. Dimensional Accuracy of Dental Models for Three-Unit Prostheses Fabricated by Various 3D Printing Technologies. *Materials* **2021**, *14*, 1550. [CrossRef]
8. Mandić, M.; Galeta, T.; Raos, P.; Jugovic, V. Dimensional accuracy of camera casing models 3D printed on Mcor IRIS: A case study. *Adv. Prod. Eng. Manag.* **2016**, *11*, 324–332. [CrossRef]
9. Moon, W.; Kim, S.; Lim, B.-S.; Park, Y.-S.; Kim, R.; Chung, S. Dimensional Accuracy Evaluation of Temporary Dental Restorations with Different 3D Printing Systems. *Materials* **2021**, *14*, 1487. [CrossRef] [PubMed]
10. Fahmy, A.R.; Becker, T.; Jekle, M. 3D printing and additive manufacturing of cereal-based materials: Quality analysis of starch-based systems using a camera-based morphological approach. *Innov. Food Sci. Emerg. Technol.* **2020**, *63*, 102384. [CrossRef]
11. Villarraga-Gómez, H. The Role of X-ray Computed Tomography in the 3D Printing Revolution. In Proceedings of the RAPID + TCT 2018 Conference, Fort Worth, TX, USA, 23–26 April 2018. [CrossRef]
12. Dorweiler, B.; Baqué, P.; Chaban, R.; Ghazy, A.; Salem, O. Quality Control in 3D Printing: Accuracy Analysis of 3D-Printed Models of Patient-Specific Anatomy. *Materials* **2021**, *14*, 1021. [CrossRef]
13. Khosravani, M.R.; Reinicke, T. On the Use of X-ray Computed Tomography in Assessment of 3D-Printed Components. *J. Nondestruct. Eval.* **2020**, *39*, 1–17. [CrossRef]
14. Rokicki, P.; Budzik, G.; Kubiak, K.; Bernaczek, J.; Dziubek, T.; Magniszewski, M.; Nowotnik, A.; Sieniawski, J.; Matysiak, H.; Cygan, R.; et al. Rapid prototyping in manufacturing of core models of aircraft engine blades. *Aircr. Eng. Aerosp. Technol.* **2014**, *86*, 323–327. [CrossRef]
15. Magerramova, L.; Vasilyev, B.; Kinzburskiy, V. Novel Designs of Turbine Blades for Additive Manufacturing. In *Proceedings of the Volume 5C: Heat Transfer*; ASME International Seoul: Seoul, Korea, 2016.
16. Rokicki, P.; Budzik, G.; Kubiak, K.; Dziubek, T.; Zaborniak, M.; Kozik, B.; Bernaczek, J.; Przeszlowski, L.; Nowotnik, A. The assessment of geometric accuracy of aircraft engine blades with the use of an optical coordinate scanner. *Aircr. Eng. Aerosp. Technol.* **2016**, *88*, 374–381. [CrossRef]
17. Liu, M.; Zhang, Q.; Shao, Y.; Liu, C.; Zhao, Y. Research of a Novel 3D Printed Strain Gauge Type Force Sensor. *Micromachines* **2018**, *10*, 20. [CrossRef] [PubMed]
18. Wu, H.-C.; Chen, T.-C.T. Quality control issues in 3D-printing manufacturing: A review. *Rapid Prototyp. J.* **2018**, *24*, 607–614. [CrossRef]
19. Guo, B.; Xu, Z.; Luo, X.; Bai, J. A detailed evaluation of surface, thermal, and flammable properties of polyamide 12/glass beads composites fabricated by multi jet fusion. *Virtual Phys. Prototyp.* **2021**, 1–14. [CrossRef]

20. Siemiński, P.; Budzik, G. *Additive Techniques. 3D Printing. 3D Printers*; Oficyna Wydawnicza PW: Warszawa, Poland, 2015; pp. 18–23.
21. Badiru, A.B.; Valencia, V.V.; Liu, D. *Additive Manufacturing Handbook: Product Development for the Defense Industry*; CRC Press: Boca Raton, FL, USA, 2017; pp. 162–164.
22. Gapinski, B.; Wieczorowski, M.; Grzelka, M.; Alonso, P.A.; Tome, A.B. The application of micro computed tomography to assess quality of parts manufactured by means of rapid prototyping. *Polimery* **2017**, *62*, 53–59. [CrossRef]
23. PN/EN ISO/ASTM 52900:2017-06 Additive Manufacturing—General Principles—Terminology. Available online: https://www.pkn.pl/ (accessed on 16 February 2021).
24. PN/EN ISO/ASTM 52901:2019-01 Additive Manufacturing—General Principles—Requirements Relevant to Parts Manufactured by Means of Additive Manufacturing (AM) Processes. Available online: https://www.pkn.pl/ (accessed on 16 February 2021).
25. PN/EN ISO 17296-2:2016-10 Additive Manufacturing—General Principles—Part 2: Overview of Process Categories and Feedstock. Available online: https://www.pkn.pl/ (accessed on 16 February 2021).
26. PN/EN ISO 17296-3:2016-10 Additive Manufacturing—General Principles—Part 3: Principal Characteristics and Appropriate Research Methods. Available online: https://www.pkn.pl/ (accessed on 16 February 2021).
27. PN/EN ISO 17296-4:2016-10 Additive Manufacturing—General Principles—Part 4: Overview of Data Processing. Available online: https://standards.iteh.ai/catalog/standards/cen/5b10b8ed-c472-49d7-a4fd-335eb8de2e11/en-iso-17296-4-2016 (accessed on 16 February 2021).
28. ISO/ASTM 52902:2019 Additive Manufacturing—Test Artifacts—Geometric Capability Assessment of Additive Manufacturing Systems. Available online: https://www.iso.org/ (accessed on 16 February 2021).
29. ISO/ASTM TR 52912:2020 Additive Manufacturing—Design—Functionally Graded Additive Manufacturing. Available online: https://www.iso.org/ (accessed on 16 February 2021).
30. ISO/ASTM 52911-1:2019 Additive Manufacturing—Design—Part 1: Laser-Based Powder Bed Fusion of Metals. Available online: https://www.iso.org/ (accessed on 16 February 2021).
31. ISO/ASTM 52911-2:2019 Additive Manufacturing—Design—Part 2: Laser-Based Powder Bed Fusion of Polymers. Available online: https://www.iso.org/ (accessed on 16 February 2021).
32. Kopsacheilis, C.; Charalampous, P.; Kostavelis, I.; Tzovaras, D. In Situ Visual Quality Control in 3D Printing. In *Proceedings of the 15th International Joint Conference on Computer Vision, Imaging and Computer Graphics Theory and Applications*; SCITEPRESS—Science and Technology Publications: Setúbal, Portugal, 2020; pp. 317–324.
33. Mitutoyo. Compendium of Metrology for Precision Measuring Instruments. Available online: https://www.mitutoyo.pl/application/files/1215/5888/6942/Mitutoyo_kompendium_metrologii_2013_WWW_opt_2.pdf (accessed on 30 March 2020).
34. Vagovský, J.; Buranský, I.; Görög, A. Evaluation of Measuring Capability of the Optical 3D Scanner. *Procedia Eng.* **2015**, *100*, 1198–1206. [CrossRef]
35. Dziubek, T.; Oleksy, M. Application of ATOS II optical system in the techniques of rapid prototyping of epoxy resin-based gear models. *Polimery* **2017**, *62*, 44–52. [CrossRef]
36. Siraj, I.; Bharti, P.S. Reliability analysis of a 3D Printing process. *Procedia Comput. Sci.* **2020**, *173*, 191–200. [CrossRef]
37. Rayna, T.; Striukova, L. From rapid prototyping to home fabrication: How 3D printing is changing business model innovation. *Technol. Forecast. Soc. Chang.* **2016**, *102*, 214–224. [CrossRef]
38. Paszkiewicz, A.; Bolanowski, M.; Budzik, G.; Przeszłowski, Ł.; Oleksy, M. Process of Creating an Integrated Design and Manufacturing Environment as Part of the Structure of Industry 4.0. *Processes* **2020**, *8*, 1019. [CrossRef]
39. Dilberoglu, U.M.; Gharehpapagh, B.; Yaman, U.; Dolen, M. The Role of Additive Manufacturing in the Era of Industry 4.0. *Procedia Manuf.* **2017**, *11*, 545–554. [CrossRef]
40. Paraskevoudis, K.; Karayannis, P.; Koumoulos, E.P. Real-Time 3D Printing Remote Defect Detection (Stringing) with Computer Vision and Artificial Intelligence. *Processes* **2020**, *8*, 1464. [CrossRef]
41. Octoprint. Available online: www.octoprint.org (accessed on 30 March 2020).
42. Matzkanin, G.A. Selecting a nondestructive testing method: Visual inspection. *Adv. Mater. Manuf. Test. Inf. Anal. Cent.* **2006**, *1*, 7–10.

Article

Nanoparticle Additivation Effects on Laser Powder Bed Fusion of Metals and Polymers—A Theoretical Concept for an Inter-Laboratory Study Design All Along the Process Chain, Including Research Data Management

Ihsan Murat Kusoglu [1], Florian Huber [2], Carlos Doñate-Buendía [1,3], Anna Rosa Ziefuss [1], Bilal Gökce [1,3], Jan T. Sehrt [4], Arno Kwade [5], Michael Schmidt [2] and Stephan Barcikowski [1,*]

1. Technical Chemistry I, Center for Nanointegration Duisburg-Essen (CENIDE), University of Duisburg Essen, 45141 Essen, Germany; ihsan.kusoglu@uni-due.de (I.M.K.); carlos.donate-buendia@uni-due.de (C.D.-B.); anna.ziefuss@uni-due.de (A.R.Z.); goekce@uni-wuppertal.de (B.G.)
2. Institute of Photonic Technology, Friedrich-Alexander-University Erlangen-Nürnberg, 91052 Erlangen, Germany; Florian.Huber@lpt.uni-erlangen.de (F.H.); michael.schmidt@lpt.uni-erlangen.de (M.S.)
3. Materials Science and Additive Manufacturing, School of Mechanical Engineering and Safety Engineering, University of Wuppertal, 42119 Wuppertal, Germany
4. Department of Hybrid Additive Manufacturing, Ruhr University of Bochum, 44801 Bochum, Germany; Jan.Sehrt@ruhr-uni-bochum.de
5. Institute for Particle Technology, Technical University of Braunschweig, 38104 Braunschweig, Germany; a.kwade@tu-braunschweig.de
* Correspondence: stephan.barcikowski@uni-due.de

Citation: Kusoglu, I.M.; Huber, F.; Doñate-Buendía, C.; Rosa Ziefuss, A.; Gökce, B.; T. Sehrt, J.; Kwade, A.; Schmidt, M.; Barcikowski, S. Nanoparticle Additivation Effects on Laser Powder Bed Fusion of Metals and Polymers—A Theoretical Concept for an Inter-Laboratory Study Design All Along the Process Chain, Including Research Data Management. *Materials* 2021, 14, 4892. https://doi.org/10.3390/ma14174892

Academic Editors: Mika Salmi and Young-Hag Koh

Received: 15 July 2021
Accepted: 24 August 2021
Published: 27 August 2021

Publisher's Note: MDPI stays neutral with regard to jurisdictional claims in published maps and institutional affiliations.

Abstract: In recent years, the application field of laser powder bed fusion of metals and polymers extends through an increasing variability of powder compositions in the market. New powder formulations such as nanoparticle (NP) additivated powder feedstocks are available today. Interestingly, they behave differently along with the entire laser powder bed fusion (PBF-LB) process chain, from flowability over absorbance and microstructure formation to processability and final part properties. Recent studies show that supporting NPs on metal and polymer powder feedstocks enhances processability, avoids crack formation, refines grain size, increases functionality, and improves as-built part properties. Although several inter-laboratory studies (ILSs) on metal and polymer PBF-LB exist, they mainly focus on mechanical properties and primarily ignore nano-additivated feedstocks or standardized assessment of powder feedstock properties. However, those studies must obtain reliable data to validate each property metric's repeatability and reproducibility limits related to the PBF-LB process chain. We herein propose the design of a large-scale ILS to quantify the effect of nanoparticle additivation on powder characteristics, process behavior, microstructure, and part properties in PBF-LB. Besides the work and sample flow to organize the ILS, the test methods to measure the NP-additivated metal and polymer powder feedstock properties and resulting part properties are defined. A research data management (RDM) plan is designed to extract scientific results from the vast amount of material, process, and part data. The RDM focuses not only on the repeatability and reproducibility of a metric but also on the FAIR principle to include findable, accessible, interoperable, and reusable data/meta-data in additive manufacturing. The proposed ILS design gives access to principal component analysis (PCA) to compute the correlations between the material–process–microstructure–part properties.

Keywords: laser melting; laser sintering; 3D printing; AlSi10Mg; PA12; Round-Robin

1. Introduction

Powder bed fusion using laser beam (PBF-LB) [1] is a sub-class production technique of additive manufacturing (AM). Highly complex 3D structures and tailor-made designs can be produced by melting metal and sintering polymer powder feedstocks. Scientific studies in AM showed an exponential growth within the last decade, and studies on PBF-LB of metal and polymer powder feedstocks followed the growth trend [2,3]. Besides the scientific contributions, progress in the industrialization of PBF-LB can be seen in the increasing number of machine suppliers [4–7]. In addition, PBF-LB plays an essential role as a critical technology in the context of industry 4.0 as parts and products can be tracked. Therefore, the production chain is flexible and can be automated, leading to communication between machines, parts, and products [8] and understanding the metallurgical phenomenon of the PBF-LB process together with mechanistic models and machine learning opening a path to control the process by machines further and increase the reproducibility of the as-built parts with the repeatable microstructure and part properties [9,10]. However, current industrialization efforts for the PBF-LB technology need standardization in testing powder feedstocks, processability, and as-built part properties. Increasing repeatability and reproducibility in PBF-LB highly depends on these standardization activities. The ISO committee (ISO/TC 261) and ASTM committee (F42) work in cooperation to develop and publish new standards on PBF-LB and AM technologies [11,12].

Recent PBF-LB progress focuses on increasing metal and polymer powder compositions' variability, developing processability, and enhancing part properties [13–24]. In order to overcome material-related limitations and expand the application fields of PBF-LB technology, recent studies address the nano-additivation of metal and polymer powder feedstocks [2,3,25–42]. Several material types of nanoparticles (NPs) can be supported on the surface of the metal and polymer powder feedstocks [2,3,25], leading to enhanced material properties, such as flowability, laser absorbance, and thermal conductivity. Furthermore, NPs are known to enhance the processability [26,27], microstructural properties by avoiding grain-growth [43–46], and as-built part properties [3,25,30,33,36,42], as well as broaden the process window [28,37–41].

Several reviews [2,3,47–49] have been published to understand the process–structure–property relationships in the PBF-LB process but a well-designed inter laboratory study (ILS) to test nano-additivated commercial powder feedstocks can further consolidate industrialization, standardization, robustness, and fundamental understanding along the entire process chain of PBF-LB. ILSs allow the statistical evaluation of measurable metrics' repeatability and reproducibility linked to material, process, or as-built part properties. Participants' repeatability and reproducibility of each measurable metric are valuable seed data to develop and standardize test procedures in the PBF-LB process.

Several ILSs are designed and organized for PBF-LB of metal and polymer powder feedstocks [50–60]. Previous ILSs on PBF-LB focused on quantifying the variability in an as-produced and used state of metal [50,51] and polymer [52,53] powder feedstock properties without focusing on processability and as-built part properties. They included measurements of powder properties such as powder size distribution (PSD), shape, chemical composition, crystallographic phases, material rheology, flowability, and thermal behavior [50–53]. However, they showed that the statistical evaluation of powder properties is affected by the experimental setup, calibration in measurements, and powder modification during handling. Previous ILSs on PBF-LB also statistically evaluated the deviations in process parameter sets and as-built part properties of commercially available metal and polymer powder feedstocks [50–53]. However, these studies did not include a statement on powder properties that affect processability, microstructural formation, and as-built part property. Additionally, it could be demonstrated that the variability of as-build part properties between-participant is higher than within-participant [54–60], resulting in a strict recommendation to follow a manufacturing plan.

To overcome these limitations, the proposed design herein includes the measurement of as-produced and used powder properties and evaluates process parameter sets and as-built part properties, including microstructural formation with at least 20 PBF-LB process participants (10 polymers, 10 metals). Besides testing as-produced powder polymer and metal properties along the entire process chain, nano-additivated feedstocks are included in the ILS design. The design is instructive for non-profit organizations, industry, and national or international standardization organizations conducting ILS to test new metal or polymer powder feedstocks developed for the PBF-LB process. The design includes an organizational workflow, sample flow, and data flow that evaluate the entire process chain based on decentralized PBF-LB processing, but centralized testing. In addition, this design focuses on the research data management that further evaluates the generated data by PCA. Reusable data opens a path to obtain relevant seed data that data analysts can further analyze for additive manufacturing and powder material design. The proposed ILS herein includes evaluating 27 powder property metrics, 12 process property metrics, 26 part property metrics for three metals (unmodified and two nano-additivated), and three polymers (unmodified and two nano-additivated) powder feedstocks along the entire PBF-LB process chain.

2. Design of Interlaboratory Study

The ILS organization has to be well-structured to trace work and material flows through processing and testing. Moylan et al. [61] briefly defined an organizational structure, which they recommend following in an ILS. Additionally, Maier et al. [62] confirmed that such an ILS's primary limitation is an organization that needs particular, high scientific and management knowledge. A general organizational workflow of ILS specific to PBF-LB is given in Figure 1.

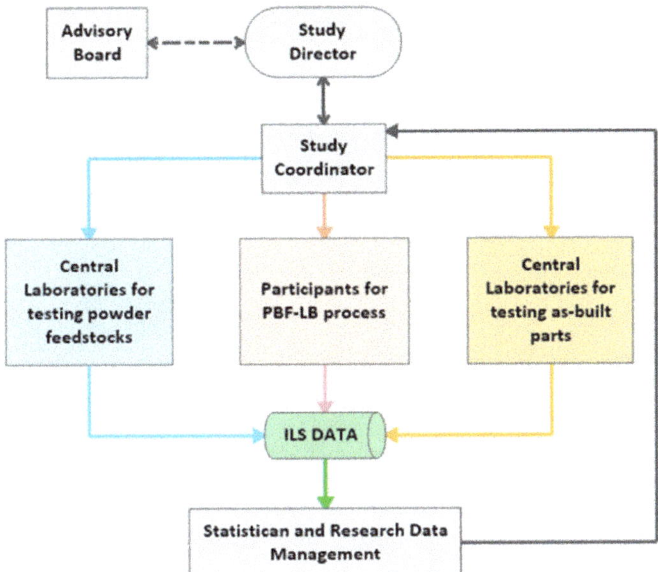

Figure 1. The organizational workflow of inter-laboratory study (ILS) design for powder bed fusion using laser beam, including decentralized processing and centralized testing.

The advisory board has to be selected from research institutes and the AM industry. Important requirements are long-term experience in powder feedstock production, PBF-LB of metal and polymer powder feedstocks, material characterization techniques, and conducting ILS. However, it is good to include a specialist working on developing international standards for AM technology. The study director should have excellent managing skills and closely communicate with the study coordinator. The latter has a critical position in organizing, tracking, and evaluating the workflows' progress with scientific research data management experience. The organization includes coordinating workflows in ILS to statistically evaluate the repeatability and reproducibility of NPs additivated powder feedstock and as-built part properties.

Note that previous ILS [50–59] started introducing commercially available powder feedstocks followed by measuring one or two properties of the as-built parts for statistical evaluation of the measured metrics' repeatability and reproducibility between different machine users. However, the ILS proposed herein includes testing NPs additivated metal and polymer powder feedstocks; hence, the ILS design needs to have additional workflows. These workflows have to test both unmodified and additivated powder feedstocks to measure NPs' effect on powder material properties. Further, it includes testing NPs' effect on microstructural formations and mechanical properties after the PBF-LB process, as shown in Figure 2. In that way, traceability along the entire process chain is enabled and the ILS will be completed in six steps, with each step being discussed in the following sections.

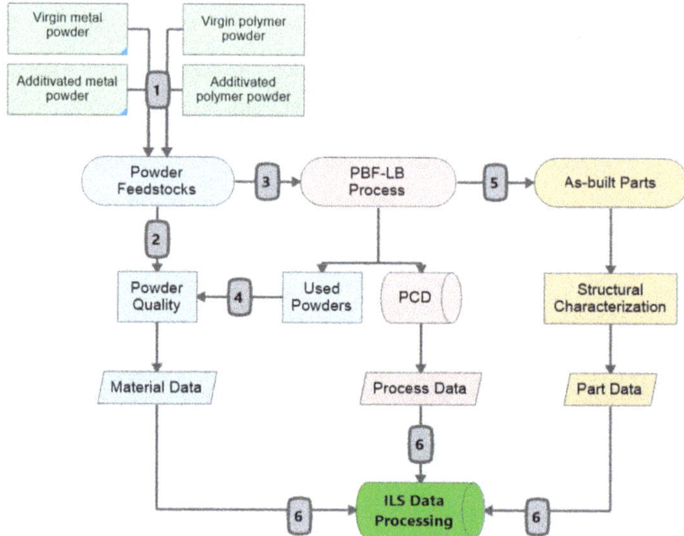

Figure 2. Workflow of inter-laboratory study (ILS) for powder bed fusion using laser beam (PBF-LB) of unmodified and NP-additivated metal and polymer powder feedstocks. The workflow consists of six steps: (1) nano-additivated powder production, (2) powder quality of unmodified and nanoparticle additivated powders, (3) PBF-LB of unmodified and additivated powders, (4) powder quality of used powders after the PBF-LB process, (5) testing as-built parts, (6) powder, process, part data generation, and ILS data processing. Topics in blue, pink, and orange represent the workflows to obtain the powder, process, and part data, respectively.

In each sub-section, the amount of acquired datasets delivered by the respective ILS activity is estimated. These datasets will be tagged with the user, time, SOP, and may mathematically consist of one-dimensional (e.g., flowability value), pairs (e.g., meting-crystallization temperature), or multidimensional data (e.g., spectra or histograms). The term "data" will be used through the manuscript and counted as one for one tagged dataset, independent of its dimensionality.

2.1. Additivation of NPs on As-Produced Metal and Polymer Powder Feedstocks

Before starting additivation of commercially available powder feedstocks, highly reproducible NP-additivated powder feedstocks are required to minimize material-dependant deviations in the ILS statistical evaluation. While as-produced metal and polymer powder feedstocks show good reproducibility and can be commercially produced on a ton scale, this is not the case for powder feedstocks in the development phase. In addition, up-scaling limitations in NP production and feasibility of additivation techniques can limit producing high amounts of NP additivated powder feedstocks for the PBF-LB process. The proposed ILS herein will require less than 700 kg feedstock material (187.5 kg polymer, 502.5 kg metal) for 20 PBF-LB processing participants (10 PBF-LB/P and 10 PBF-LB). There is a base amount of powder needed (1 kg polymer, 3 kg metal) for property assessment activity (see Figure 3) before the powders can be sent out to processing partners; after that, the required powder mass scales linearly with the number of participants that run PBF-LB (details on the build chamber volumes are given in chapter 2.1.2.). For example, 10 powder recipients each for PBF-LB/P and PBF-LB/M would require a total of 30 and 90 kg of polymer (3 different compositions) and metal powder (3 different compositions), and 20 each for PBF-LB/P and PBF-LB/M requires 247.5 and 592.5 kg of polymer and metal powder for the execution of the whole ILS.

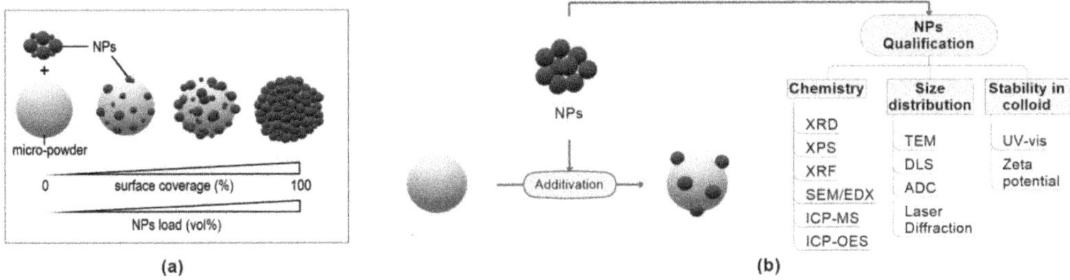

Figure 3. (a) Illustration of the surface coverage by micro powders with increasing nanoparticle loading (vol%) and (b) the most relevant nanoparticle properties for the additivation process and recommended nanoparticles characterization techniques.

As mentioned above, the inclusion of nanoparticles in the powder feedstock represents a new challenge that has been ignored to date in ILSs and takes a central point in this ILS. The production of nano-additivated powder feedstocks requires the supporting of NPs on the surface of commercially available metal and polymer powder feedstocks. Hence, knowing the NP properties before and after supporting is crucial and needs a proper analysis beforehand. The main requirement is a high reproducibility of the whole synthesis process, including the NP synthesis route and the additivation process. As shown in Figure 3a, the surface coverage and the distribution of NPs on the micro-powder's surface can vary depending on the NP load and surface additivation technique. Additionally, NPs may be present as a dry powder [41] or in a colloidal solution [28,30]. These variations may affect powder feedstock material properties relevant for PBF-LB.

Lüddecke et al. [40] studied the powder characteristics of several metal powders coated with different NP material types, sizes, and supporting amounts (<1 vol%) in fluidized bed coatings. They showed a comparison of the flowability behavior of nano-additivated metal powders by measuring bulk density, unconfined yield strength using ring shear tests, and dynamic angle of repose within a rotating drum. They found that the flowability of powder feedstocks can be improved by the optimum NP size, NP material type, and loading amounts. Hupfeld et al. [28,29,32] presented a simplified scaling diagram for NP supporting on micro powders, link in NP particle size, surface coverage (surf%), and vol% as a prerequisite to creating very well defined, highly dispersed submonolayer NP coatings on PA12 and TPU, including the comparison between wet and dry coating. Pannitz et al. [39] showed that coating SiC NPs or few-layer graphene (FLG) NPs on stainless steel powder feedstocks decreases the laser reflectivity of the powder feedstock and increases the thermal conductivity of the material during processing, leading to a rapid heat dissipation into the already solidified underlying layers during the melting of next powder layer. These different studies show the importance of measuring the physical and chemical properties of NPs before producing the additivated metal and polymer powder feedstocks to probe correlations between NP property, supported powder property, and the behavior along the process chain during ILS. Hence, before and after supporting NPs on the ILS´s powders, complementary measurement techniques are recommended to determine the physical and chemical properties of NPs, summarized in Figure 3.

It has been shown that the chemical composition of NPs does not change during the supporting process, but may change during processing, where the NPs see high temperatures and may undergo aggregation in the melt during processing [33,36,63]. Different material types of NPs, such as metals, oxides, non-oxides, and compounds, can be used as additives in powder feedstock formulations [3,25]. The crystalline phases of NPs should be identified by X-ray diffraction (XRD), a common material characterization technique. X-ray fluorescence (XRF) is the recommended, non-destructive characterization technique to determine the elemental composition of NPs and trace the chemical formations and elemental compositions in as-built parts after the PBF-LB process [64]. Therefore, within the ILS, XRD and XRF methodologies are recommended to determine the NP's chemical structure before starting the additivation process.

The size distribution of NPs is a vital factor in tracing the reproducibility of NPs. Further, NP size distribution must be known to calculate the amount of NPs needed for an intended surface coverage level on the micro-powders [29]. It is known that the NP´s size influences the powder feedstock flowability [40,41] as attractive forces between micro-powders are reduced by roughening the surface with NPs [65,66]. However, a statistical investigation is still missing in the literature. Following Gaertner et al. [41], the feedstock's flowability behavior will be enhanced by optimum surface coverage level and optimum NPs size. As discussed in references [67,68], transmission electron microscope (TEM), dynamic light scattering (DLS), and analytical disc centrifugation (ADC) characterization techniques are recommended methods within an ILS to measure dry and colloidal NP size distributions.

A widespread approach to measure dry NP size distribution is TEM, but careful sample preparation is needed, and depending on the NP size, hundreds to thousands of particles must be measured for statistical relevance [67]. Moreover, TEM may create drying artifacts and lack good size statistics. Instead, widely employed techniques to measure the hydrodynamic size distribution of NPs are DLS and ADC. While DLS lacks in determining multimodal size distributions, ADC is capable of measuring polydisperse colloids. Detailed information on the comparison between TEM, DLS, and ADC can be found elsewhere [67,68], including the comparison of five analytical methods for particle diameter differentiation and bimodality identification [69]. Please note that the particle size can alter during the supporting process, mainly forced by agglomeration-induced effects [41]. Therefore, we advise determining both supported and colloidal NP size distribution.

Additivation Process

The surface additivation of the metal and polymer powders can be performed by dry mixing or wet coating and subsequent or parallel drying [28,33,37,40,41]. The choice of coatingtechnique highly depend on the availability of NPs. Dry powder mixing techniques such as three-dimensional free-fall shaker [41] and rotary mixing drum [37] can be used. Then, the mixing parameters such as duration and rotating speeds need optimization for homogeneous dispersion of NPs on the surface of powder feedstocks. Gärtner et al. [41] studied the effect of dry-coating SiO_2 NPs (<1 vol%) on the flowability behavior of several particle size fractions of CoCrFeNi alloy powder. They showed that short dry mixing durations avoided NP agglomeration on the metal powder surface and increased the powder flowability. Homogeneous NP distribution on the metal and polymer powder surface [28,33,37,40,41] will result in reproducible powder feedstocks and as-built part properties.

Supporting colloidal NPs requires a wet coating technique followed by additional drying [27]. Fluidized bed drying is a suitable method where the powders can be wet mixed and dried in the same unit with a relatively short processing time, facilitating minimum cross-contamination [26,40]. Another wet mixing technique while wet mixing the materials is dielectrophoretic deposition [28–30]. It depends on the stirring medium's pH, the concentration of the materials, isoelectric points of NPs, and powder feedstocks in the medium [67]. Key to the dielectrophoretic deposition is the particle's surface charge density, which promotes this process. For a progressive wet surface additivation, NPs and micro-powders should be oppositely charged. For this purpose, the isoelectric points of both materials should be measured under different pH values before starting the additivation process. Further information on this process can be found elsewhere [19,32,42,67]. The resulting mixture must be sieved and dried before using the PBF-LB process [28–33]. This ILS is designed to test six different powder composition configurations as unmodified and 2 different NP additivated metal as well as unmodified, and 2 different NP additivated polymer powder feedstocks.

The total amount of powder production is a crucial point for a complete run in ILS. It depends critically on the height of the build job, the build area of each participant's machine, and the number of participants in PBF-LB processing. Participating PBF-LB machines can have different specifications and build areas. The build job area and height must be minimized to avoid excessive powder consumption in the ILS. The build job height has to be defined before starting the ILS, and is defined by the longest dimension of the Z-direction specimen. In detail, a build area of 35 cm × 35 cm and height of 10 cm will have a volume of 12.25 cm^3 per build job. Polymer powder such as polyamide-12 (PA12) with a theoretical density of almost 1 g/cm^3 and a relative powder bed density of 50% requires 6.25 kg of powder feedstocks per build job. Likewise, metal powder such as AlSi10Mg (2.68 g/cm^3) having a relative powder density of 50% will need 16.75 kg of powder feedstocks per build job. An ILS design testing three PA12 powder compositions with 10 participants requires 187.5 kg of powder feedstocks, and three AlSi10Mg powder compositions with 10 participants need 502.5 kg of powder feedstocks. Some PBF-LB participants may have a smaller build area, e.g., 25 cm × 25 cm, in their machine, which will reduce the total amount of powder requirement to run the ILS.

Further, it should be noted that the height of the build volume for polymer PBF-LB will include an additional base layer for thermal decoupling from the metal platform with top layers on top of the finished parts. In contrast, the height of the build volume for metal PBF-LB will include an additional support structure. Therefore, some PBF-LB machines can have a smaller build area than the planned build job, not allowing participation in the designed ILS. In this case, splitting build jobs can enable participation. In addition to the participant's process failures, each machine's recoating mechanism must be considered as well in calculating the total amount of powder required for the ILS. The volume of the powder conveyor can be different from machine to machine. However, additional powder for recoating the first layers will be needed, and overdosing in powder recoating is required to spread a good powder layer during processing. Supplying a double amount of powder feedstock to each participant per build job will further accelerate ILS progress (e.g., 200 kg of PA12 and 500 kg of AlSi10Mg powders in total for 10 participants and 3 powder variants processing a build volume of 6250 cm^3 per build job). After delivery by the ILS study coordinator, participants must store the delivered powder feedstocks containers according to the provided SOP conditions.

2.2. Powder Quality of Unmodified and NP-Additivated Powder Feedstocks

Since feedstocks' material properties can differ from company to company, all unmodified metal or polymer powder feedstocks should be supplied from the same metal and polymer powder producer. The powder chemical composition, size distribution, shape, flowability, thermal behavior, and laser interaction of metal or polymer powder feedstocks are the most reported material properties affecting the laser processability microstructure, porosity, and build parts' mechanical properties [2,3]. Depending on the physicochemical properties and volumetric loadings of NPs, additivated powder feedstock material properties will be differentiated from unmodified powder feedstock material properties. In addition to the material properties given above, surface occupation density (surface coverage %) and interparticle distance between NPs covering the surface must be measured to differentiate volume loading from surface dispersion effects. Within the proposed ILS, each participant's feedstock batch must be measured before starting the PBF-LB process, which will later be correlated with statistical deviations in each participant's as-built part properties and microstructural analysis. The material properties recommended to be analyzed within this ILS design are discussed in the following, and the recommended techniques to measure those material properties (Figure 4) are given in the following sub-sections. The number of data generated for each property metric is shown in Figure 4 and evaluated in the material data generation (Section 2.6). Therefore, in the second step of the ILS (Figure 2), both unmodified and additivated powder feedstocks' material properties as chemical composition, shape, size distribution, flowability, thermal behavior, laser interaction, moisture content, surface occupation density of NP, and interparticle distance of NP on the micro powder surface will be measured.

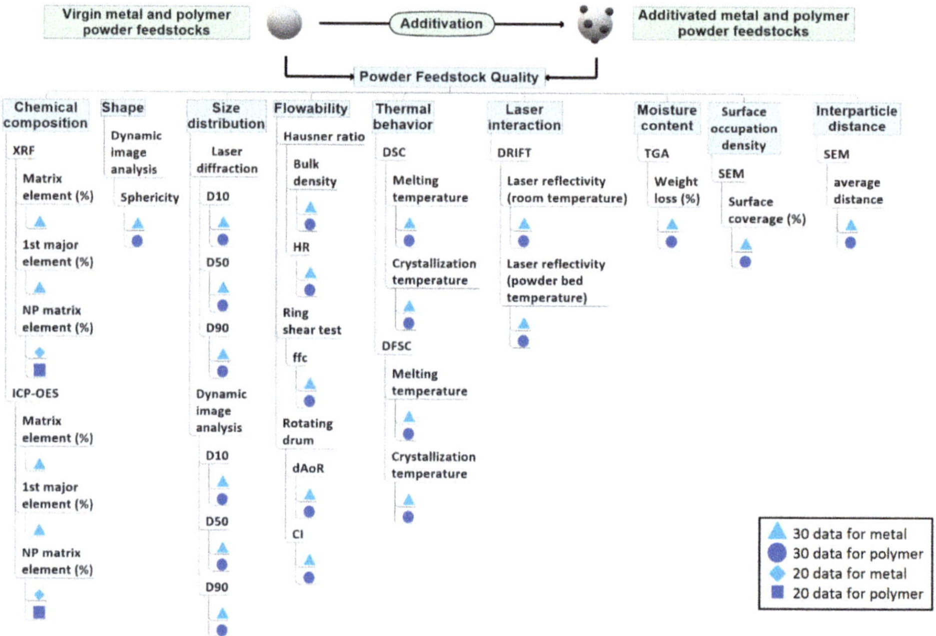

Figure 4. The material properties of powder feedstocks affected by nanoparticles recommended test methods to measure each material property for 30 metal powder batches and 30 polymer powder batches, and the number of data generated for each property metric. Explanation for abbreviations are as follow: X-ray fluorescence (XRF); inductively coupled plasma-optical emission spectroscopy (ICP-OES); the portion of particles with diameters smaller than this value is 10% (D10); the median diameter (D50); the portion of particles with diameters below this value is 90% (D90); hausner ratio (HR); consolidation stress/unconfined yield strength (ffc); dynamic angle of repose (dAoR); cohesive index (CI); differential scanning calorimetry (DSC), differential fast scanning calorimetry (DFSC); diffuse reflectance infrared fourier transform (DRIFT); thermogravimetric analysis (TGA); scanning electron microscope (SEM).

A corresponding characterization of the unmodified metal and polymer powder feedstocks must reference the additivated powder's data in this context. The repeatability of measurements and reusability of generated data (see Section 2.6) needs standardization in sample preparation and testing conditions, which will be described in Section 2.5.

2.2.1. Chemical Composition

The chemical composition of an unmodified metal or polymer powder feedstock directly affects the laser interaction, process window, microstructural formations, and as-built part properties [2,3,35]. These properties can be further affected by NPs' chemical composition, supported amount, and surface coverage levels [29–33,38,39]. In this ILS, the chemical composition of powder feedstock is recommended to be measured by XRF and ICP-OES. Measuring the weight % of the matrix element (i.e., wt% Al in AlSi10Mg alloy), first major element (i.e., wt% Si in AlSi10Mg alloy), and NP matrix element (the highest wt% metallic element of NP phase) will generate a total of 160 items of data for metals and 40 items of data for polymers (Figure 4). Note that XRF cannot measure the polymer's elemental composition but the amount of NPs in the polymer matrix. Alternatively to XRF, ICP-OES can be used within the ILS, but the polymer digestion method must be carefully executed for ICP-OES analysis of composites to avoid matrix cross-effect [70]. Measuring chemical composition of feedstocks by XRF and ICP-OES will generate 160 data points for metals and 40 data points for polymers (Figure 4).

2.2.2. Powder Shape and Powder Size Distribution

Powder bed density is directly linked to powder shape and size distribution during the PBF-LB process [45]. Additionally, different powder size fractions using the same process parameters influence the part quality [2,3,71–73]. Hence, in the ILS, the shape of the powder feedstocks has to be measured by dynamic image analysis. Measuring the sphericity of powder feedstocks will generate a total of 30 data points for metals and 30 data points for polymers (Figure 4).

The particle size distribution (D10, D50, and D90) will be measured by statistically relevant laser diffraction and dynamic image analysis [18]. Thus, a total of 180 data points for both metals and polymers will be generated as material data in ILS (Figure 4).

2.2.3. Flowability

PBF-LB is a layer-by-layer powder processing technique, and in each layer, the recoater spreads powder feedstocks on the building platform. During spreading, powder feedstocks must have excellent flowability to increase powder bed density and facilitate recoating a homogeneous and defect-free powder layer [74]. Since there is no international standard to measure PBF-LB powder feedstocks' flowability, a combination of several techniques such as density measurement (Hausner ratio, HR), ring shear test, and the rotating drum has to be used to measure the unmodified and additivated powder feedstocks' flowability. More details on the mentioned techniques can be found elsewhere [37,40,41,60,61,75].

In short, measuring the HR is a simple technique defined by the ratio between tap density and bulk density of powders [52]. Values smaller than 1.25 describe free-flowing characteristics [53,75]. Besides measuring the bulk density of powders, HR test conditions do not mimic the dynamic powder deposition conditions of PBF-LB. For this purpose, a ring shear test and the rotating drum test have to be used within the ILS to understand the dynamic behavior of powders. Note that the ring shear test measures the so-called ffc ratio (consolidation stress/unconfined yield strength) of quasi static (low shear rates) and the powders and classify those from non-flowing to free-flowing (with values > 10 [40]), and the rotating drum test is another technique to determine the flowability and cohesion of powders within dynamic movement [76] in which the dynamic angle of repose (dAoR) and cohesive index (CI) is linked to the flowability of powders. The measurements at different rotation speeds are appropriate to determine the recoating speed limitations during the PBF-LB process [37,40]. Avalanche angles smaller than 45° describe excellent to good flowability [41], and a cohesive index greater than 0 describes good flowability [40]. In this ILS, the flowability measurements by the three methods will generate 150 data points for metals and 150 for polymers (Figure 4).

2.2.4. Thermal Behavior

Depending on the material types, powder feedstocks have different melting and crystallization temperatures during the heating and cooling stages of PBF-LB. These temperatures can be shifted by introducing NPs [27,28,38]. Metal powders entirely melt during the PBF-LB process. The as-built metal parts can result in over 99.5% of theoretical density after processing [2]; however, this is often not the case for polymer powder feedstocks. By increasing the laser energy densities, polymer feedstocks will partially melt to the point where sintering and elastoplastic flow mechanisms between powders occur. Kusoglu et al. [3] showed that over 99% of densification in the as-built part has rarely been achieved for most thermoplastics. The processing window, the difference between the onset melting and crystallization temperature, determines the required build chamber temperature to process polymer powder feedstocks. The processing window of nano-additivated powders depends on the NP material type and volume loadings, affecting melting–crystallization temperatures [28,59]. Differential scanning calorimetry (DSC) is a well-known approach to measure the melting and crystallization temperatures of unmodified and additivated powder feedstocks to predict the processing window of PBF-LB/P. Additionally, powder bed temperatures can be set according to the process window to delay rapid crystallization during the PBF-LB process. The DSC heating rate is too low compared to the PBF-LB process, but the cooling rate can be well represented by DSC between 0.2 and 20 K/min in PBF-LB/P. However, there are different cooling rates throughout the powder bed volume, which is difficult to mimic for DSC. A new approach of differential fast scanning calorimetry (DFSC) with very rapid heating and cooling rates can represent the PBF-LB process conditions better, to understand the rapid heating and cooling behavior of metal and polymer powder feedstocks [27,28,62]. Derived from DSC and DFCS, a total of 120 data points for metals and 120 data points for polymers will be generated by measuring the melting and crystallization temperature of the metal and polymer powders (Figure 4).

2.2.5. Laser Reflectivity

PBF-LB is a process where powders are heated up to melting temperatures of the powder material. The laser absorbance or reflectance of powders is known to influence the required PBF process parameters and melt pool dynamics. Laser interaction with NPs, unmodified metal powders, unmodified polymer powders, and additivated powders will differ depending on powder crystalline structure and surface chemistry (affecting the NP's absorbance at the laser wavelength). As seen in the inset of Figure 3a, by increasing the NP load, the surface coverage differentiates. Additionally, depending on the degree of surface coverage, the laser reflectivity of metal and polymer powder feedstocks will be different after NPs additivation [40]. Diffuse reflectance infrared Fourier transform spectroscopy (DRIFTS) measures the reflectivity (scattering) of powder feedstocks for a wide spectral range depending. Ideally, the DRIFTS is equipped with a heating unit to heat the powder feedstocks to the PBF bed's temperatures [76]. The reflectivity measurements should be performed under inert gas atmosphere (Ar for metals and nitrogen for polymers) to mimic the PBF-LB process conditions. The reflectivity at a wavenumber of 943 cm^{-1} corresponds to wavelength 10.6 μm. This characteristic laser wavelength of carbondioxide (CO_2) lasers is mainly employed in the PBF-LB of polymer powder feedstocks. The reflectivity at a wavenumber of 9398 cm^{-1} corresponds to a wavelength of 1064 nm (emission of the Nd:YAG lasers), primarily used in PBF-LB of metal powder feedstocks. DRIFT measurements will generate a total of 60 data points for metals and 60 data points for polymers at room temperature and powder bed temperature (Figure 4).

2.2.6. Moisture Content

The moisture content is another property that adversely affects flowability, processability, and pore formation during PBF-LB. Therefore, if a wet additivation is used, the powder feedstocks must be dried until weight loss is completed in a condition avoiding degradation or oxidation of powders. Note that moisture absorption from the atmosphere can lead to gas pores forming within the as-built part during metal or polymer powder feedstocks processing. Therefore, it is recommended to dry the powders under vacuum followed by storage under inert gas. The moisture content of the metal and polymer powders will be measured by thermo gravimetric analysis (TGA), and a total of 30 data points for the metal and 30 data points for the polymer will be generated (Figure 4).

2.2.7. Surface Coverage of Micro-Powders by NPs and Interparticle Distance between NPs

It is known that powder size distribution, flowability, powder bed density, the powder's laser interaction properties, and the additivation technique are influenced by the surface coverage value of NPs on the micro powders [39–41]. This surface coverage is highly dependent on NP size distribution, NP load, and additivation technique [28]. SEM has to be employed in the ILS to measure the surface occupation density of NPs and interparticle distance between NPs on the surface of metal and polymer powder feedstocks. It is a time-consuming image analysis technique, and tens to hundreds of additivated micro-powder particles must be imaged and measured for statistical relevance [76]. A total of 60 data points for the metal and 60 data points for the polymer will be generated by measuring surface coverage % and the average distance between NPs on the surface of micro powders (Figure 4).

Several techniques to measure unmodified and NPs additivated metal and polymer powder feedstocks' material properties were described in Section 2.2. During ILS, it is strongly recommended to measure each property by the same operator using the same equipment in a central laboratory (Figure 1). The study coordinator has to deliver standard operational procedures (SOPs) for each measurement technique that central laboratories must follow. Available international standards for conducting related tests must be embedded in these SOPs (see Section 2.6).

2.3. PBF-LB Process

After material properties of unmodified and additivated metal and polymer powder feedstocks are tested and embedded in the RDM, the powder feedstocks are sent out to the processing participants in the third step of ILS (Figure 2). It should include the manufacturing plan SOPs, process control documents (PCDs), technical drawings, standard triangle language (STL) files, and, if possible, building plates for metal parts. Every participant must use the central STL and additive manufacturing file format, which defines the whole geometry, specimen placement, and build job support structures. For example, every test specimen must be labeled, e.g., with numbers to indicate its position in the build job. Since tens of specimens will be built in each build job, all samples with built-in unique numbers will be formatted in STL. In doing so, participants will not have to mark each built specimen after PBF-LB, facilitating traceability. The position in the building envelope and parts orientated in different directions will be traceable even during post-processing. STL files have to be sent to participants at least one month before the start of ILS. CAD drawings will be converted to STL files by a central entity. Later, STL files will be distributed to each process participant to ensure that everyone uses the duplicate files. It is known that CAD to .stl conversion can be challenging in very complex parts. Our ILS design mainly addresses the powder material qualification and processability question rather than creating complex 3D parts. Hence, printing is based on simple geometries (i.e., cubes, bars, and cylinders) where the CAD to .stl conversion is not challenging. Each process participant will use their build processor for data preparation that fits their PBF-LB machine. Various machines and software versions of participants can have problems introducing the STL of the build job [61]. In this case, the study coordinator must be informed immediately to decide further participation in the ILS. Machine users must carefully follow the PBF-LB process SOPs, including allowed parameter corridors, e.g., laser power, scanning speed, hatch distance, and powder layer thickness.

Several strategies can be followed in processing powder feedstocks, while it is highly dependent on each participant's PBF-LB machine type, specifications, and experiences.

Previous ILSs on processing unmodified powder feedstocks showed that a strict manufacturing plan with a narrow corridor range of process parameters (shown in Figure 5) could be supported if each participant's machine type is identical, which is often not the case. Otherwise, between-participant deviations in as-built part property can be high [54,61]. Another strategy can be, selecting experienced institutions in processing unmodified metal or polymer powder type of ILS. In this case, proposed for the present ILS, each participant can set their best practice processing parameters within the corridor range given in the manufacturing plan [57]. As a limitation to best practice process parameters of participants, achieving a certain level of relative density and/or UTS in unmodified as-built parts is requested from the participants, e.g., relative density (RD) > 99%, ultimate tensile strength (UTS) > 400 MPa for AlSi10Mg [2], and RD > 90%, UTS > 48 MPa for PA12 [3]. As a result of the second strategy, participants will individually set their best-effort process parameters shown in Figure 5. Process parameters will include laser power, scanning speed, hatch spacing, powder layer thickness, laser beam diameter, powder bed temperature, contour parameters, powder recoater speed, chamber atmosphere, and laser wavelength. Each participant must report the process parameter sets shown in Figure 5 into PCDs and send them to the study coordinator after processing. PCDs will be used as a meta-data source during ILS data evaluation. Each participant must process the metal build job on the building plates supplied by the study coordinator and send it to the coordination office without separating as-built parts. A total of 360 data points both for metals and polymers will be generated as the process data (Figure 5).

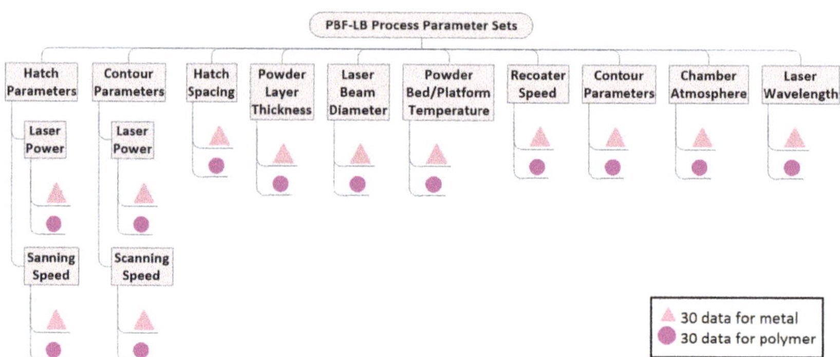

Figure 5. Participant-dependent PBF-LB process parameter sets and number of data generated for each property.

2.4. Powder Quality of Used Powder Feedstocks

At the end of the PBF-LB process, each participant must remove an aliquot of the unprocessed powder feedstocks, ideally 2.5 kg for AlSi10Mg and 1 kg for PA12 powders. These used powder aliquots must be stored in the containers without sieving and sent back to the ILS coordination office with the build jobs. As a fourth step of ILS (Figure 2), each participant's used unmodified and NP-additivated metal and polymer powder feedstock properties will be analyzed with the same test methods stated in the second step (see Section 2.2) to draw conclusions on the powder´s reusability and property changes. PBF-LB of metal powders may cause spatters and agglomerates, e.g., direct ejections from the molten pool, the dragging of particles by the gas stream over the process zone. These spattered powders might be sintered and have been reported to deviate from the chemical composition of the as-produced feedstock powders and degradation due to heat, oxidation, and change in the powders surface morphology have been observed during PBF-LB [51]. Used powder quality is another interest in ILS to determine the material properties of used metal and polymer powder feedstocks for reusability and statistical interlaboratory dependence in the PBF-LB process. Therefore, it is highly recommended to test the material properties of used powder feedstocks after sieving. The sieve must match the measured biggest powder size of the as-produced metal and polymer powders, e.g., a sieve size > DV, 99 of metal and polymer powder feedstocks.

2.5. As-Built Parts

All metal and polymer build job parts must be delivered to the coordination office before proceeding to the ILS's next step. The coordination office will organize preparing specimens and tests at the central laboratories in the fifth step (Figure 2). As-built polymer parts will be built on the powder layers without support structures, and after the dimensional accuracy of each part is measured, polymer specimens will be sent to the assigned central laboratories for testing.

Compared to PBF-LB/P, building a metal part needs a support structure in the PBF-LB/M process. These support structures are being manufactured along with the parts on the building plate. As-built parts must be separated with a cutting machine such as electrical discharge machining (EDM).

Metal alloy PBF-LB parts are usually heat-treated to reduce residual stresses in the as-built parts or rearrange microstructural formations affecting the mechanical properties of as-built parts. If heat-treatment is of interest in ILS, as-built parts must be heat-treated before measuring part properties as given in the next section of the ILS plan. The effect of heat-treatment conditions on mechanical properties of the AlSi10Mg alloy is given in ASTM F3318-18 standard [77]. Depending on the heat-treatment conditions, microstructural formations in the metal matrix, residual stress in the as-built parts, tensile strength, and tensile elongation is improved compared to the as-built state [78]. As mentioned in the introduction, NPs may have a strong impact on microstructural formations during processing that affect the mechanical properties of as-built parts, requiring a different heat-treatment condition. If heat treatment is included in the ILS, unmodified and NP-additivated specimens should be heat-treated in a single central lab using the same furnace and heat treatment conditions. The coordination office must provide related SOPs to central labs for metal part separation and heat-treatment. However, this ILS is designed to test directly as-built conditions in order to minimize the required powder amount and specimens to be characterized (that would double in case of heat treatment is included, as understanding of the material response would requires analysis of both the heat-treated and as-built specimen). In this case, after cutting, removing support structures, and measuring the dimensional accuracy of the specimens, the as-built parts are sent to assigned central laboratories to test as-built part properties.

Following sample preparation, testing the effect of NPs on the as-built part properties of each participant's build job will be carried out in the fifth step (Figure 2). Both powder properties and process parameter sets affect the microstructural formations, porosity, and mechanical properties of as-built metal and polymer parts [2,3,13,17,18,30,34,36]. Therefore, the effect of NPs on microstructural and mechanical properties has to be investigated with the test methods shown in Figure 6. Composition, relative density, porosity, NP imaging, and tensile properties will be measured for both metals and polymers. Matrix grain size and crystal orientation will be measured separately for metal parts; molecular weight and crystallinity will be separately measured for polymer parts to assess the parts' microstructural formations.

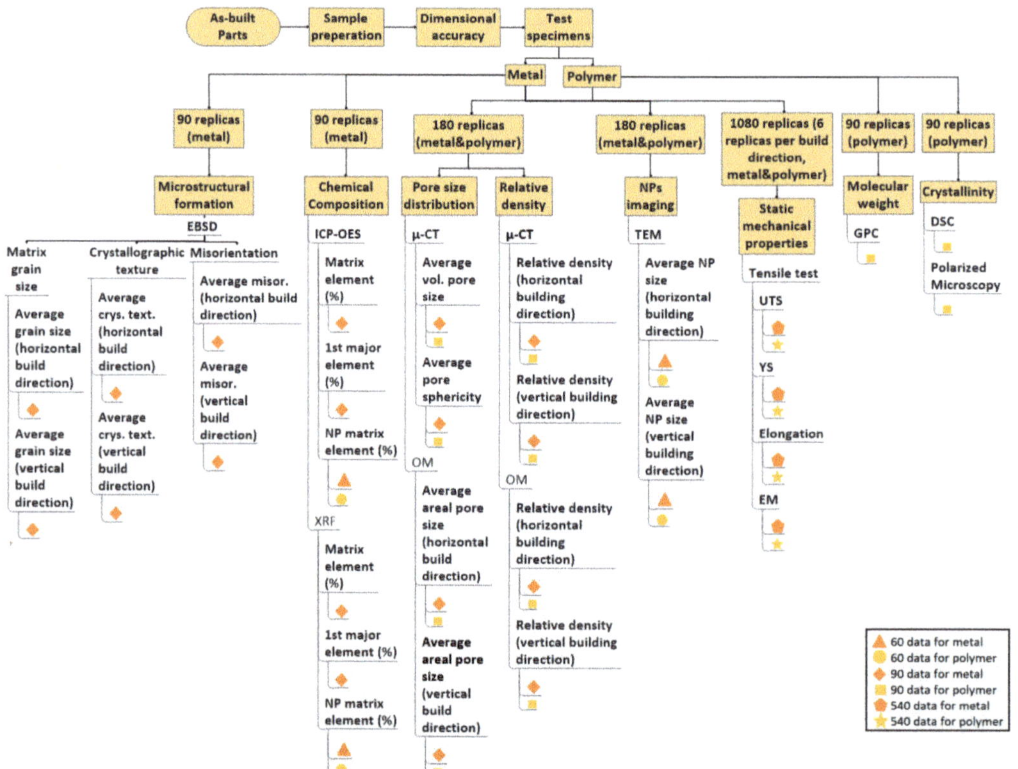

Figure 6. As-built part property analysis with related measurement techniques for nanoparticle-additivated metal and polymer parts, number of replicas for each property measurements per build job, and number of data generated for each property metric. Electron backscattered diffraction (EBSD), inductively coupled plasma-optical emission spectroscopy (ICP-OES); X-ray fluorescence (XRF); micro-computer tomography (μ-CT); optical microscope (OM); transmission electron microscope (TEM); ultimate tensile strength (UTS); yield strength (YS); elastic modulus (EM); gel permeation chromatography (GPC); differential scanning calorimetry (DSC).

2.5.1. Microstructural Formations

NPs are known to rearrange microstructural formation in both metal and polymer parts [2,3,25,26,28–30,33,36]. Since metal and polymer are different material types, investigation techniques to understand the effect of NPs on microstructural formations need different analytic techniques. The grain size of metal parts is one of the measurable properties to understand the effect of NPs on microstructural formation. NPs can act as nucleation sites during the cooling of the melt pool, which fines the grain size in the microstructure, and a decrease in grain size can be linked to an increase in tensile strength by the Hall–Petch equation [36]. SEM will observe metallographically prepared surfaces/cross-sections of the as-built parts. Measurements have to be carried out both from the vertical and horizontal building directions of the sample according to the ASTM E112 [79] standard.

To investigate the NP effects on the crystallographic orientation during the melt pool's solidification, electron backscattered diffraction (EBSD) measurements in the as-built metal parts [36] are implemented in the ILS. At least 3 replicas should be used for EBSD analysis that generates 540 data points for metals (Figure 6). The effect of NPs on the polymer parts' microstructure will be determined by measuring the parts' molecular weight and crystallinity. The molecular weight of the as-built polymer parts will be measured by gel permeation chromatography (GPC). GPC analysis showed that depending on the process parameter sets, e.g., different build chamber temperatures affect the polymer's average molecular weight, and a change in chain length alters the crystallization temperature of the polymer [80]. The crystallinity degree of as-built polymers will be analyzed by polarized light microscopy and DSC [80,81]. Crystallinity measurements will be performed on 3 replicas and generate 270 data points for polymers (Figure 6).

2.5.2. Chemical Composition

A defined number and type of NPs are supported on the feedstocks, and the chemical composition of additivated metal and polymer powder feedstocks are measured in the second step of ILS. However, it is unknown if all additivated NPs will be transferred to the as-built metal and polymer parts during processing, as losses during powder handling or laser melting cannot be excluded. Furthermore, some metal alloy elements can be vaporized during laser melting that affects the solidification microstructure and part quality [43]. As discussed above (Section 2.2.1), XRF [27,33,40] and/or ICP-OES [36] will measure the as-built polymer and metal parts' chemical composition. Measurements by XRF have to be carried out on the polished surface of metal and polymer cubes. 3 replicas will be measured by non-destructive XRF, and the same replicas will be sent to relevant central laboratory to conduct ICP-OES measurements. Measurements by XRF and ICP-OES will generate 480 data points for metals and 120 data points for polymers (Figure 6). The transferability of the amount of additivated NPs along the process chain will be determined by comparing the chemical composition (nanoparticle vol%) of unmodified and processed, as well as unmodified and additivated powders with as-built parts.

2.5.3. Relative Density and Pore Size Distribution

Porosity in as-built metal and polymer parts significantly impacts the density, mechanical, and functional properties [57–60]. The microstructural formations and porosity distributions of the as-built parts can be affected by the dissolution degree, lattice misfit degree, high-temperature chemical stability, wetting degree, interface reactions, and laser absorptivity of NPs during PBF-LB processing [36,82–86]. Further, the parts' relative density and porosity content can affect as-built parts' mechanical and functional properties [2,3]. After the PBF-LB process, as-built parts can contain pores, and the effect of NPs on the final density and pore size distribution must be determined. Volume μ-CT and cross-sectional optical microscope (OM) measurements must be performed to measure the relative density, volumetric porosity content, and volumetric pore size distributions of as-built polymer and metal parts [58–60]. 3 replicas will be measured for each material. Relative density and pore size distribution will generate 720 data points both for metals and polymers (Figure 6).

2.5.4. NP Imaging

Measuring the distribution and sizes of NPs in the metal or polymer matrix is essential for evaluating the as-built part properties. Depending on the material properties of NPs, different reaction mechanisms can occur with metal melt and polymer melt that affect the microstructural formation in the as-built parts [36,38]. For example, NPs can dissolve in metal melt and, during the cooling stage of PBF, can form nano- or microprecipitates with a different crystalline phase than starting NPs structure [33,36]. On the other hand, undissolved NPs in the melt pool can rearrange in the metal matrix during the cooling stage of the PBF-LB process. As illustrated in Figure 7, laser power or energy density variations can affect the melt pool's undissolved the NPs rearrangement in the melt pool. Marangoni convection significantly impacts NP rearrangement in the metal melt pool [63,86,87]. Depending on the laser power or energy density, surface tension gradients between melt and solid drive a fluid flow in the melt pool. Increasing laser power or energy density will increase the Marangoni convection and melt pool depth during processing. In addition, there can be other factors in melt pool dynamics: evaporation in the melt pool, recoil pressure, and inverted Marangoni flow direction due to inhomogeneous melt pool chemistry [36,63,75,78,86,87].

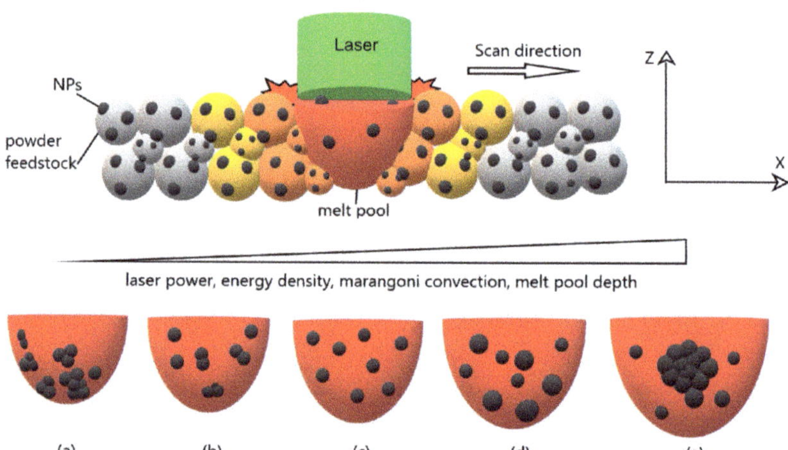

Figure 7. Illustration of nanoparticle distribution in the melt pool during laser powder bed fusion process. Depending on the process conditions, within the melt pool volume, the undissolved nanoparticles can be (**a**) poorly mixed and aggregated at the bottom, (**b**) loosely accumulated with improved distribution, (**c**) homogeneously distributed without accumulation, (**d**) well distributed but coarsened, and (**e**) highly aggregated to (sub)micrometer scale.

Conducted ILSs showed that deviations in the as-built part property are high between participants rather than within-participant properties [54,61]. As stated in Section 2.3, participants can set different laser energy densities within a given range to process the powders. Depending on the process parameter sets, the melt pool dynamics can vary between participants, and NP rearrangement in the as-built parts can vary, affecting microstructural formation. Therefore, the final status of NPs in each participants' as-built part must be investigated. Thin cross-sections will be prepared from 3 exemplary replicas (randomly selected each for polymer and metal) by microtome for polymer parts and focused ion beam (FIB) for metal parts before conducting TEM measurement. Thin cross-sections of both vertical and horizontal directions will be investigated by TEM to measure the NPs sizes in the polymer or metal matrix [28,36]. TEM measurements will generate 120 data points both for metals and polymers (Figure 6). Since there is no standard test method for this investigation, an SOP must be prepared to follow the same experimental and analysis procedure in cross-section TEM investigations.

2.5.5. Static Mechanical Properties

Previous ILSs showed that the mechanical properties of the unmodified parts could vary between participants [54,56,57]. The tensile test is the most common method that can quantify the effect of NPs on as-built parts and can be used both for polymer and metals. Depending on the building direction within the build job volume, parts can exhibit different tensile behavior [55,57]. It is highly recommended to build test specimens in three directions: vertical, horizontal, and diagonal to the building direction. The dimensions, geometry, and testing conditions have a high impact on the tensile test results. There is no international standard yet precisely defining tensile testing of as-built metal and polymer parts processed by PBF-LB. As stated previously, minimizing the tensile specimen's length will decrease the building height of the build job, so the total amount of powder requirement in the ILS run will be reduced. By conducting tensile tests, the effect of NPs on the ultimate tensile strength (UTS), yield strength (YS), elastic modulus (EM), and tensile elongation

(TE) of metal and polymer parts can be measured. Therefore, 6 replicas will be built in vertical, diagonal, and horizontal build directions for each polymer and metal build jobs. It is recommended to test metal tensile specimens with a height/length of 60 mm (Form B, DIN 50125:2016-12) and polymer tensile specimens with a height/length of 49 mm (Form C, DIN 50125:2016-12) within this ILS. Tensile specimens will be tested according to DIN EN ISO 6892-1 standard for metals [88] and ISO 527-2:2012 standard for molding and extrusion plastics [89]. Tensile tests will generate a total of 2160 data points both for metals and polymers (Figure 6).

Overall, the effect of NPs on volumetric relative density, microstructural formation, and mechanical properties of the as-built polymer and metal parts (Figure 6) will be measured for each participant's build job in the fifth step of ILS (Figure 2).

2.6. Research Data Management (RDM) of ILS

Since this is the first time a proposed PBF-LB ILS evaluates a comprehensive set of powder material, process, microstructure, and part properties, there will be a vast data source at the end of the ILS. On the one hand, this rich data pool is an ideal basis for statistical assessments and correlation function extraction along the full process chain, from the powder to the part. On the other hand, solid statements and robust data processing, in particular correlation analysis, is only possible if data and meta-data management is strictly embedded into the ILS concept. Therefore, it is essential to set up an RDM plan beforehand and follow FAIR principles [90]. In this manner, transparent, reproducible, and reusable data can be provided by the available research process with all documented components. Here, data registration plays a critical role in the success of the ILS RDM. Each central laboratory will document the sampling techniques and test conditions while measuring each property metric, collecting metadata, and generating datasets of related property metrics to improve data registration. All datasets with their corresponding documentation and SOPs will be stored in a central repository created by the coordination office. A long-term accessible repository will give open access to validated data for future data-mining interests in the field of AM. Data reporting should be performed according to ASTM F2971-13 [91]. In addition, ASTM is developing a new standard (ASTM WK73978 [92]) to comprise actions for data registration which will draw a route map for high-quality data registration under FAIR principles in the field of AM. Then, all the data can be further analyzed by data miners or linked to the further ILSs to advance discovery in material selection and design of AM.

Furthermore, this traceable data forming and structuring will set the base for functional correlations in PBF-LB via PCA. An exemplary data flow is prepared to show the complexity of ILS design and the importance of research data management to further evaluate the generated data before starting the ILS. As shown in Figure 8, the dataflow matrix systemically shows the data generation for the selected process variables in the ILS design. The dataflow matrix shown in Figure 8 is designed for ten polymer and ten metal PBF-LB process participants testing unmodified and two different NPs additivated metal and polymer powder feedstocks.

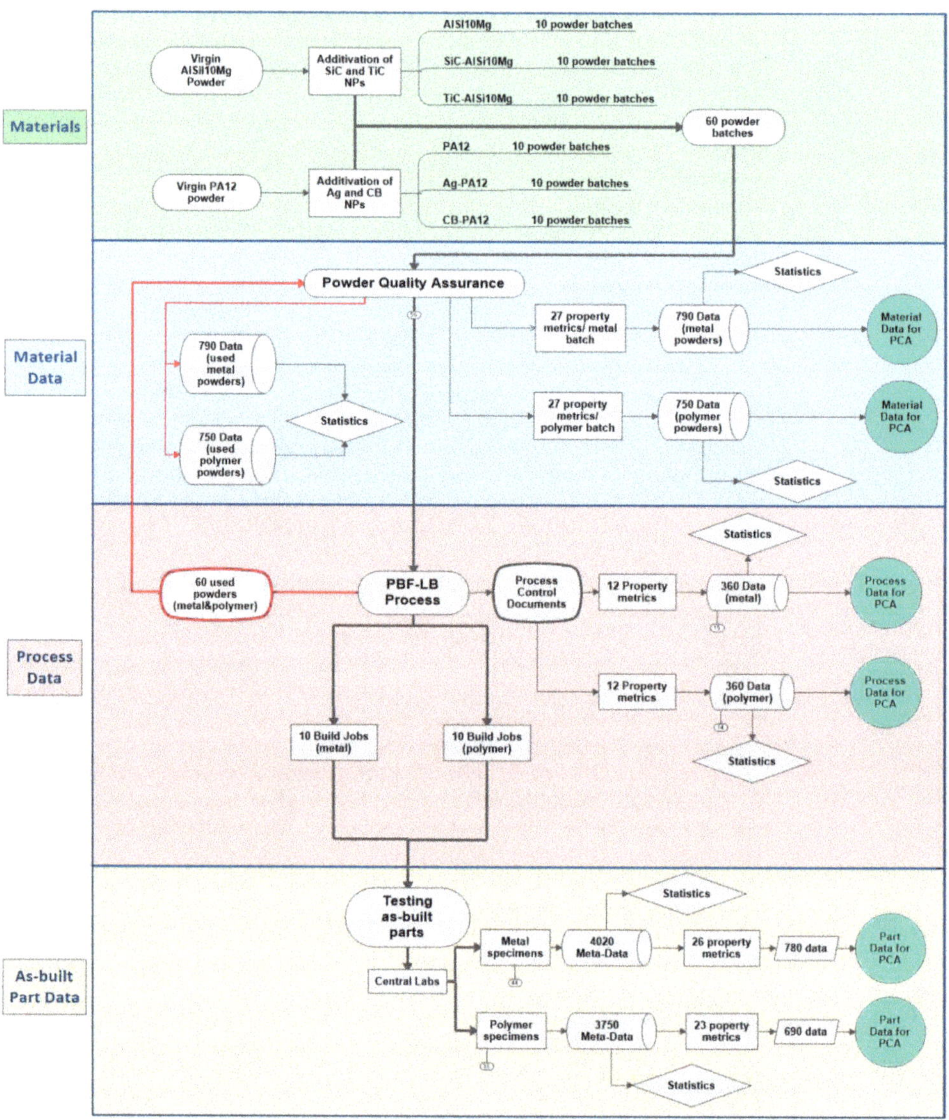

Figure 8. Material, process, part data flow chart of inter-laboratory study, and the entire process chain. Numbers are based on testing 1 unmodified and 2 nanoparticle-additivated powders, each for metal and polymer powder with 10 processing participants (each for polymer and metal) plus central lab participants.

In the sixth step of the ILS, the ILS design's data generation is classified into three sections along the process chain: material, process, and parts. The material data of as-produced (second step, Section 2.2) and used (fourth step, Section 2.4) unmodified and NPs additivated powder feedstocks per participant have to be evaluated to compare the properties of as-produced and used powder feedstocks in the sixth step (Figure 2). As shown in Figures 5 and 8, in total of 9 variables (chemical composition, powder size distribution, powder shape, flowability, thermal behavior, laser interaction, moisture content, surface occupation density, and interparticle distance between the NPs on the surface of micro powders) will be evaluated by measuring a total of 52 property metrics. A total of 790 data points for metals and 750 data points for polymers will be evaluated to compare the as-produced and used powder properties. The material data generated from as-produced, unmodified, and NP- additivated powder feedstocks will be correlated with the generated process and part data by PCA (Figure 8).

A total of 20 PCDs filled by participants will be sent back to the study coordinator after PBF-LB process. By doing so, each participant's 12 PBF-LB process properties (Figure 5) will be extracted to generate a total of 360 data points both for metals and polymers (Figure 8), and further will be correlated with the microstructure and part properties to determine the effect of 20 participants' process parameter variation on the unmodified and nanoadditivated metal and polymer as-built part properties.

The part data generated from as-built part properties of unmodified and NP-additivated powder feedstocks per participant are statistically evaluated in the sixth step (Figure 2). As shown in Figures 6 and 8, in total 5 variables (composition, density, pore size distribution, average NP size within the as-built parts, the tensile properties of both metals and polymers), an additional 3 variables (grain size, crystallographic texture, and misorientation) for metals, and 2 variables (molecular weight, crystallinity) for polymers will be evaluated by measuring a total of 49 property metrics.

As shown in Figure 8, a total of 65 property metrics in PBF-LB of unmodified and NP-additivated metal and polymer powder feedstocks will be statistically evaluated in ILS. Following the FAIR principle in data generation, subsequent PCA allows a correlation between material–process–part properties to understand the PBF-LB process's complexity. For example, Kusoglu et al. [2,3] conducted a PCA of a small data subset by extracting the reported powder material, process, and part properties data from the most-cited 100 SCI-Expanded articles published within the last decade for PBF-LB of Al alloys and polymers. A data matrix was generated in that study from 139 Al powder compositions [2] and 257 polymer powder compositions [3] with the corresponding material, process, and as-built part properties. Unfortunately, only 33 Al powder compositions and eight polymer powder compositions could be successfully processed by PCA because of missing information parts in most published work. As a result, only one powder property, six process properties, and three as-built properties were successfully correlated with these limited datasets. These studies [2,3] showed the importance of reporting the same properties, using international standards in test methodologies, building the same test specimen geometries, and following FAIR principles.

The statistical evaluation of ILS data will focus on calculating the repeatability and reproducibility of each property metric for between-participant and within-participant test results. ASTM E691-20 [93] described how to conduct an ILS to determine a test method's precision. Luping and Schouenborg [94] showed the equations to calculate the repeatability/reproducibility of the test results. These equations should be used in the proposed ILS's statistical evaluation. Both repeatability and reproducibility conditions strictly rely on using the same test methods to measure identical material within a short interval.

Additionally, repeatability conditions benefit from the same operator's tests at the same laboratory using the same equipment, which is embedded in the ILS design by the concept of central testing laboratories. With SOPs, reproducibility conditions depend less on the tests at different laboratories by different operators using different equipment. Repeatability and reproducibility limits will be allowed to a probability of 95% between two test results and can be obtained by multiplying sr or sR by a factor of 2.8 as a rule of thumb. Since the reproducibility depends on conducting the tests under the same experimental conditions in different laboratories, reproducibility can be calculated for the same build jobs processed in different PBF-LB machines. Finally, the consistency of the test results is assessed by graphical techniques as Mandel's k, which is a within-participant consistency statistic, and Mandel's h, which is a between-participant consistency statistic [93,94].

A total of 68 property metrics will be evaluated for unmodified and additivated metal and polymer powder feedstocks in the suggested ILS design. Internationally accepted standards should be used to define each property's testing conditions. By doing so, significant variations in the test results can be minimized. As a result, each property metric's repeatability and reproducibility in original data can be increased. For this purpose, it is strongly recommended to prepare SOPs for each test method or property. Published and developing standards considered in the ILS design for the SOPs are given in Table 1.

Table 1. Published and developing international standards on the characterization of laser powder bed fusion of metal and polymer powder feedstocks and as-built parts, as basis for the standard operational procedures of the inter-laboratory study.

Material	International Standards	Content
Metal Powder	ISO/ASTM 52907-19 [95]	Technical specification of as-produced and used feedstocks
	ASTM F3049-14 [96]	Test methods for powder size, morphology, chemistry, flowability, and density
	ASTM B527-20 [97]	Test method for tap density
	ASTM B822-20 [98]	Particle size distribution by light scattering
	ISO 13320:2020 [99]	Particle size distribution by laser diffraction
	WK66030 [100]	Quality assessment guidelines for powder reusability
	WK55610 [101]	Powder dynamic flow properties
	WK74905 [102]	Particle shape analysis to identify agglomerates/satellites
Polymer Powder	ISO 13320:2020 [99]	Particle size distribution by laser diffraction
	WK55610 [101]	Powder dynamic flow properties
As-built Metal Part	ASTM F3122-14 [103]	Evaluating mechanical properties
	ASTM E572-13 [104]	Measuring chemical composition by wide wavelength XRF
	ASTM E8/E8M-16ae1 [105]	Tensile test
	WK49229 [106]	Orientation and location-dependent mechanical properties
	ASTM E3166-20 [107]	Non-destructive examination of as-built parts
As-built Polymer Part	ISO 527-1:2012 [108]	General principles of tensile test
	WK66029 [109]	Tensile test

Besides testing the properties of additivated metal and polymer powders, it is recommended to prepare additional SOP for NP characterization. Available standards as ASTM F1877-16 (characterization of the morphology, number, size, and size distribution of particles) [110], ASTM E2834-12(2018) (particle size distribution of nanomaterials in suspension) [111], ASTM E3247-20 (measuring the size of nanoparticles in aqueous media using DLS) [112], ISO 17200:2020 (characteristics and measurements of nanoparticles in powder form) [113], ISO/TR 14187:2020 (surface chemical analysis of nanostructured materials) [114] have to be used to prepare SOPs to test the properties of NPs shown in Figure 4.

3. Recommended Implementation Procedure

Due to the complexity of the design testing new powder feedstocks in PBF-LB, it is not recommended to start directly with the full width of the ILS before pre-testing the properties of unmodified and NP additivated powder feedstocks, defining the range of various PBF-LB process parameters, and testing the sample flow and workflow of the intensive structural characterization in central labs. The basic workflow should be tested in a minimalized pre-test with 2 processing partners. The following control on the workflows, data generation, and cycle duration should be traced in pilot run (2–4 processing partners, central labs) before running the full-run ILS design (200 processing partners, central labs, research data management).

In this manner, ILS must be completed in three parts, (i) pre-test, (ii) pilot run, and (iii) full run. The purpose of each part is given in Figure 9.

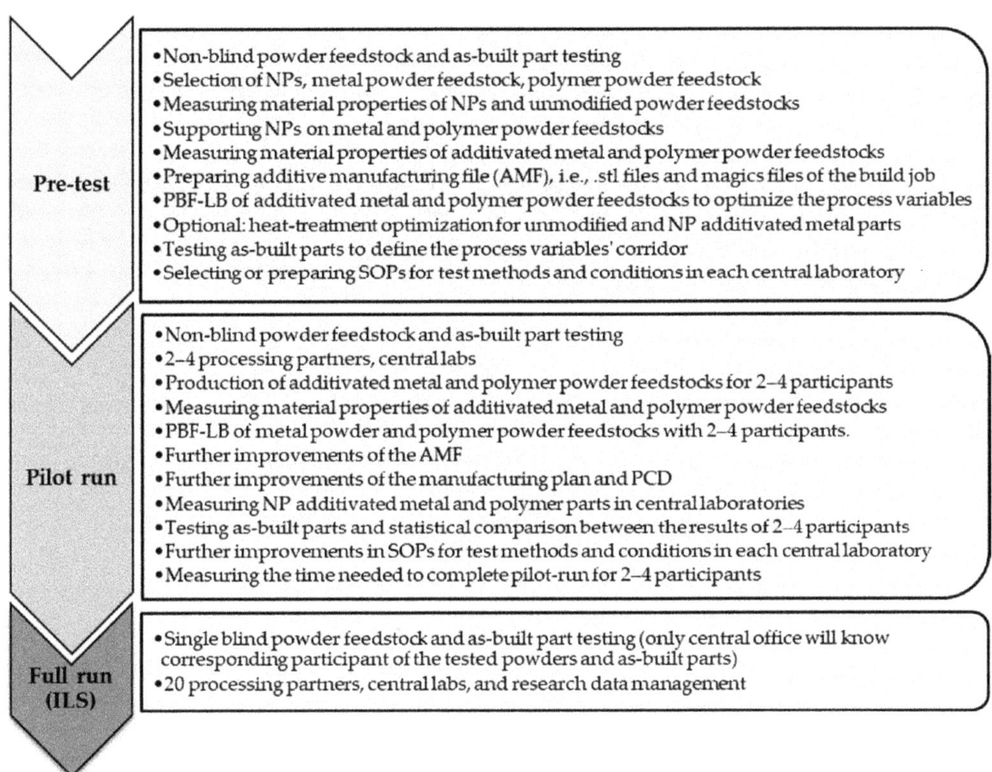

Figure 9. Design of the stepwise inter-laboratory study implementation via execution of pre-test, pilot run, and full ILS run.

Besides planning workflows, the study director and study coordinator must trace the logistics' progress and timing already during the pilot run. Controlling logistics between the workflows will help draw flexible time planning for the full run and avoid ILS logistic-based delays.

4. Conclusions

The implementation of interlaboratory studies in additive manufacturing is indispensable for further developing new materials for 3D printing methods such as PBF-LB. Here, a well-designed ILS can consolidate progress in industrialization, standardization, robustness, and fundamental understanding along the entire process chain of PBF-LB. Further, it is essential to accurately evaluate measurable property metrics and calculate property's repeatability and reproducibility between ILS participants.

Although several ILS exist, no study describes the complete process chain based on different feedstock materials. We herein designed such a large-scale interdisciplinary ILS, ensuring a comprehensive analytical characterization along the process chain. In existing ILSs, a focus often lies on (e.g., part property) repeatability to manufacture specific parts (to see, e.g., machine-dependences) or to see inter-laboratory effects of general LPBF manufacturing of a material class (e.g., steel powder from varying sources). Such designs do not make sure that a process chain parameter can later be firmly attributed to another, not yet known one, via PCA. Additionally, the strong focus on a) powder feedstock properties and b) RDM is unique. To provide a more concrete picture of the general theoretical concept and ILS design, and provide exemplary numbers to the proposed RDM concept, the presented exemplary design statistically evaluates the entire process chain by measuring 27 powder property metrics, 12 process property metrics, and 26 part property metrics. Of course, an extension or reduction in the number of varied material properties is easily possible based on the presented general ILS concept. Further, the effect of nano-additives on the commercially available metal and polymer powder feedstocks and the repeatability of the modifications by NP can be measured for a better understanding of the PBF-LB process chain. Hence, this ILS design may be a blueprint for future ILS on LPBF, in particular in those ILS where powder feedstock effects shall be understood, and/or robust statistical analysis and PCA data processing is intended, where a good RDM design before ILS execution is key. The implementation of nano-additivated powders is not required, but is recommended, as it allows an expansion of the existing feedstock materials towards enhanced processability and shows further advantages in terms of avoiding crack formation, rearranging grain structure, and adding functionality to the as-built parts.

The ILS includes the experimental procedure and an RDM which is essential in a study of this scale. It enables an in-depth PCA based on FAIR principles in research data management. We highly recommend dividing the ILS into three stages—pre-test, pilot run, and full run—to successfully establish the latter for pure and nano-additivated feedstock materials. While the pre-test and pilot run of this ILS mainly focus on understanding the effect of powder properties, processability, and as-built properties during the PBF-LB process, the full run is required to prove its feasibility.

Testing new powder feedstocks within an ILS performed with a statistically relevant number of participants will guarantee reproducibility for both metal as well as polymer processing. Further, the ILS will set the data for determining correlations along the entire process chain and help to evaluate the available characterization techniques used for quality control of feedstocks and as-built parts in the field of additive manufacturing.

Author Contributions: Conceptualization, I.M.K., A.R.Z. and S.B.; methodology, I.M.K.; data curation, I.M.K.; writing—original draft preparation, I.M.K., A.R.Z., C.D.-B. and F.H.; writing—review and editing, F.H., C.D.-B., B.G., J.T.S., A.K., M.S., S.B.; visualization, I.M.K.; supervision, S.B.; funding acquisition, B.G., J.T.S., A.K., M.S. and S.B. All authors have read and agreed to the published version of the manuscript.

Funding: This research was funded by the Deutsche Forschungsgemeinschaft (DFG, German Research Foundation) under the Priority Program "Materials for Additive Manufacturing" (SPP 2122, project BA 3580/28-1 and BA 3580/27-1, SCHM 2115/78-1, KW 9/32-1 and SE 2935/1-1) and under the Heisenberg Programme (GO 2566/10-1, Project-ID 445127149).

Institutional Review Board Statement: Not applicable.

Informed Consent Statement: Not applicable.

Data Availability Statement: All the data is available within the manuscript.

Conflicts of Interest: The authors declare no conflict of interest.

References

1. ISO/ASTM 52910, Additive Manufacturing–Design–Requirements, Guidelines and Recommendations. Available online: https://www.iso.org/standard/67289.html (accessed on 14 February 2020).
2. Kusoglu, I.M.; Gökce, B.; Barcikowski, S. Research trends in Laser Powder Bed Fusion of Al alloys within the last decade. *Addit. Manuf.* **2020**, *36*, 101489. [CrossRef]
3. Kusoglu, I.M.; Doñate-Buendía, C.; Barcikowski, S.; Gökce, B. Laser Powder Bed Fusion of Polymers: Quantitative Research Direction Indices. *Materials* **2021**, *14*, 1169. [CrossRef]
4. EOS, 3D Printing for Metals. Available online: https://www.eos.info/en/additive-manufacturing/3d-printing-metal (accessed on 3 April 2021).
5. SLM Solutions, Industrial Metal Additive Manufacturing Machines. Available online: https://www.slm-solutions.com/products-and-solutions/machines/ (accessed on 3 April 2021).
6. 3D systems, Selective Laser Sintering. Available online: https://www.3dsystems.com/resources/information-guides/selective-laser-sintering/sls (accessed on 3 April 2021).
7. EOS, 3D Printing with Plastics. Available online: https://www.eos.info/en/additive-manufacturing/3d-printing-plastic (accessed on 3 April 2021).

8. Paul, C.P.; Jinoop, A.N.; Nayak, S.K.; Paul, A.C. Laser Additive Manufacturing in Industry 4.0: Overview, Applications, and Scenario in Developing Economies. In *Additive Manufacturing Applications for Metals and Composites*; Balasubramanian, K., Senthilkumar, V., Eds.; IGI Global: Hershey, PA, USA, 2020; pp. 271–295.

9. DebRoy, T.; Mukherjee, T.; Wei, H.L.; Elmer, J.W.; Milewski, J.O. Metallurgy, mechanistic models and machine learning in metal printing. *Nat. Rev. Mater.* **2021**, *6*, 48–68. [CrossRef]

10. Wei, H.L.; Mukherjee, T.; Zhang, W.; Zuback, J.S.; Knapp, G.L.; De, A.; DebRoy, T. Mechanistic models for additive manufacturing of metallic components. *Prog. Mater. Sci.* **2021**, *116*, 100703. [CrossRef]

11. International Standard Organization, Committee ISO/TC261 on Additive Manufacturing. Available online: https://www.iso.org/committee/629086.html (accessed on 3 April 2021).

12. ASTM International, Committee F42 on Additive Manufacturing Technologies. Available online: https://www.astm.org/COMMIT/SUBCOMMIT/F42.htm (accessed on 3 April 2021).

13. Pannitz, O.; Sehrt, J.T. Transferability of Process Parameters in Laser Powder Bed Fusion Processes for an Energy and Cost Efficient Manufacturing. *Sustainability* **2020**, *12*, 1565. [CrossRef]

14. Dechet, M.A.; Demina, A.; Römling, L.; Gómez Bonilla, J.S.; Lanyi, F.J.; Schubert, D.W.; Bück, A.; Peukert, W.; Schmidt, J. Development of poly(L-lactide) (PLLA) microspheres precipitated from triacetin for application in powder bed fusion of polymers. *Addit. Manuf.* **2020**, *32*, 100966. [CrossRef]

15. Dechet, M.A.; Baumeister, I.; Schmidt, J. Development of Polyoxymethylene Particles via the Solution-Dissolution Process and Application to the Powder Bed Fusion of Polymers. *Spec. Polym. Addit. Manuf.* **2020**, *13*, 1535. [CrossRef]

16. Dechet, M.A.; Schmidt, J. On the Development of Polymer Particles for Laser Powder Bed Fusion via Precipitation. *Procedia CIRP* **2020**, *94*, 95–99. [CrossRef]

17. Bierwisch, C.; Mohseni-Mofidi, S.; Dietemann, B.; Kraft, T.; Rudloff, J.; Baumann, S.; Popp, K.; Lang, M. Particle-based Simulation and Dimensional Analysis of Laser Powder Bed Fusion for Polymers and Metals. *Procedia CIRP* **2020**, *94*, 74–79. [CrossRef]

18. Sommereyns, A.; Hupfeld, T.; Gökce, B.; Barcikowski, S.; Schmidt, M. Evaluation of essential powder properties through complementary particle size analysis methods for laser powder bed fusion of polymers. *Procedia CIRP* **2020**, *94*, 116–121. [CrossRef]

19. Hupfeld, T.; Doñate-Buendía, C.; Krause, M.; Sommereyns, A.; Wegner, A.; Sinnemann, T.; Schmidt, M.; Gökce, B.; Barcikowski, S. Scaling up colloidal surface additivation of polymer powders for laser powder bed fusion. *Procedia CIRP* **2020**, *94*, 110–115. [CrossRef]

20. Bierwisch, C.; Mohseni-Mofidi, S.; Dietemann, B.; Kraft, T.; Rudloff, J.; Lang, M. Particle-based simulation, dimensional analysis and experimental validation of laser absorption and thermo-viscous flow during sintering of polymers. *Procedia CIRP* **2020**, *94*, 74–79. [CrossRef]

21. Taruttis, A.; Hardes, C.; Röttger, A.; Uhlenwinkel, V.; Chehreh, A.B.; Theisen, W.; Walther, F.; Zoch, H.W. Laser additive manufacturing of hot work tool steel by means of a starting powder containing partly spherical pure elements and ferroalloys. *Procedia CIRP* **2020**, *94*, 46–51. [CrossRef]

22. Abel, A.; Wessarges, Y.; Julmi, S.; Hoff, C.; Hermsdorf, J.; Klose, C.; Maier, H.J.; Kaierle, S.; Overmeyer, L. Laser powder bed fusion of WE43 in hydrogen-argon-gas atmosphere. *Procedia CIRP* **2020**, *94*, 21–24. [CrossRef]

23. Döring, M.; Boussinot, G.; Hagen, J.F.; Apel, M.; Kohl, S.; Schmidt, M. Scaling melt pool geometry over a wide range of laser scanning speeds during laser-based Powder Bed Fusion. *Procedia CIRP* **2020**, *94*, 58–63. [CrossRef]

24. Rudloff, J.; Lang, M.; Mohseni-Mofidi, S.; Bierwisch, C. Experimental investigations for improved modelling of the laser sintering process of polymers. *Procedia CIRP* **2020**, *94*, 80–84. [CrossRef]

25. Kusoglu, I.M.; Gökce, B.; Barcikowski, S. Use of (Nano-)Additives in Laser Powder Bed Fusion of Al powder feedstocks: Research directions within the last decade. *Procedia CIRP* **2020**, *94*, 11–16. [CrossRef]

26. Pannitz, O.; Lüddecke, A.; Sehrt, J.T.; Kwade, A. Improvement of the laser powder bed fusion process by surface tailored metal powders. In Proceedings of the Fraunhofer Direct Digital Manufacturing Conference (DDMC) 2020, Berlin, Germany, 18–19 March 2019.

27. Vieth, P.; Voigt, M.; Ebberta, C.; Milkereit, B.; Zhuravlev, E.; Yang, B.; Keßler, O.; Grundmeier, G. Surface inoculation of aluminium powders for additive manufacturing of Al-7075 alloys. *Procedia CIRP* **2020**, *94*, 17–20. [CrossRef]

28. Hupfeld, T.; Blasczyk, A.; Schuffenhauer, T.; Zhuravlev, E.; Krebs, M.; Gann, S.; Keßler, O.; Schmidt, M.; Gökce, B.; Barcikowski, S. How colloidal surface additivation of polyamide 12 powders with well-dispersed silver nanoparticles influences the crystallization already at low 0.01 vol%. *Addit. Manuf.* **2020**, *36*, 101419. [CrossRef]

29. Hupfeld, T.; Sommereyns, A.; Riahi, F.; Doñate-Buendía, C.; Gann, S.; Schmidt, M.; Gökce, B.; Barcikowski, S. Analysis of the nanoparticle dispersion and its effect on the crystalline microstructure in carbon-additivated PA12 feedstock material for laser powder bed fusion. *Materials* **2020**, *13*, 3312. [CrossRef] [PubMed]

30. Doñate-Buendia, C.; Streubel, R.; Kürnsteiner, P.; Wilms, M.B.; Stern, F.; Tenkamp, J.; Bruder, E.; Barcikowski, S.; Gault, B.; Durst, K.; et al. Effect of nanoparticle additivation on the microstructure and microhardness of oxide dispersion strengthened steels produced by laser powder bed fusion and directed energy deposition. *Procedia CIRP* **2020**, *94*, 41–45. [CrossRef]

31. Hupfeld, T.; Salamon, S.; Landers, J.; Sommereyns, A.; Doñate-Buendía, C.; Schmidt, J.; Wende, H.; Schmidt, M.; Barcikowski, S.; Gökce, B. 3D Printing of magnetic parts by Laser Powder Bed Fusion of iron oxide nanoparticle functionalized polyamide powders. *J. Mater. Chem. C* **2020**, *8*, 12204–12217. [CrossRef]

32. Hupfeld, T.; Wegner, A.; Blanke, M.; Doñate-Buendía, C.; Sharov, V.; Nieskens, S.; Piechotta, M.; Giese, M.; Barcikowski, S.; Gökce, B. Plasmonic seasoning: Giving color to desktop laser 3D-printed polymers by highly dispersed nanoparticles. *Adv. Opt. Mater.* **2020**, *8*, 2000473. [CrossRef]

33. Doñate-Buendia, C.; Kürnsteiner, P.; Stern, F.; Wilms, M.B.; Streubel, R.; Kusoglu, I.M.; Tenkamp, J.; Bruder, E.; Pirch, N.; Barcikowski, S.; et al. Microstructure formation and mechanical properties of ODS steels built by Laser Additive Manufacturing of nanoparticle coated iron-chromium powders. *Acta Mater.* **2021**, *26*, 116566. [CrossRef]

34. Sonawane, A.; Roux, G.; Blandin, J.-J.; Despres, A.; Martin, G. Cracking mechanism and its sensitivity to processing conditions during laser powder bed fusion of a structural aluminum alloy. *Materiala* **2021**, *15*, 100976. [CrossRef]

35. Aboulkhair, N.T.; Simonelli, M.; Parry, L.; Ashcroft, I.; Tuck, C.; Hague, R. 3D printing of Aluminium alloys: Additive Manufacturing of Aluminium alloys using selective laser melting. *Prog. Mater. Sci.* **2019**, *106*, 100578. [CrossRef]

36. Martin, J.H.; Yahata, B.; Mayer, J.; Mone, R.; Stonkevitch, E.; Miller, J.; O'Masta, M.R.; Schaedler, T.; Hundley, J.; Callahan, P.; et al. Grain refinement mechanisms in additively manufactured nano-functionalized aluminum. *Acta Mater.* **2020**, *200*, 1022–1037. [CrossRef]

37. Kleijnen, R.G.; Schmid, M.; Wegener, K. Impact of Flow Aid on the Flowability and Coalescence of Polymer Laser Sintering Powder, Solid Freeform Fabrication 2019. In Proceedings of the 30th Annual International Solid Freeform Fabrication Symposium—An Additive Manufacturing Conference 2019, Austin, TX, USA, 12–14 August 2019; pp. 806–817.

38. Sommereyns, A.; Hupfeld, T.; Gann, S.; Wang, T.; Wu, C.; Zhuravlev, E.; Lüddecke, A.; Baumann, S.; Rudloff, J.; Lang, M.; et al. Influence of sub-monolayer quantities of carbon nanoparticles on the melting and crystallization behavior of polyamide 12 powders for additive manufacturing. *Mater. Des.* **2021**, *201*, 109487. [CrossRef]

39. Pannitz, O.; Lüddecke, A.; Kwade, A.; Sehrt, J.T. Investigation of the in situ thermal conductivity and absorption behavior of nanocomposite powder materials in laser powder bed fusion processes. *Mater. Des.* **2021**, *201*. [CrossRef]

40. Lüddecke, A.; Pannitz, O.; Zetzener, H.; Sehrt, J.T.; Kwade, A. Powder properties and flowability measurements of tailored nanocomposites for powder bed fusion applications. *Mater. Des.* **2021**, *202*, 109536. [CrossRef]

41. Gärtner, E.; Jung, H.Y.; Peter, N.J.; Dehm, G.; Jägle, E.A.; Uhlenwinkel, V.; Mädler, L. Reducing cohesion of metal powders for additive manufacturing by nanoparticle dry-coating. *Powder Technol.* **2021**, *379*, 585–595. [CrossRef]

42. Doñate-Buendía, C.; Frömel, F.; Wilms, M.B.; Streubel, R.; Tenkamp, J.; Hupfeld, T.; Nachev, M.; Gökce, E.; Weisheit, A.; Barcikowski, S.; et al. Oxide dispersion-strengthened alloys generated by laser metal deposition of laser-generated nanoparticle-metal powder composites. *Mater. Des.* **2018**, *154*, 360–369. [CrossRef]

43. Martin, J.H.; Yahata, B.D.; Hundley, J.M.; Mayer, J.; Schaedler, T.A.; Pollock, T.M. 3D printing of high-strength aluminium alloys. *Nature* **2017**, *549*, 365–369. [CrossRef]

44. Ho, I.-T.; Chen, Y.-T.; Yeh, A.-C.; Chen, C.-P.; Jen, K.-K. Microstructure evolution induced by inoculants during the selective laser melting of IN718. *Addit. Manuf.* **2018**, *21*, 465–471. [CrossRef]

45. Gao, C.; Wang, Z.; Xiao, Z.; You, D.; Wong, K.; Akbarzadeh, A.H. Selective laser melting of TiN nanoparticle-reinforced AlSi10Mg composite: Microstructural, interfacial, and mechanical properties. *J. Mater. Process. Technol.* **2020**, *281*, 116618. [CrossRef]

46. Mangour, B.A.; Kim, Y.-K.; Grzesiak, D.; Lee, K.-A. Novel TiB2-reinforced 316L stainless steel nanocomposites with excellent room- and high-temperature yield strength developed by additive manufacturing. *Compos. Part B Eng.* **2019**, *156*, 51–63. [CrossRef]

47. DebRoy, T.; Wei, H.L.; Zuback, J.S.; Mukherjee, T.; Elmer, J.W.; Milewski, J.O.; Beese, A.M.; Wilson-Heid, A.; De, A.; Zhang, W. Additive manufacturing of metallic components–Process, structure and properties. *Prog. Mater. Sci.* **2018**, *92*, 112–224. [CrossRef]

48. Yakout, M.; Cadamuro, A.; Elbestawi, M.A.; Veldhuiset, S.C. The selection of process parameters in additive manufacturing for aerospace alloys. *Int. J. Adv. Manuf. Technol.* **2017**, *92*, 2081–2098. [CrossRef]

49. Dowling, L.; Kennedy, J.; O'Shaughnessy, S.; Trimble, D. A review of critical repeatability and reproducibility issues in powder bed fusion. *Mater. Des.* **2020**, *186*, 108346. [CrossRef]

50. Lefebvre, L.P.; Whiting, J.; Nijikovsky, B.; Brika, S.E.; Fayazfar, H.; Lyckfeldt, O. Assessing the robustness of powder rheology and permeability measurements. *Addit. Manuf.* **2020**, *35*, 101203. [CrossRef]

51. Slotwinski, J.A.; Garboczi, E.J.; Stutzman, P.E.; Ferraris, C.F.; Watson, S.S.; Peltz, M.A. Characterization of metal powders used for additive manufacturing. *J. Res. Natl. Inst. Stand. Technol.* **2014**, *119*, 460–493. [CrossRef]

52. Schmid, M.; Amado, F.; Levy, G.; Wegener, K. Flowability of powders for selective laser sintering (SLS) investigated by round robin test. In *High Value Manufacturing, Proceedings of the 6th International Conference on Advanced Research in Virtual and Rapid Prototyping, Leiria, Portugal, 1–5 October 2013*; Taylor & Francis: Abingdon, UK, 2013; pp. 95–99. [CrossRef]

53. Affolter, S.; Ritter, A.; Schmid, M. Interlaboratory tests on polymers by Differential Scanning Calorimetry (DSC): Determination of glass transition temperature (Tg). *Macromol. Mater. Eng.* **2001**, *286*, 605–610. [CrossRef]

54. Brown, C.U.; Jacob, G.; Stoudt, M.; Moylan, S.; Slotwinski, J.; Donmez, A. Interlaboratory Study for Nickel Alloy 625 Made by Laser Powder Bed Fusion to Quantify Mechanical Property Variability. *J. Mater. Eng. Perform.* **2016**, *25*, 3390–3397. [CrossRef]

55. Ahuja, B.; Schaub, A.; Junker, D.; Karg, M.; Tenner, F.; Plettke, R.; Merklein, M.; Schmidt, M. A round robin study for laser beam melting in a metal powder. *S. Afr. J. Ind. Eng.* **2016**, *27*, 30–42. [CrossRef]

56. Slotwinski, J.; Luecke, W.; Lass, E.; Possolo, A. Interlaboratory mechanical-property study for Cobalt-Chromium alloy made by laser powder-bed-fusion additive manufacturing. *J. Res. NIST* **2018**. [CrossRef]

57. Stichel, T.; Frick, T.; Laumer, T.; Tenner, F.; Hausotte, T.; Merklein, M.; Schmidt, M. A round robin study for selective laser sintering of polyamide 12: Microstructural origin of the mechanical properties. *Opt. Laser Technol.* **2017**, *89*, 31–40. [CrossRef]

58. Stichel, T.; Frick, T.; Laumer, T.; Tenner, F.; Hausotte, T.; Merklein, M.; Schmidt, M. A round robin study for selective laser sintering of polyamide 12: Back tracing of the pore morphology to the process parameters. *J. Mater. Process. Tech.* **2018**, *252*, 537–545. [CrossRef]

59. du Plessis, A.; le Roux, S.G. Standardized X-ray tomography testing of additively manufactured parts: A round robin test. *Addit. Manuf.* **2018**, *24*, 125–136. [CrossRef]

60. Townsend, A.; Racasan, R.; Leach, R.; Senin, N.; Thompson, A.; Ramsey, A.; Bate, D.; Woolliams, P.; Brown, S.; Blunt, L. An interlaboratory comparison of X-ray computed tomography measurement for texture and dimensional characterization of additively manufactured parts. *Addit. Manuf.* **2018**, *23*, 422–432.

61. Moylan, S.; Brown, C.U.; Slotwinski, J. Recommended protocol for Round-Robin studies in Additive Manufacturing. *J. Test. Eval.* **2016**, *44*, 1009–1018. [CrossRef]

62. Maier, E.A. *Techniques and Instrumentation in Analytical Chemistry, Chapter 12 Interlaboratory Studies*; Elsevier: Amsterdam, The Netherlands, 1999; Volume 22, pp. 481–535.

63. Yuan, P.; Gu, D.; Dai, D. Particulate migration behavior and its mechanism during selective laser melting of TiC reinforced Al matrix nanocomposites. *Mater. Des.* **2015**, *82*, 46–55. [CrossRef]

64. Garmay, A.V.; Oskolok, K.V.; Monogarova, O.V. Improved Accuracy of Multicomponent Samples Analysis by X-ray Fluorescence Using Relative Intensities and Scattered Radiation: A Review. *Anal. Lett.* **2020**, *53*, 2685–2699. [CrossRef]

65. Rumpf, H. Die Wissenschaft des Agglomerierens. *Chem. Ing. Tech.* **1974**, *1*, 1–46. [CrossRef]

66. Meyer, K.; Zimmermann, I. Effect of glidants in binary powder mixtures. *Powder Technol.* **2004**, *139*, 40–54. [CrossRef]

67. Mourdikoudis, S.; Pallares, R.M.; Thanh, N.T.K. Characterization techniques for nanoparticles: Comparison and complementarity upon studying nanoparticle properties. *Nanoscale* **2018**, *10*, 12871–12934. [CrossRef]

68. Teulon, J.-M.; Godon, C.; Chantalat, L.; Moriscot, C.; Cambedouzou, J.; Odorico, M.; Ravaux, J.; Podor, R.; Gerdil, A.; Habert, A.; et al. On the Operational Aspects of Measuring Nano-particle Sizes. *Nanomaterials* **2019**, *9*, 18. [CrossRef]

69. Letzel, A.; Gökce, B.; Menzel, A.; Plech, A.; Barcikowski, S. Primary particle diameter differentiation and bimodality identification by five analytical methods using gold nanoparticle size distributions synthesized by pulsed laser ablation in liquids. *Appl. Surf. Sci.* **2018**, *435*, 743–751. [CrossRef]

70. Pereira, J.S.F.; Knorr, C.L.; Pereira, L.S.F.; Moraes, D.P.; Paniz, J.N.G.; Flores, E.M.M.; Knapp, G. Evaluation of sample preparation methods for polymer digestion and trace elements determination by ICPMS and ICPOES. *J. Anal. At. Spectrom.* **2011**, *26*, 1849–1857. [CrossRef]

71. Bonesso, M.; Rebesan, P.; Gennari, C.; Dima, R.; Pepato, A.; Calliari, I. Effect of particle size distribution on laser powder bed fusion manufacturability of copper. *Berg Huettenmaenn Mon.* **2021**, *166*, 256–262. [CrossRef]

72. Brika, S.E.; Letenneur, M.; Dion, C.A.; Brailovski, V. Influence of particle morphology and size distribution on the powder flowability and laser powder bed fusion manufacturability of Ti-6Al-4V alloy. *Addit. Manuf.* **2020**, *31*, 100929. [CrossRef]

73. Kuznetsov, P.A.; Shakirov, I.V.; Zukov, A.S.; Bobyr', V.V.; Starytsin, M.V. Effect of particle size distribution on the structure and mechanical properties in the process of laser powder bed fusion. *J. Phys. Conf. Ser.* **2021**, *1758*, 1–9. [CrossRef]

74. Beitz, S.; Uerlich, R.; Bokelmann, T.; Diener, A.; Vietor, T.; Kwade, A. Influence of Powder Deposition on Powder Bed and Specimen Properties. *Materials* **2019**, *12*, 297. [CrossRef]

75. Tan, J.H.; Wong, W.L.E.; Dalgarno, K.W. An overview of powder granulometry on feedstock and part performance in the selective laser melting process. *Addit. Manuf.* **2017**, *18*, 228–255. [CrossRef]

76. Fissan, H.; Ristig, S.; Kaminski, H.; Asbach, C.; Epple, M. Comparison of different characterization methods for nanoparticle dispersions before and after aerosolization. *Anal. Methods* **2014**, *6*, 7324–7334. [CrossRef]

77. ASTM F3318-18: Standard for Additive Manufacturing, Finished Part Properties, Specification for AlSi10Mg with Powder Bed Fusion-Laser Beam. ASTM International: West Conshohocken, PA, USA, 2018. Available online: www.astm.org (accessed on 3 April 2021).

78. Khairallah, S.A.; Anderson, A.T.; Rubenchik, A.; King, W.E. Laser powder-bed fusion additive manufacturing: Physics of complex melt flow and formation mechanisms of pores, spatter, and denudation zones. *Acta Mater.* **2016**, *108*, 36–45. [CrossRef]

79. ASTM E112-13: Standard Test Methods for Determining Average Grain Size. ASTM International: West Conshohocken, PA, USA, 2018. Available online: www.astm.org (accessed on 3 April 2021).

80. Wudy, K.; Drummer, D. Aging effects of polyamide 12 in selective laser sintering: Molecular weight distribution and thermal properties. *Addit. Manuf.* **2019**, *25*, 1–9. [CrossRef]

81. Zhao, M.; Wudy, K.; Drummer, D. Crystallization Kinetics of Polyamide 12 during Selective Laser Sintering. *Polymers* **2018**, *10*, 168. [CrossRef]

82. Sehrt, J.T.; Kleszczynski, S.; Notthoff, C. Nanoparticle improved metal materials for additive manufacturing. *Prog. Addit. Manuf.* **2017**, *2*, 179–191. [CrossRef]

83. Wang, Y.; Shi, J.; Lu, S.; Xiao, W. Investigation of Porosity and Mechanical Properties of Graphene Nanoplatelets-Reinforced AlSi10Mg by Selective Laser Melting. *J. Micro Nano-Manuf. March* **2018**, *6*. [CrossRef]

84. Sastri, V.R. 5-Polymer Additives Used to Enhance Material Properties for Medical Device Applications. In *Plastics in Medical Devices*, 2nd ed.; Sastri, V.R., Ed.; William Andrew Publishing: Norwich, NY, USA, 2014; pp. 55–72.

85. Wagner, H.D.; Vaia, R.A. Nanocomposites: Issues at the interface. *Mater. Today* **2004**, *7*, 38–42. [CrossRef]

86. Yang, Y.; Doñate-Buendía, C.; Oyedeji, T.D.; Gökce, B.; Xu, B.-X. Nanoparticle Tracing during Laser Powder Bed Fusion of Oxide Dispersion Strengthened Steels. *Materials* **2021**, *14*, 3463. [CrossRef]

87. Gu, D. Nanoscale TiC particle-reinforced AlSi10Mg bulk-form nanocomposites by selective laser melting (SLM) additive manufacturing (AM): Tailored microstructures and enhanced properties. In *Laser Additive Manufacturing of High-Performance Materials*; Springer: Berlin/Heidelberg, Germany, 2015.

88. DIN EN ISO 6892-1:2020-06: Metallic Materials—Tensile Testing—Part 1: Method of Test at Room Temperature, DIN Standards. Available online: https://www.beuth.de/de/norm/din-en-iso-6892-1/317931281 (accessed on 3 April 2021).

89. ISO 527-2:2012, Plastics-Determination of tensile properties—Part 2: Test Conditions for Moulding and Extrusion Plastics. Available online: https://www.iso.org/standard/56046.html (accessed on 3 April 2021).

90. Wilkinson, M.D.; Dumontier, M.; Aalbersberg, I.J.; Appleton, G.; Axton, M.; Baak, A.; Blomberg, N.; Boiten, J.-W.; da Silva Santos, L.B.; Bourne, P.E.; et al. The FAIR Guiding Principles for scientific data management and stewardship. *Sci. Data* **2016**, *3*, 160018. [CrossRef]

91. *ASTM F2971-13: Standard Practice for Reporting Data for Test Specimens Prepared by Additive Manufacturing*; ASTM International: West Conshohocken, PA, USA, 2013; Available online: www.astm.org (accessed on 3 April 2021).

92. ASTM WK73978, New Specification for Additive Manufacturing—Data Registration. Available online: https://www.astm.org/DATABASE.CART/WORKITEMS/WK73978.htm (accessed on 3 April 2021).

93. *ASTM E691-20: Standard Practice for Conducting an Interlaboratory Study to Determine the Precision of a Test Method*; ASTM International: West Conshohocken, PA, USA, 2019; Available online: www.astm.org (accessed on 3 April 2021).

94. Luping, T.; Schouenborg, B. *Methodology of Inter-Comparison Tests and Statistical Analysis of Test Results, SP Report*; SP Swedish National Testing and Research Institute: Boras, Sweden, 2000; Volume 35, Available online: http://www.nordtest.info/wp/2001/10/28/methodology-of-inter-comparison-tests-and-statistical-analysis-of-test-results-nt-tr-482/ (accessed on 3 April 2021).

95. ISO/ASTM 52907-19: Additive Manufacturing, Feedstock Materials, Methods to Characterize Metallic Powders. Available online: https://www.iso.org/standard/73565.html (accessed on 3 April 2021).

96. *ASTM F3049-14: Standard Guide for Characterizing Properties of Metal Powders Used for Additive Manufacturing Processes*; ASTM International: West Conshohocken, PA, USA, 2014; Available online: www.astm.org (accessed on 3 April 2021).

97. *ASTM B527-20: Standard Test Method for Tap Density of Metal Powders and Compounds by Fluorescence Spectrometry*; ASTM International: West Conshohocken, PA, USA, 2020; Available online: www.astm.org (accessed on 3 April 2021).

98. *ASTM B822-20: Standard Test Method for Particle Size Distribution of Metal Powders and Related Compounds by Light Scattering*; ASTM International: West Conshohocken, PA, USA, 2020; Available online: www.astm.org (accessed on 3 April 2021).

99. ISO 13320:2020 Particle Size Analysis, Laser Diffraction Methods. Available online: https://www.iso.org/standard/69111.html (accessed on 3 April 2021).

100. WK66030: Quality Assessment of Metal Powder Feedstock Characterization Data for Additive Manufacturing. Available online: https://www.astm.org/DATABASE.CART/WORKITEMS/WK66030.htm (accessed on 3 April 2021).

101. WK55610: The Characterization of Powder Flow Properties for Additive Manufacturing Applications. Available online: https://www.astm.org/DATABASE.CART/WORKITEMS/WK55610.htm (accessed on 3 April 2021).

102. WK74905: Additive Manufacturing-Feedstock-Particle Shape Analysis to Identify Agglomerates/Satellites in Feedstock. Available online: https://www.astm.org/DATABASE.CART/WORKITEMS/WK74905.htm (accessed on 3 April 2021).

103. *ASTM F3122-14: Standard Guide for Evaluating Mechanical Properties of Metal Materials Made via Additive Manufacturing Processes*; ASTM International: West Conshohocken, PA, USA, 2014; Available online: www.astm.org (accessed on 3 April 2021).

104. *ASTM E572-13: Standard Test Method for Analysis of Stainless and Alloy Steels by Wavelength Dispersive X-Ray Fluorescence Spectrometry*; ASTM International: West Conshohocken, PA, USA, 2013; Available online: www.astm.org (accessed on 3 April 2021).

105. *ASTM E8/E8M-16ae1: Standard Test Methods for Tension Testing of Metallic Materials*; ASTM International: West Conshohocken, PA, USA, 2016; Available online: www.astm.org (accessed on 3 April 2021).

106. WK49229: Orientation and Location Dependence Mechanical Properties for Metal Additive Manufacturing. Available online: https://www.astm.org/DATABASE.CART/WORKITEMS/WK49229.htm (accessed on 3 April 2021).

107. *ASTM E3166-20: Standard Guide for Nondestructive Examination of Metal Additively Manufactured Aerospace Parts after Build*; ASTM International: West Conshohocken, PA, USA, 2020; Available online: www.astm.org (accessed on 3 April 2021).

108. ISO 527-1:2012, Plastics-Determination of Tensile Properties—Part 1: General Principles. Available online: https://www.iso.org/standard/56045.html (accessed on 3 April 2021).

109. WK66029: Mechanical Testing of Polymer Additively Manufactured Materials. Available online: https://www.astm.org/DATABASE.CART/WORKITEMS/WK66029.htm (accessed on 3 April 2021).

110. *ASTM F1877-16: Standard Practice for Characterization of Particles*; ASTM International: West Conshohocken, PA, USA, 2016; Available online: www.astm.org (accessed on 3 April 2021).

111. *ASTM E2834-12(2018) Standard Guide for Measurement of Particle Size Distribution of Nanomaterials in Suspension by Nanoparticle Tracking Analysis (NTA)*; ASTM International: West Conshohocken, PA, USA, 2018; Available online: www.astm.org (accessed on 3 April 2021).

112. *ASTM E3247-20 Standard Test Method for Measuring the Size of Nano-particles in Aqueous Media Using Dynamic Light Scattering*; ASTM International: West Conshohocken, PA, USA, 2020; Available online: www.astm.org (accessed on 3 April 2021).

113. ISO 17200:2020(en), Nanotechnology, Nanoparticles in Powder Form, Characteristics and Measurements. Available online: https://www.iso.org/obp/ui/#iso:std:iso:17200:ed-1:v1:en (accessed on 3 April 2021).
114. ISO/TR 14187:2020(en), Surface Chemical Analysis, Characterization of Nanostructured Materials. Available online: https://www.iso.org/obp/ui/#iso:std:iso:tr:14187:ed-2:v1:en (accessed on 3 April 2021).

materials

Article

Towards Automatic Detection of Precipitates in Inconel 625 Superalloy Additively Manufactured by the L-PBF Method

Piotr Macioł [1,*], **Jan Falkus** [1], **Paulina Indyka** [2] **and Beata Dubiel** [1]

1 Faculty of Metals Engineering and Industrial Computer Science, AGH University of Science and Technology, Czarnowiejska 66, 30-054 Kraków, Poland; jfalkus@agh.edu.pl (J.F.); bdubiel@agh.edu.pl (B.D.)
2 Solaris National Synchrotron Radiation Centre, Faculty of Chemistry, Jagiellonian University, Czerwone Maki 98, 30-392 Kraków, Poland; paulina.indyka@uj.edu.pl
* Correspondence: pmaciol@agh.edu.pl; Tel.: +48-(12)-617-45-25

Abstract: In our study, the comparison of the automatically detected precipitates in L-PBF Inconel 625, with experimentally detected phases and with the results of the thermodynamic modeling was used to test their compliance. The combination of the complementary electron microscopy techniques with the microanalysis of chemical composition allowed us to examine the structure and chemical composition of related features. The possibility of automatic detection and identification of precipitated phases based on the STEM-EDS data was presented and discussed. The automatic segmentation of images and identifying of distinguishing regions are based on the processing of STEM-EDS data as multispectral images. Image processing methods and statistical tools are applied to maximize an information gain from data with low signal-to-noise ratio, keeping human interactions on a minimal level. The proposed algorithm allowed for automatic detection of precipitates and identification of interesting regions in the Inconel 625, while significantly reducing the processing time with acceptable quality of results.

Keywords: inconel 625; additive manufacturing; EDS microanalysis; automatic image analysis

Citation: Macioł, P.; Falkus, J.; Indyka, P.; Dubiel, B. Towards Automatic Detection of Precipitates in Inconel 625 Superalloy Additively Manufactured by the L-PBF Method. *Materials* **2021**, *14*, 4507. https://doi.org/10.3390/ma14164507

Academic Editor: Mika Salmi

Received: 29 June 2021
Accepted: 6 August 2021
Published: 11 August 2021

Publisher's Note: MDPI stays neutral with regard to jurisdictional claims in published maps and institutional affiliations.

1. Introduction

Recent advances in additive manufacturing (AM) of metals and alloys require intensification of the research of their microstructure, which should be precisely controlled to assure the good quality of the manufactured products. Moreover, the evolution of the microstructure in the working conditions should be recognized. It is especially important due to the growing application of AM technology to produce parts made of superalloys for aerospace, energy, automotive, and chemical industries. From the perspective of materials science, significant challenges remain in the field of understanding the effect of the build process from metal powder experiencing rapid melting and solidification on the microstructure and its development during high temperature exposure.

Due to the elimination of difficulties with subtractive machining and the possibility of making parts with complex geometry, the additive manufacturing of nickel-based superalloy components by laser powder fusion (L-PBF) has recently attracted great interest [1]. One of the superalloys most widely applied for L-PBF manufacturing is Inconel 625, which exhibits high temperature strength and exceptional corrosion resistance in harmful environments. High cooling rate and repeated heating and cooling cycles during the L-PBF process shape a distinctly different microstructure of Inconel 625 than in other methods. In as-built condition, it consists of paths resulting from the bonding of powder particles by the laser beam, in which the rapidly solidified melt pools are distinguished. Inside them, fine grains with a very fine cellular-dendritic structure are present. The non-equilibrium solidification conditions promote the microsegregation of alloying elements and formation of nano-precipitates, as well as the substructure of dislocation cells [2–7].

One of the important phenomena that require careful study is the examination of the precipitates present in L-PBF Inconel 625 and their evolution after prolonged exposure at high temperature.

1.1. Microstructural and Compositional Characterization of Precipitates in Additively Manufactured Inconel 625 Superalloy

The literature data concerning nano-sized precipitates in L-PBF Inconel 625 are mostly related to as-built and stress-relief annealed condition [2–6], in which the occurrence of the γ'' phase and δ phase, as well as MC, M_6C, and $M_{23}C_6$ carbides, was established. The presence of the Laves phase and P phase is also reported [8]. In addition, nanometric oxide inclusions, which are formed by the oxidation of the feed powder and/or phenomena occurring during additive manufacturing, are observed [9,10]. These particles were detected on the basis of different methods, such as SEM and TEM imaging, X-ray, and electron diffraction, as well as EDS microanalysis of chemical composition. However, there is limited data concerning the evolution of precipitates after high temperature exposure of L-PBF Inconel 625. Examination of the evolution of the microstructure and the local changes in chemical composition caused by annealing can be performed by scanning transmission electron microscopy (STEM) imaging, combined with energy dispersive X-ray (EDS) elemental mapping, which allows for both microstructure observation and determination of elemental distribution within nanoareas. This technique requires the use of thin specimens transparent to the electron beam. In the STEM mode, a focused electron beam scans the sample. Due to the interaction of electrons with the sample, different signals are excited and collected at each scanned point by appropriate detectors. The collected signals can be used to create various types of images. High-angle annular dark-field (HAADF) images show the intensity distribution of electrons scattered near the atomic nuclei of the atoms in the sample. The intensity of the images depends on the square of the atomic number (Z) and the thickness of the sample, so a mass-thickness contrast is obtained. The contrast in STEM-HAADF mode is also influenced by the deformation fields around structural defects, e.g., dislocations [11]. In turn, EDS spectral images, showing maps of the distribution of chemical elements, are obtained by collecting EDS spectra at each point of the scanned sample (Figure 1). The information obtained from the compositional maps is influenced by many factors related to the instrument configuration, the specimen, and the settings of the data analysis software [12].

1.2. Automatic Detection of Particles in Metal Alloys Produced by Additive Manufacturing

Interpretation of micrographs, including EDS spectral images, is a tedious and time-consuming work. It is even more challenging if the results do not offer high quality (meaning sufficient number of X-ray counts per pixel). Analyzing of acquired data requires precise processing by an experience human researcher. Since this work is frequently based on trial-and-error approach, mainly, while adjusting the background level, it is not only time-consuming but may lead to incorrect or only partially correct results. Automatic analysis of images cannot replace a skilled human researcher (at least not today) but might save a lot of most cumbersome work and provide a good starting point for in-depth manual investigation.

There are several domains, where automatic image/data processing could be useful, from the most typical (denoising) to the ones requiring domain-specific knowledge, such as supporting phases identification. The phases' identification is in fact a form of clustering of data, and many types of data science are applicable.

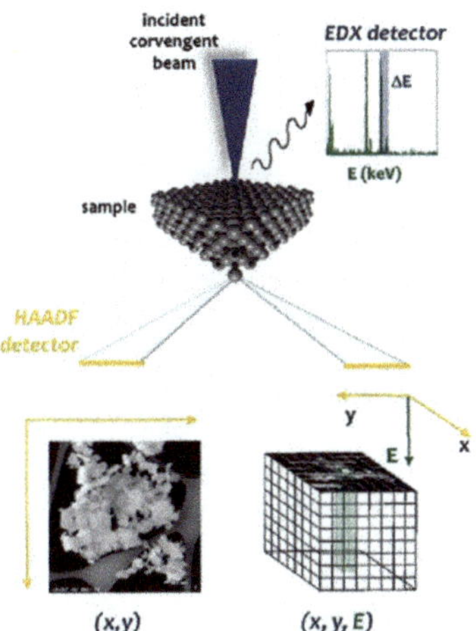

Figure 1. Schematic layout of the Spectrum Imaging (SI) method in STEM mode. The STEM probe is rastered over the specimen (x, y), and local information is retrieved by mapping the HAADF and EDX signals against to the position of the probe (x, y, E). The EDX spectrum shows the characteristic EDX peaks above the background signal. The obtained STEM-HAADF structural images can be complemented by elemental information extracted slice-by-slice, while sampling the EDS spectrum energy space I with an energy corresponding to the ΔE range of the EDS characteristic peaks.

1.3. Denoising

STEM imaging and EDS mapping require considerable time to acquire a single image. The need of reducing that time introduces artifacts, which are generally described as a noise. Many techniques have been developed to deal with this issue. Some are general methods of image processing; others are focused on processing microscope images. Often, simple general purpose filters, such as a Gaussian blur, median filter, frequency space low-pass, or Wiener filters, are applied [13–15]. More advanced techniques consider specificity of imaged materials character/features. In metal alloys, a periodicity of a structure is usually used. An example of employing this approach for high-resolution STEM is presented, i.e., in Reference [16]. The most successful image denoising algorithm employing recognition of similar features rely on non-local detection and averaging of self-similar image regions. The first algorithm based on this strategy is the non-local means filter (NLM) in Reference [17]. Due to the richness in self-similarity of electron micrographs of crystals, NLM is, in principle, very well suited for denoising such micrographs [18,19]. Recently, Mevenkamp et al. [20] proposed a multi-modal and multi-scale non-local means (M3S-NLM) method to extract atomically resolved spectroscopic maps. Their approach joins NLM approach with a multimodality based on searching on similarities not only between different features recorded with a single method and spread spatially on image but also considering several bound signals. The method is limited to periodic structures. Results presented by Mevenkamp et al. are based on a synthetic structure, generated from a single real image. Such approaches based on a cyclicity and NLM works very well if an imaged object exhibits cyclic structure or at least features on the image are strongly similar. However, none of these requirements is fulfilled in case of imaging of cellular microstructure in L-PBF Inconel 625.

The most common, classical, and widely used tools for automatic denoising are binning and smearing methods. The binning is based on replacing of a set (usually a rectangle or a square) of pixels with a single one, with a signal being a sum of signals from replaced pixels. The main disadvantage of this approach is significant loss of spatial resolution. The smearing is based on averaging (unweighted, linear or nonlinear weighted) of adjacent pixels. This approach reduces noise and might be very efficient; however, it usually requires manual fitting of weighting function parameters. Improperly chosen values lead to ineffective denoising or losing important information. Since classical denoising methods have reduced application in automatic processing of highly noised and non-uniform images, recently, in material science, most attention is paid to Machine Learning techniques. Ede presented a neural network trained to remove noise from micrographs. The presented case studies show significant improve of image quality [21]. The author claims that once trained network can be successfully used for a wide range of types of images, including STEM images. However, the level of noise present on the exemplary images was never high enough to hinder automatic identification of features. Other examples of ML algorithms for noise removal in microscopy are presented in References [22,23].

While ML techniques can be effective, they are burdened with two important drawbacks. At first, the most of ML techniques is based on a black-box approach. Hence, while it is possible to get an answer for the question of what is the prediction, it is not possible to explain the prediction (black-box approach). Second, ML techniques require manually labeled data (supervised ML) or huge datasets (unsupervised ML). Due to those important issues, other techniques useful in automatic analysis of micrographs are also investigated. One of the most promising is Principal Component Analysis (PCA). Potapov et al. and Potapov and Lubk [24,25] presented an application of PCA to enhance quality of highly noised EDS images. Combination of several classical filtering methods with PCA are presented. Processed images have much better quality, but several manually fitted coefficients were necessary.

Some reviews describing various advanced denoising methods of microscope images (e.g., modified Rudin-Osher-Fatemi model, the adaptive Total variation method or Poisson Unbiased Risk Estimation—Linear Expansion of Thresholds) might be found also in References [26,27].

1.4. Segmentation

Other relatively common use of image processing techniques in micrographs' analysis is segmentation. One of the primary purposes of microstructural analysis of metal alloys is examination of grain size and morphology, as well as identification of structural compounds, phases, and other features. Traditionally, this job is performed by a human researcher on the ground of his knowledge of investigated material and with the use of microscopy techniques. If a metal alloy is fabricated using additive manufacturing, this task gets somewhat complicated because of a combinatorial explosion of possible microstructural features to identify. Hence, any support which might be provided by computational techniques is very welcome. The first step in computer aided image analysis is distinguishing of present features. The problem of segmentation of images is well known and important in many disciplines, such as in i.a. autonomous driving. A review of segmentation techniques for general applications including traditional and modern ML-based approaches might be found in Reference [28].

Segmentation of microscope images has been a subject of many publications [29–31]. Recently, a consistent framework for segmentation was proposed in the study of Reference [32], which presents a thorough review of possible techniques. However, rapid progress in this discipline meets some important obstacles against spreading into aiding of STEM images and EDS compositional maps analysis. One obstacle is that ML-based approaches require large sets of labeled training data, while, in conventional STEM-EDS imaging, only several SI can be collected within a reasonable period of experimental time, mainly due to relatively low efficiency for X-ray signal collection. Examples of successful segmentation of images are presented by Uusimaeki et al. [33]. The software called AutoEM is presented with some examples of particles identifying from TEM images. Several well-established image processing methods (for filtering and segmentation) are used. Excellent results were obtained; however, it has to be emphasized that the input images had also excellent quality (very high contrast and noise-to-signal ratio). In daily routine, STEM-EDS images have much worse quality, and such traditional approaches, based on filtering and thresholding, suffer from very low signal-to-noise ratio.

An alternative approach is based on employing primary additional knowledge, rather than only image processing algorithms. One feasible solution is an image decomposing on the ground of local features. Jany et al. [34] proposed application of FFT within local window moving over a micrograph and later application of blind decomposition of the data via Non-Negative Matrix Factorization. Their approach is sensitive to the local features of the image such as symmetry, orientation, and characteristic spacing. It is worth mentioning is that the method does not require a learning process. The other approach to include a domain-knowledge was presented in Reference [35], based on including an expert knowledge in the algorithm. Mirzaei and Rafsanjani [36] presented an application of image processing methods, including nonlinear denoising, edge sharpening filtering, and segmentation to identify nanoparticles from TEM images. The algorithm is dedicated for particles with circular cross-section. With this assumption, computational efficiency and particles' identification quality were significantly improved, at the expense of loss of generality.

The common drawback of all methods presented above is neglecting a part of initially available specimen knowledge, e.g., chemical composition and repartition of elements. Combination of STEM imaging with compositional imaging by simultaneous acquisition of EDS signal gives the possibility to use information obtained from these complementary methods. Mevenkamp et al. [20] applied this approach into filtering of an image, but only in the context of NLM filtering. Furthermore, a synthetic EDS dataset was used. In this paper, we present an approach extending this concept and applying a multispectral signal analysis for the whole process of automated recognition of cellular structure from denoising to segmentation.

1.5. Multispectral Images

If data includes more than one type of information (multi- or hyperspectral images), the problem of distinguishing specific regions of a sample is a clustering problem. There are several approaches available, such as Gaussian mixture clustering [37], SVM [38], the k-means [35], or decision-tree [39]. Very recently, Georget et al. presented a paper focused on identifying phases with combination of BSE and SEM-EDS images [40]. They presented a comprehensive approach to process multispectral data consist of denoising, researcher-defined clustering and visualization. The results are very promising; however, a share of a manual work is significant.

Automation of phases identification with SEM images is a core of Automated Mineralogy—a range of analytical solutions for quantitative analysis of minerals, rocks, and ceramic materials. Schulz et al. presented applied summation of low-counts spectra characteristic for predefined minerals to increase image quality. It has been proved that this approach is very promising for investigated rare earth element ores [41]. Iglesias et al. presented an application of image processing algorithms for automated identification of phases in iron ores. A complex scripts were applied, with several manually fitted coefficients [42]. Presented results are promising; however, it should be emphasized that the processed samples were relatively easier to analyze than typical STEM images of additively manufactured Inconel 625 samples. Moreover, effectiveness of the method was presented only for limited number of cases

1.6. Metal Alloys and Additive Manufacturing

Automatization of microscopic images of additively manufactured metal alloys are rather rare. Snow et al. presented an application of Neural Network for detecting flaws in layerwise images of a build. Analyzed images were acquired with X-ray computed tomography (XCT). Obtained accuracy is high but, as other NN-based works, large number of labeled training examples were necessary [43].

Applications of ML for analyzing of microstructures of additively manufactured materials are reviewed by Wang et al. [44]. Several important works are discussed, but the most important conclusion is that character ML algorithms require labeled data or, in the case of algorithms with ability to extract representative features automatically, large number of training images (thousands). Johnson et al. presented another wide review of applications of ML in image analysis application in AM materials [45]. Examples include classification of grain structures, measurements of grain size, pore size calculations, and more. Miyazaki et al. [46] presented an application of a complex set of ML techniques of image processing to classify a and b phases in Ti–6Al–4V AM alloy, basing on SEM data. The presented results confirm ability to automatically distinguish these two phases.

Gupta et al. [47] presented an approach to identify phases in steels, based on SEM images processing. Local binary pattern, random forest, and Otsu operators are used for extracting features, classification, and segmentation of microstructure images. The SEM images had been denoised with Gabor filter with coefficient manually fitted. Denoising had been followed with histogram equalization. The Local Binary Pattern algorithm had been used to describe 72 SEM microstructures with a set of properties, which, in turn, had been used to develop the classifier by random forest technique. All 72 images had been manually labeled prior to the process. The images had been segmented with Otsu algorithm, and distinguished phases had been assigned to predefined ones on the ground of the type of steel. The presented results confirm high accuracy of quantifying phases fractions.

From this short state-of the-art review, several conclusions might be drawn. A lot of efforts are put on denoising, segmentation, and phase identification with ML techniques. However, the need of labeled data or abundant training sets limits their practical applications. Classical denoising and segmentation algorithms might be very effective, as well, but they generally require a significant manual fitting to particular conditions and materials. Multispectral analysis is an efficient tool in automatized processing of data, including EDS images, but its applications in identification of phases in metal alloys are very rare. Hence, the efficient at highly automatized way to identify phases in metal alloys, particularly additively manufactured samples, must be still investigated.

Phases identification process in its assumptions is straightforward. The measurements (usually an electron microscopy analysis) provide data, which is processed to maximize information gain. None of the currently available techniques can identify phases directly. Hence, additional step of data interpretation is necessary. The interpretation might be generally based on three approaches: (i) researcher's intuition and experience, (ii) computational modeling, and (iii) comparative analysis. Currently, the researcher's interpretation is inevitable—as we showed in the literature review, none of currently available methods can specify the microstructure fully automatically. On the other hand, raw measurements are far too complex to be directly interpreted by human. The goal of the presented research has been not to fully automatize the process of phase identification but to design a solution, maximally decreasing the amount of human-researcher work while keeping reliability on a such high level as it is possible. Hence, if human researcher assistance might ensure high reliability with minimal involvement required, it was decided to not replace them with a fully automatic one.

Another approach to phase analysis is to perform thermodynamic modeling for the average chemical compositions of automatically detected particles, that show the theoretically possible phases in equilibrium conditions. This allows for the creation of a reference system for real results obtained by experimental method. In addition, it enables determination of the influence of temperature on the occurrence of individual equilibrium phases. This information is valuable when interpreting the evolution of identified phases during high temperature exposure.

Therefore, the aim of this work is to demonstrate the novel approach for automated recognition of precipitates in compositional EDS maps and to verify the results with the microstructural analysis and thermodynamic modeling.

2. Material and Methods

2.1. Material

Specimens of additively manufactured Inconel 625 were delivered by Bibus Menos Sp. z o.o. (Gdańsk, Poland). The chemical composition of the alloy is given in Table 1. Specimens fabricated by Direct Metal Laser Sintering® (DMLS), the commercially available L-PBF process of EOS GmbH (Krailling, Germany), were post-build annealed at a temperature of 980 °C for 1 h and slowly cooled in an argon atmosphere. Detailed characterization of the microstructure in the stress-relieved condition can be found in Reference [7]. Subsequently, the specimens were isothermally annealed at a temperature of 800 °C for 5, 100, and 500 h, and cooled in air. This temperature was selected because it is typical for the use of Inconel 625 for prolonged exposure in aerospace, and energy industry applications.

Table 1. Nominal chemical composition of the Inconel 625 (wt. %).

Ni	Cr	Fe	Mo	Nb	C	Mn	Si	Al	Ti	Co
58.00	22.00	5.00	9.00	3.50	0.10	0.40	0.40	0.30	0.30	1.00

2.2. Microstructural Analysis

The TEM microstructural investigation was performed using JEM-2010 ARP microscope (Jeol, Tokio, Japan). Thin foils for TEM were prepared by electropolishing. STEM investigation was carried out using a Tecnai Osiris microscope (FEI, Hillsboro, OR, USA) operating at 200 kV, equipped with a ChemiSTEM™ system. STEM imaging was performed by applying the high angle annular dark-field (HAADF) mode. Microanalysis of the chemical composition was performed using energy-dispersive X-ray spectrometry (EDS) with the use of Esprit (Bruker, Billerica, MA, USA) software. The distribution of the specific chemical elements was determined using STEM images coupled with EDS maps. The beam current was low to avoid sample damage; thus, the acquisition time was long enough to achieve the signal-to-background ratio minimum of 3:1 for every peak/pixel in the EDS spectrum selected for the collection of elemental maps. The EDS spectra contained Cu peaks generated by the copper support grid of the specimens. To exclude the Cu from quantification, this element was chosen for deconvolution only and was not included in quantitative analysis, as well as in the subsequent automatic detection of microstructural features. The standardless Cliff-Lorimer quantification model was used, as it is convenient for quantitative EDS measurements of thin specimens with relatively high accuracy. EDS spectra recorded at every pixel in the elemental maps allowed us to reconstruct the quantitative EDS maps and linescans, which were used to determine the mean concentrations of chemical elements in the analyzed phases. The indexing of the electron diffraction patterns was performed with the use of JEMS software (version 4.4131U2016, JEMS-SWISS, Jongny, Switzerland) by P. Stadelmann [48]. The crystallographic data of analyzed phases available in Atom Work software (NIMS, Tsukuba, Japan) [49] were used.

2.3. Thermodynamic Modeling

The equilibrium phase stability of the Inconel 625 was calculated using the FactSage™ 7.2 program in conjunction with the SGTE 2017. The calculations were carried out in the temperature range 200–1000 °C using a 10 °C temperature step and the alloy nominal composition given in Table 1. Moreover, the relative amounts of equilibrium phases that can exist at a temperature of 800 °C for the measured compositions of the matrix and precipitates were calculated.

2.4. Automatic Detection of Precipitates

The first part of the proposed approach is processing measurements. In our research, the raw data are STEM-EDS data.

We also assume that the chemical composition of an investigated sample is known with high level of certainty. A classical way of EDS data analysis is based on prior identification of chemical elements. In our work, we choose a different approach. We treat measurements analogously to a data-science approach, without assigning a physical or chemical knowledge to data. Instead, for the majority of the data processing steps, we rely on statistical tools.

EDS data for a single 2D measurements are large 3D datasets. A raw measurement in our case would be arrays of $1024 \times 1024 \times 4096$ integer values representing a number of registered electrons of particular energy (4096 levels) for each 'pixel' of analyzed region (1024×1024). EDS maps usually suffer from low signal-to-noise ratio, which is caused by an urge to shorten a measurement duration. To maximize information's gain from data, several operations must be performed—summarization of measurements over domains for particular energy peaks (this step results in generating single, 2D map for each peak) and improving quality of obtained images. While the goal of our research is automatization, identification of energy peaks is performed manually. This is caused by significant disadvantages of automatic peaks' finding (relatively high probability of incorrectness of solution) and easiness of manual solution—since the chemical composition is known, it is straightforward to define all possible peaks. Other steps, on the other hand, is fully automatized.

The procedure starts with a raw STEM-EDS datafile readout followed by the EDS peak identification. The raw data of a Spectrum Image (SI) is represented as a 3D matrix of integral values (x, y, E), where (x, y) correspond to a 2D image pixel coordinates, and E consists of a column-by-column EDS spectral information. *HyperSpy* library was used to process EDS data [50]. With the tools provided by this library, energy ranges defined for particular peaks were converted to the ranges of counts (ΔE). For each recognized peak, all counts are summed in particular energy range of ΔE channels (keV) (separately for each (x, y) pixel). A depth of the data structure is reduced from a number of energy channels (4096 in analyzed cases) into a number of recognized peaks, keeping width and height of the peaks unchanged. For each (x, y) pixel of an image, a vector of integral numbers represents the number of X-ray counts in a particular energy range ΔE that corresponds with detected EDS characteristic peak (i) Equation (1).

$$intensities = \left[\text{intensity}_{(\Delta E)1}, \text{intensity}_{(\Delta E)2}, \ldots \text{intensity}_{(\Delta E)N} \right]. \tag{1}$$

The next step deals with a common problem with the thickness of the wedge-shaped sample in STEM, resulted from a thin foil preparation by electropolishing. The wedge shape sample thickness may lead to a different electron beam interaction volume in the sample, thus altering the emerging X-rays and subsequently detected EDS signal intensities. To compensate for the sample thickness correlated EDS peak intensities, a semi-empirical normalization procedure was applied, based on a linear regression. The detailed unwedging algorithm for 2D (x, y) image is presented in Figure 2.

```
For axis x, axis y:

    For each tenth index of axis:

        Values = image[index]

        Calculate trend coefficient for Values

    Calculate Avg(all trend coefficients)

For axis x, axis y:

    For each row in axis

        Values in row = (Values in row) / (number of  row ×
mean trend coefficient)
```

Figure 2. The unwedging algorithm for 2D (x, y) image.

Digitally unwedged images are smoothed with Gaussian filter with a small standard deviation for Gaussian kernel (default value for 1024 × 1024 image is 3) and normalized to the range (0,1). Gaussian filtering was performed with the *gaussian* method from scikit-image library [51]. Next, the correlation matrix between images is computed. First, all data is converted to the form of *DataFrame* from pandas library [52]. Then, the *corr* method from this library is applied. This method returns a correlation matrix. If any of images' pair have Pearson correlation coefficient above 0.5, the images are summed (a 'layer' with index indicated by a row number in correlation matrix is algebraically summed with a 'layer' with index indicated by a column number) and renormalized to the range (0,1). Both original layers are removed from data and replaced by the new layer. These two steps are repeated till none of the images' pairs have Pearson correlation coefficient above 0.5.

The pre-prepared image is processed with a set of operations. First, the image is binarized with threshold value computed with triangle algorithm [53]. For this purpose, a set of methods provided by *scikit-images.filters* module is used. First, each image is smoothed with gaussian method. Then, the method *thresholding.threshold_triangle* is used. Positive values ("bright pixels") indicates pixels with a number of counts higher than the threshold. Since the input images are relatively noisy, the result includes numerous defects (small dark patches in generally light areas and opposite). These "small" objects are removed. Some might be an effect of noise, and others might be a real feature, but with size significantly smaller than the size of objects to be identified. The "size" is understood as an area of directly connected pixels with the same value. This step consists of diluting (*morphology.dilute* method), binary closing (*morphology.binary_closing* method), and eroding (*morphology.erosion* method).

After the presented above operation, the 3D SEM dataset is replaced with a set of binary images, each representing regions with elevated intensity, computed for combinations of correlated maps of intensities for predefined energy ranges. The quality of images is additionally improved due to application of described above graphical filters. Further, all these images are logically summed (OR operator) to represent all regions, indicating 'uncommon' composition. For all logical operation on images, NumPy library was used.

Further steps are focused on identification of phases for each 'uncommon' region. The first step requires some human-provided knowledge. Original images (before summing correlated ones), believed to represent peaks coming from different bands of a single chemical element, are summed and renormalized. The mask defined with 'uncommon regions' indicator is applied (AND logical operator applied on mask and images for chemical components maps). From the obtained images, regions with intensities in upper quartile are identified. Resulting binary images might be described as maps of regions with high share of particular components and being uncommon. Coherence of the images is improved with some image-processing procedures (applying naming after scikit-image numerical library, dilation, binary closing, and erosion).

The last step, based on STEM-EDS data, is an attempt to bind 'uncommon' regions with particular phases. Since quantitative analysis with low signal-to-nose ratio is unreliable, we define phases in qualitative mode. A phase is constituted by combination of binary logical values, indicating "increased" or "not increased" share of particular chemical component. Furthermore, "groups of phases" might also be defined. Such a group is defined with trinary logic, as a combination of values indicating "increased", "not increased", and "not relevant" shares. The problem of attaching phases to specimen regions might be treated as a multicriteria optimization function. The objective functions depend on number of used phases, number of used groups of shapes, and area of regions not assigned to any phase or groups of phases. Unfortunately, solving such defined problem is complex, and, currently, it is not automatized. Instead, a researcher must fulfill this step with an "educated guess", supported by his/her materials science knowledge and computational procedure identifying coverage of "uncommon" regions with a given combination of available phases or phases' groups.

3. Results and Discussion

3.1. Microstructural Analysis and Thermodynamic Modeling

Equilibrium phases possible to occur for the overall composition of the alloy given in Table 1 were determined using the thermodynamic calculation by means of FactSage software and SGTE database. Figure 3 shows the graph of equilibrium phases in IN625 alloy as a function of temperature. The graph is hypothetical, assuming that the alloy with the nominal composition is under equilibrium conditions. The equilibrium phases that occur at a temperature above 700 °C are the γ solid solution (FCC_1#1 and FCC_2#1), and intermetallic phases σ (SIGM#1), δ (NI_{31}), and P (P_PH), as well as $M_{23}C_6$ carbide (not visible at this scale with the content less than 2 wt.%). These results make it possible to predict the phase composition in Inconel 625 subjected to long-term annealing.

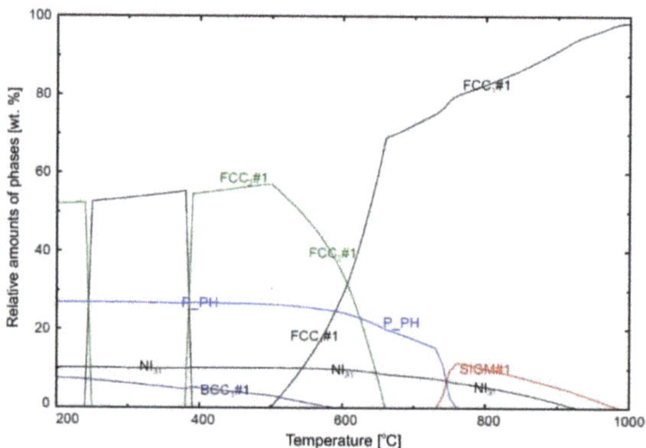

Figure 3. Calculated relative amounts of equilibrium phases in Inconel 625 alloy as a function of temperature.

Figure 4 shows the TEM, STEM-HAADF, and EDS spectral images of the L-PBF Inconel 625 post-build annealed at a temperature of 980 °C for 1 h. The micrograph shows the cellular structure with fine particles at cell wall regions, as well as inside the cells. EDS spectral images, showing distribution maps of constituent chemical elements, reveal the microsegregation of niobium, molybdenum, and silicon to the cell walls. Moreover, it can be seen that nanoparticles with dark contrast due to the content of lighter elements contain aluminum and oxygen, so they are aluminum oxide inclusions. In the further part of the study, inclusions of oxides were also observed in the samples after annealing at 80 °C, but the analysis concerned only precipitates.

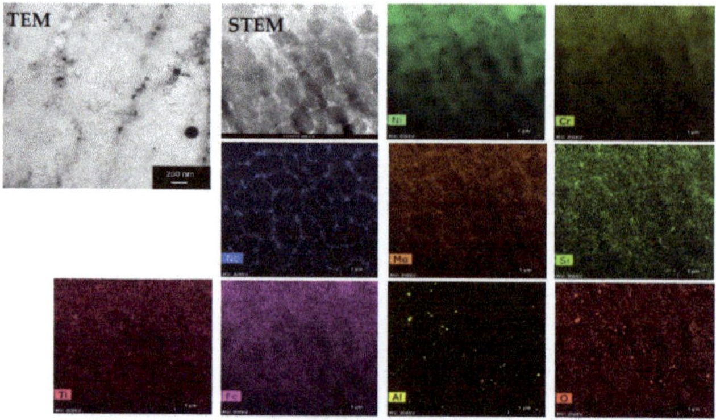

Figure 4. TEM, STEM-HAADF, and EDS spectral images showing maps of constituent chemical elements in the L-PBF Inconel 625 stress-relieve annealed at a temperature of 980 °C for 1 h.

The use of electron diffraction in TEM allowed for phase analysis of such tiny particles. Phase identification of the precipitates showed the presence of the Laves phase particles rich in Ni, Cr, Mo, Nb, and Si. Meanwhile, the particles rich in Nb were identified as MC carbides. This finding is in agreement with literature data which state that, in superalloys, Nb segregates strongly during solidification to the liquid phase and that the Nb-rich MC carbides and the Laves phase precipitates can be formed in eutectic reactions [7].

After annealing at a temperature of 800 °C for 5–500 h, the microstructure was significantly modified (Figure 5). Densely distributed plate-like precipitates were formed inside the grains and along grain boundaries. Diffraction analyses revealed that they are precipitates of the Ni_3Nb-based δ phase. Moreover, the particles with granular morphology identified as the Laves phase contained Mo, Nb, and Si. The exemplary diffraction patterns, together with their solutions for the δ phase and the Laves phase, are given in Figure 5d,e. The SAED pattern along the $[001]_\gamma$ matrix zone axis is superimposed with the [332] pattern of the δ phase (Figure 5d). In turn, the $[011]_\gamma$ pattern in Figure 5e is superimposed with the [131] pattern of the Laves phase.

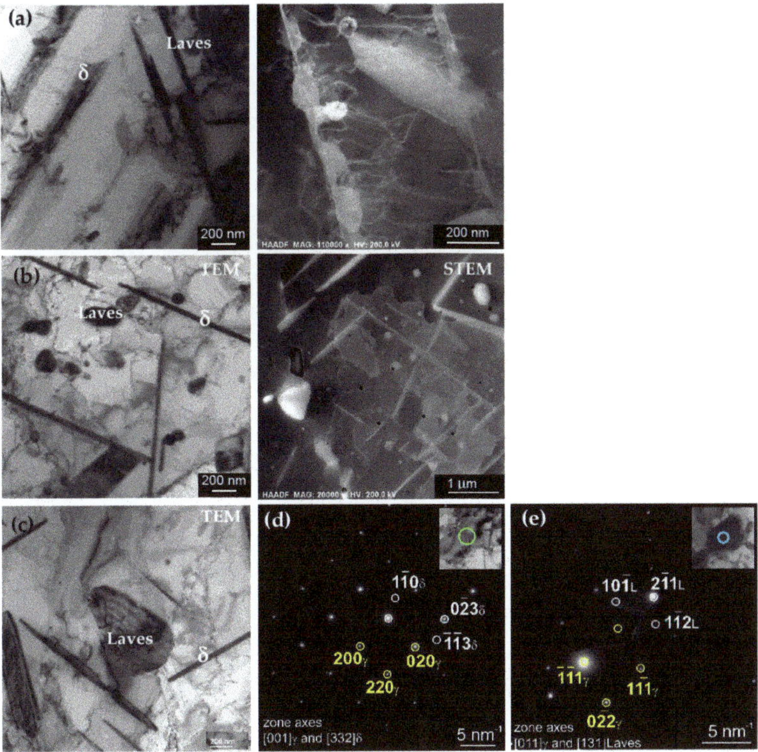

Figure 5. TEM and STEM-HAADF micrographs of the L-PBF Inconel 625 stress-relieved and subsequently annealed at (**a**) 800 °c for 5 h, (**b**) 800 °C for 100 h, and (**c**) 800 °C for 500 h, (**d,e**) SAED patterns of the exemplary δ and Laves phase particles shown in the inserts, together with their solutions; the areas covered by the selective aperture are marked by circles.

Since this work focused on the automatic analysis of the precipitates based on EDS maps, the results of the analysis of the microstructure and chemical composition were used to select groups of precipitates in terms of similar chemical composition. The results for one case, a sample annealed at 800 °C for 500 h, are provided. Figure 6 shows the STEM-HAADF image and EDS maps of the chemical elements of this specimen. Based on EDS maps, four groups of precipitates were distinguished: group 1 containing mainly Ni and Nb (δ phase), and group 2 rich in Mo and Si (Laves phase), as well as the groups of precipitates, which were not identified as different phases by electron diffraction, namely group 3 containing Ti, Ni, and Cr, and group 4 containing Ti, Cr, Ni, and C. Chemical composition of the matrix and particular groups of precipitates determined on the basis of the quantitative EDS microanalysis are given in Table 2. Subsequently, using FactSage, calculations of the mass fraction of hypothetical equilibrium phases that could be formed with such compositions were performed. The results of the thermodynamic modeling are given in Table 3.

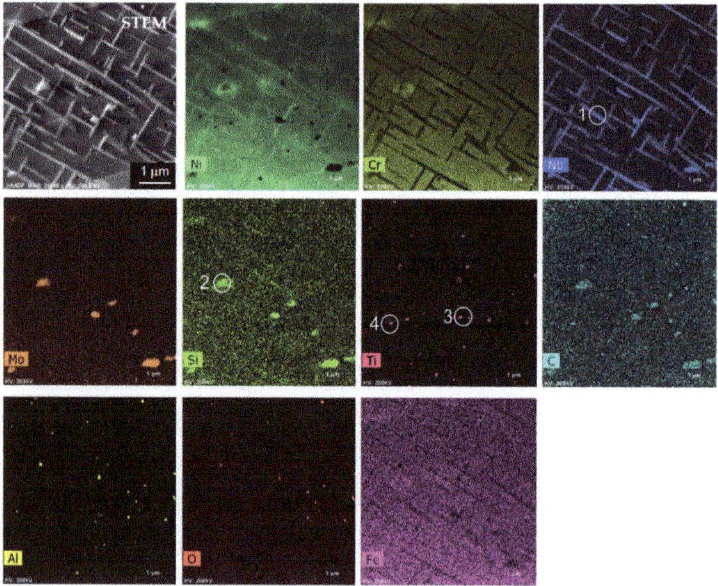

Figure 6. STEM-HAADF image together with EDS maps of chemical elements distribution in the L-PBF Inconel 625 stress-relieved and subsequently annealed at 800 °C for 500 h; exemplary precipitates of groups 1–4 are marked.

Table 2. Average chemical composition ± *standard deviation* (in wt. %) of the matrix and designated groups of precipitates in L-PBF Inconel 625 annealed at 800 °C for 500 h.

	Average Chemical Composition ± *Standard Deviation* (in wt. %)											
	O	C	Fe	Ti	Nb	Al	Cr	Mo	Ni	Si	Mn	Co
matrix	0.2	5.2	1.2	0.1	1.4	0.2	22.4	7.8	60.8	0.4	0.2	0.1
	0.2	*5.8*	*0.4*	*0.1*	*1.0*	*0.2*	*1.5*	*2.0*	*4.3*	*0.2*	*0.2*	*0.1*
Group 1 Ni, Nb	0.4	10.9	1.1	0.4	5.3	0.4	18.8	5.6	55.9	0.3	0.5	0.4
	0.4	*10.0*	*1.0*	*0.4*	*4.2*	*0.4*	*5.5*	*4.6*	*8.1*	*0.3*	*0.5*	*0.4*
Group 2 Mo, Si	0.2	9.3	0.8	0.5	6.5	0.2	17.2	19.2	44.7	0.9	0.4	0.1
	0.2	*5.2*	*0.4*	*0.5*	*5.2*	*0.2*	*3.8*	*12.4*	*14.1*	*0.7*	*0.4*	*0.1*
Group 3 Ti, Ni, Cr	1.6	3.2	0.7	5.4	7.9	1.8	15.9	12	50.6	0.7	0.1	0.1
	1.6	*3.0*	*0.4*	*2.7*	*3.0*	*1.8*	*3.0*	*9.2*	*6.6*	*0.4*	*0.1*	*0.1*
Group 4 Ti, Cr, Ni, C	0.3	11.3	0.9	3.1	3.4	0.3	17.6	7.5	54.2	0.3	0.2	0.9
	0.3	*11.2*	*0.9*	*3.0*	*3.2*	*0.3*	*2.6*	*5.1*	*7.3*	*0.3*	*0.2*	*0.9*

Table 3. Calculated relative amounts of phases that can be formed at equilibrium conditions at temperature 800 °C for the chemical compositions of the matrix and designated groups of precipitates in the specimen annealed at 800 °C for 500 h given in Table 2.

SGTE Notation	FCC A1#1	FCC A1#2	HCP A3#1	FCC L12#1	M23C6	M7C3	C_ Graphite
phase	γ phase		ε phase	γ' phase	$M_{23}C_6$	M_7C_3	graphite
Relative amount (wt. %)							
Matrix	68.2		29.4				2.4
Group 1 Ni, Nb	63.4	5.7	19.4			3.5	8.0
Group 2 Mo, Si	49.5	9.0	36.3				5.2
Group 3 Ti, Ni, Cr	23.2	15.8	10.8	45.5	4.7		
Group 4 Ti, Cr, Ni, C	61.7	10.5	20.1				7.7

For the compositions of the matrix and distinguished groups of precipitates, the discrepancies between the observed and predicted phases were noticed. In particular, for the composition of matrix, the predicted to be in equilibrium is the γ phase and the ε phase, being the M_2C carbide, with a minor fraction of graphite. For group 1, which was identified as the δ phase, calculated equilibrium phases are the γ, ε, M_7C_3, and graphite. For the measured composition of group 2, identified as the Laves phase, as well as for group 4, the predicted phases are the same as for the composition of the matrix, whereas, for the composition of group 3, the equilibrium phase that is predicted to be present at the highest mass fraction is the Ni_3 (Ti, Al)-based intermetallic γ' phase, and the remaining possible phases are γ, ε, and $M_{23}C_6$. Generally, the results of the thermodynamic modeling confirm that, for compositions given for calculations, a reasonable agreement between existing and predicted phases was only achieved for the matrix. However, such results were expected since the microstructure of the L-PBF Inconel 625 was far from equilibrium, and such a state was not achieved after annealing at 800 °C for 500 h. Another reason for the differences in the thermodynamically observed and predicted phase composition may be the fact that the chemical composition was determined by means of STEM-EDS using specimens in the form of thin foils, in which the precipitates are surrounded by a matrix. This causes inaccuracy in determining the composition; therefore, the method used does not allow for automatic detection and recognition of phases. The STEM-EDS method was purposefully selected to obtain information about the chemical composition in the form of a dataset from the analyzed area of the sample by means of automated measurement. However, the results showed that the accuracy is unsatisfactory, and, to obtain greater precision, EDS point analysis should be performed. However, in many EDS setups, this method is not automatic and is performed interactively in the areas selected by the researcher while performing the experiment. To be able to extract the EDS point analysis during the data post-processing, it has to be acquired in a Spectrum Imaging (SI) mode at the first place.

3.2. Application of the Elaborated Algorithm for the Automatic Detection of Precipitates in L-PBF Inconel 625

The algorithm presented in the Section 2.4 was verified with analysis of L-PBF Inconel 625 sample annealed at 800 °C for 500 h. STEM-HAADF image of the microstructure together with acquired EDS sum spectrum (number of X-ray counts summed over all 2D image pixels), and cropped EDS spectrum with background removed are shown in Figure 7. Defined peaks are 'OKα', 'NiLα', 'CrKα', 'CrKβ', 'NiKα', 'NiKβ', 'NbLα', 'MoLα', 'NbKα', 'MoKα', 'SiKα', 'TiKα', 'AlKα', 'NKα', with energy ranges, respectively, (0.3759, 0.676), (0.7015, 1.0015), (5.2647, 5.5647), (5.7967, 6.0967), (7.3109, 7.6109), (8.1146, 8.4146), (2.07,

2.25), (2.25, 2.55), (16.465, 16.7650), (17.330, 17.63), (1.59, 1.89), (4.362, 4.662), (1.33, 1.63), (0.292, 0.402) (all values in keV).

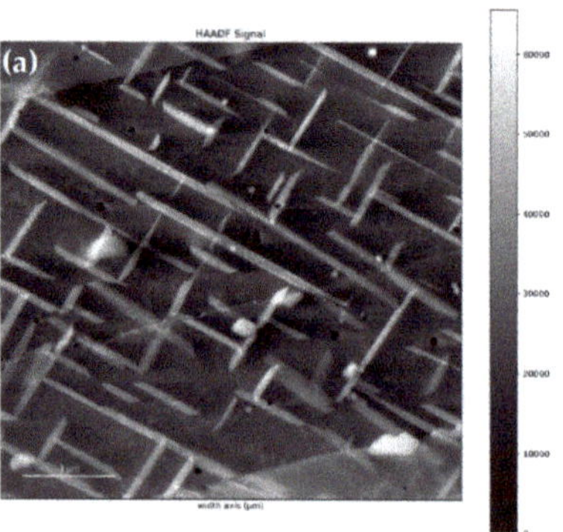

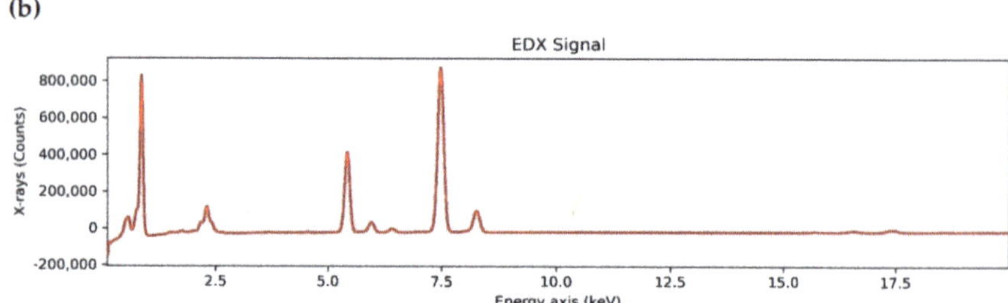

Figure 7. STEM-HAADF image of a microstructure of the L-PBF Inconel 625 post-build annealed at a temperature of 800 °C for 500 h (**a**), and EDS sum spectrum with removed background (**b**).

Three-dimensional results of EDS are flattened with summing in ranges connected to particular peaks, transforming a $1024 \times 1024 \times 4096$ matrix to a $1024 \times 1024 \times 14$ one. Exemplary results are shown in Figure 8. Further, correction of wedging and Gaussian smoothing are applied. Results are shown in Figure 9. In the next step, correlation table is computed for the data treated as 14 separate images. As it was described in the previous section, images with correlation coefficient above 0.5 are summed, and these steps are repeated till none of the coefficients are above this threshold. The first and last correlation tables are shown in Table 4.

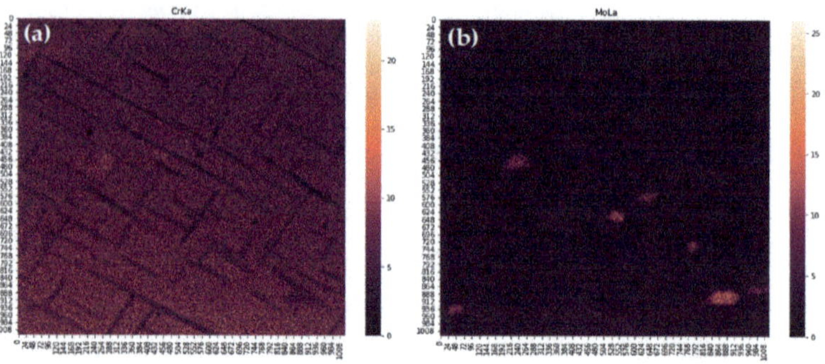

Figure 8. Unprocessed intensities summed over CrKα (**a**) and MoLα (**b**) peaks.

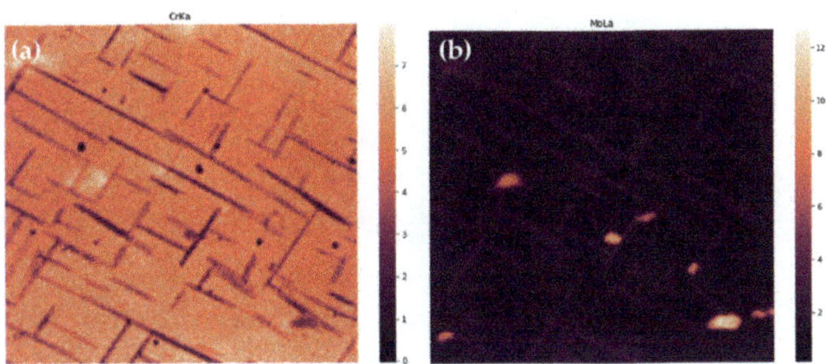

Figure 9. Unwedged and smoothed intensities summed over CrKα (**a**) and MoLα (**b**) peaks.

Table 4. Correlation tables for initial (**a**) and final (**b**) set of images.

(a)

1.000	−0.070	0.439	0.324	−0.024	−0.014	−0.470	−0.271	−0.455	−0.154	−0.168	0.120	0.477	0.297
−0.070	1.000	0.071	0.103	0.711	0.561	0.066	−0.323	0.124	−0.350	−0.058	−0.305	−0.225	−0.146
0.439	0.071	1.000	0.709	0.214	0.170	−0.722	−0.243	−0.745	−0.052	−0.172	−0.235	−0.145	0.016
0.324	0.103	0.709	1.000	0.231	0.184	−0.515	−0.160	−0.535	−0.018	−0.123	−0.207	−0.109	−0.001
−0.024	0.711	0.214	0.231	1.000	0.720	0.131	−0.151	0.175	−0.177	0.051	−0.251	−0.191	−0.097
−0.014	0.561	0.170	0.184	0.720	1.000	0.120	−0.101	0.153	−0.124	0.050	−0.189	−0.141	−0.071
−0.470	0.066	−0.722	−0.515	0.131	0.120	1.000	0.642	0.906	0.417	0.468	0.181	0.093	0.005
−0.271	−0.323	−0.243	−0.160	−0.151	−0.101	0.642	1.000	0.490	0.833	0.587	0.069	0.054	0.053
−0.455	0.124	−0.745	−0.535	0.175	0.153	0.906	0.490	1.000	0.274	0.379	0.177	0.086	−0.014
−0.154	−0.350	−0.052	−0.018	−0.177	−0.124	0.417	0.833	0.274	1.000	0.495	0.035	0.035	0.064
−0.168	−0.058	−0.172	−0.123	0.051	0.050	0.468	0.587	0.379	0.495	1.000	0.024	0.071	0.048
0.120	−0.305	−0.235	−0.207	−0.251	−0.189	0.181	0.069	0.177	0.035	0.024	1.000	0.386	0.355
0.477	−0.225	−0.145	−0.109	−0.191	−0.141	0.093	0.054	0.086	0.035	0.071	0.386	1.000	0.302
0.297	−0.146	0.016	−0.001	−0.097	−0.071	0.005	0.053	−0.014	0.064	0.048	0.355	0.302	1.000

(b)

1.000	−0.041	0.484	−0.226	0.120	0.477	0.297
−0.041	1.000	0.025	−0.156	−0.281	−0.211	−0.119
0.484	0.025	1.000	−0.386	−0.229	−0.125	0.006
−0.226	−0.156	−0.386	1.000	0.048	0.063	0.063
0.120	−0.281	−0.229	0.048	1.000	0.386	0.355
0.477	−0.211	−0.125	0.063	0.386	1.000	0.302
0.297	−0.119	0.006	0.063	0.355	0.302	1.000

From this step, images have no binding to real physical values; they must be treated as a form of abstract representation of inhomogeneity of a sample microstructure. Normalized results are shown in Figure 10. Identifying of regions, where values are within the fourth quartile for any of images (7 in the presented case), and a map of 'distinctive' regions is obtained (shown in Figure 11). These regions are compared to images showing intensities for chemical components, present in the analyzed sample (if more than one spectrum is measured for a particular component, images are summarized). Exemplary images showing increased values of O, N, Cr, and Al are shown in Figure 12.

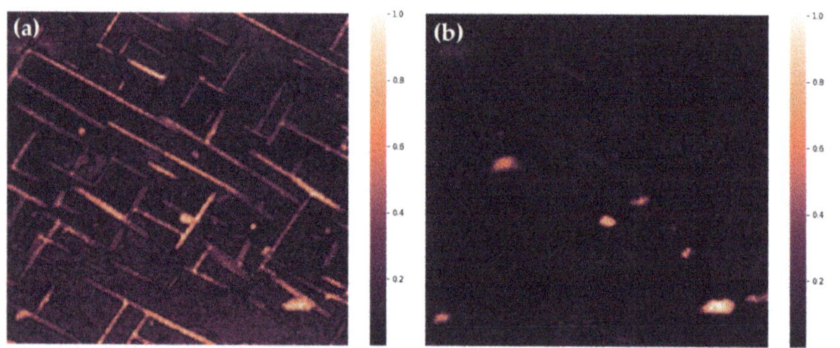

Figure 10. Summed and normalized intensities for images 3rd (**a**) and 4th (**b**) of final set of images.

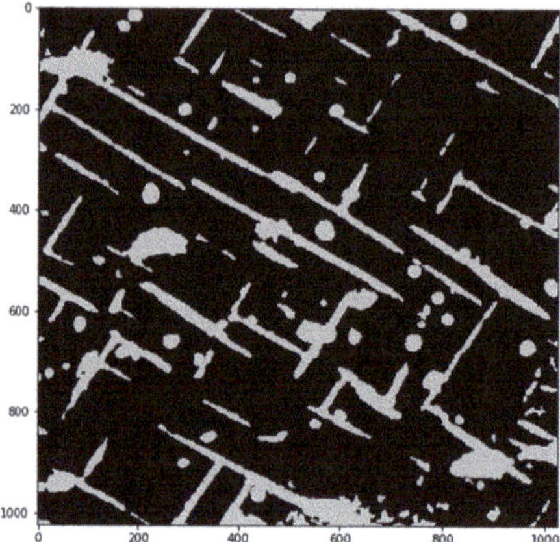

Figure 11. Regions identified as "uncommon".

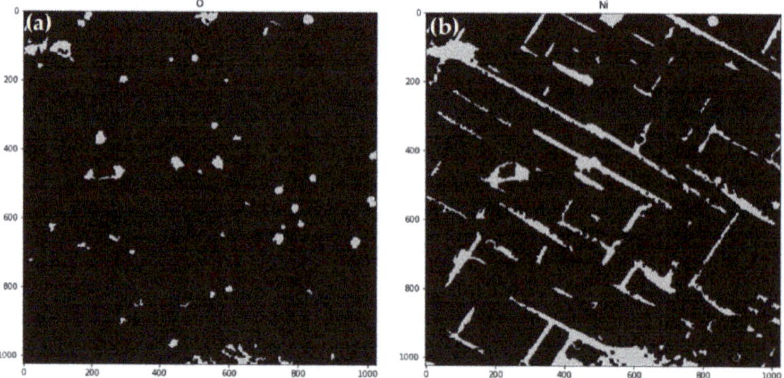

Figure 12. *Cont.*

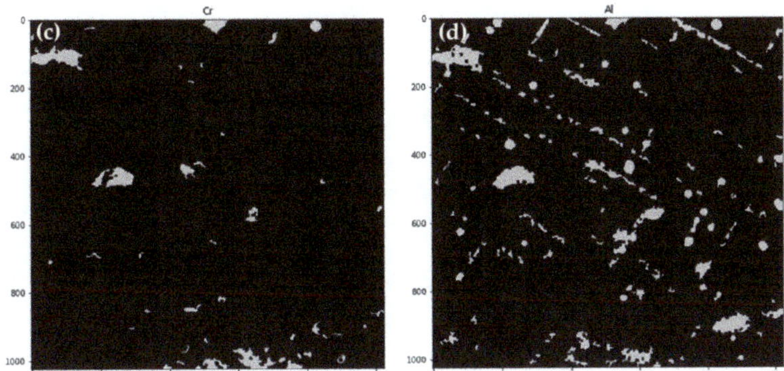

Figure 12. Regions identified as "uncommon" with components share in upper quartile, respectively, O (**a**), Ni (**b**), Cr (**c**) and Al (**d**).

As it was discussed in the previous section, automatic procedure for phases identification is still to be developed. Manual investigation led to identifying of 5 phases groups, presented in Table 5 and in Figure 13.

Table 5. Phases groups. Sign '-' indicates "not increased", no sign indicates "increased", and '*' sign indicates "not relevant".

Phase 1	-O	Ni	-Cr	Nb	Mo	Si	*Ti	*Al	*N
Phase 2	-O	Ni	-Cr	Nb	Mo	-Si	*Ti	*Al	*N
Phase 3	*O	*Ni	Cr	Nb	Mo	Si	*Ti	Al	*N
Phase 4	*O	*Ni	-Cr	*Nb	Mo	Si	*Ti	Al	*N
Phase 5	O	-Ni	-Cr	*Nb	-Mo	-Si	Ti	Al	N

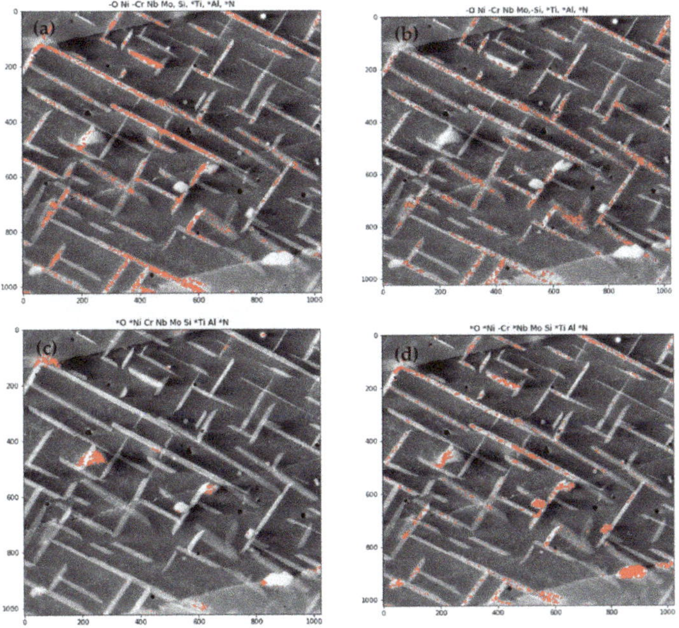

Figure 13. *Cont.*

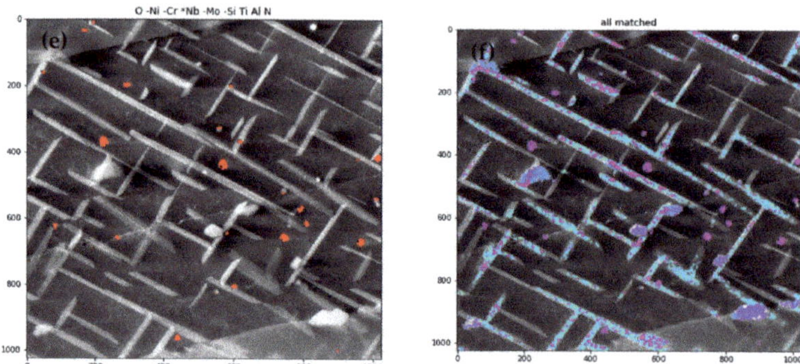

Figure 13. The overlay images of STEM-HAADF and regions identified as phases groups 1–5 marked in red (**a–e**), and regions identified as any phase group marked in violet-blue (**f**).

4. Summary

The microstructural and compositional analysis by means of TEM, STEM, and EDS were employed to investigate the precipitates in the L-PBF Inconel 625 subjected to high temperature annealing. The thermodynamic modeling was used to calculate the stable phases for compositions of experimentally detected phases. The discrepancy between the predicted and observed phases showed that the microstructure of the L-PBF Inconel 625 after annealing at 800 °C for 500 h is far from equilibrium conditions.

The application of the elaborated algorithm for the automatic detection of precipitates in L-PBF Inconel 625 provides results clearly worse than those which might be provided by a human researcher. However, the goal of the presented work was not to replace a human but to significantly decrease the amount of work needed to analyze images. The presented algorithm is able to identify interesting ("uncommon") regions within minutes (on a personal computer), proving digital masks of those domains. From that point, a researcher might start a more thorough analysis. Analysis of a coverage of precipitates with manually defined phases' groups (presented above) is one of the possible paths. Other choice is manual fitting of the algorithm's parameters to improve the segmentation quality.

Author Contributions: Conceptualization, P.M., B.D.; J.F.; methodology, P.M., J.F., P.I., and B.D.; investigation, P.M., J.F., P.I., and B.D.; writing—original draft preparation, P.M., J.F., P.I., and B.D.; writing—review and editing, P.M., J.F., P.I., and B.D.; visualization, P.M., J.F., P.I., and B.D.; supervision, B.D. All authors have read and agreed to the published version of the manuscript.

Funding: This research was funded by the National Science Centre, Poland, grant no 2017/27/B/ST8/02244.

Institutional Review Board Statement: Not applicable.

Informed Consent Statement: Not applicable.

Data Availability Statement: The data presented in this study are available at the Corresponding author.

Acknowledgments: The authors thank W. Brzegowy and K. Gola (AGH UST) for TEM specimen preparation.

Conflicts of Interest: The authors declare no conflict of interest.

References

1. Gonzalez, J.A.; Mireles, J.; Stafford, S.W.; Perez, M.A.; Terrazas, C.A.; Wicker, R.B. Characterization of Inconel 625 fabricated using powder-bed-based additive manufacturing technologies. *J. Mater. Process. Technol.* **2019**, *264*, 200–210. [CrossRef]
2. Lass, E.A.; Stoudt, M.R.; Williams, M.E.; Katz, M.B.; Levine, L.E.; Phan, T.Q.; Gnaeupel-Herold, T.H.; Ng, D.S. Formation of the Ni3Nb δ-Phase in Stress-Relieved Inconel 625 Produced via Laser Powder-Bed Fusion Additive Manufacturing. *Metall. Mater. Trans. A* **2017**, *48*, 5547–5558. [CrossRef]

3. Marchese, G.; Lorusso, M.; Parizia, S.; Bassini, E.; Lee, J.W.; Calignano, F.; Manfredi, D.; Terner, M.; Hong, H.U.; Ugues, D.; et al. Influence of heat treatments on microstructure evolution and mechanical properties of Inconel 625 processed by laser powder bed fusion. *Mater. Sci. Eng. A* **2018**, *729*, 64–75. [CrossRef]
4. Zhang, F.; Levine, L.E.; Allen, A.J.; Stoudt, M.R.; Lindwall, G.; Lass, E.A.; Williams, M.E.; Idell, Y.; Campbell, C.E. Effect of heat treatment on the microstructural evolution of a nickel-based superalloy additive-manufactured by laser powder bed fusion. *Acta Mater.* **2018**, *152*, 200–214. [CrossRef] [PubMed]
5. Hu, Y.L.; Lin, X.; Zhang, S.Y.; Jiang, Y.M.; Lu, X.F.; Yang, H.O.; Huang, W.D. Effect of solution heat treatment on the microstructure and mechanical properties of Inconel 625 superalloy fabricated by laser solid forming. *J. Alloys Compd.* **2018**, *767*, 330–344. [CrossRef]
6. Marchese, G.; Parizia, S.; Rashidi, M.; Saboori, A.; Manfredi, D.; Ugues, D.; Lombardi, M.; Hryha, E.; Biamino, S. The role of texturing and microstructure evolution on the tensile behavior of heat-treated Inconel 625 produced via laser powder bed fusion. *Mater. Sci. Eng. A* **2020**, *769*, 138500. [CrossRef]
7. Gola, K.; Dubiel, B.; Kalemba-Rec, I. Microstructural Changes in Inconel 625 Alloy Fabricated by Laser-Based Powder Bed Fusion Process and Subjected to High-Temperature Annealing. *J. Mater. Eng. Perform.* **2020**, *29*, 1528–1534. [CrossRef]
8. Keller, T.; Lindwall, G.; Ghosh, S.; Ma, L.; Lane, B.M.; Zhang, F.; Kattner, U.R.; Lass, E.A.; Heigel, J.C.; Idell, Y.; et al. Application of finite element, phase-field, and CALPHAD-based methods to additive manufacturing of Ni-based superalloys. *Acta Mater.* **2017**, *139*, 244–253. [CrossRef]
9. Li, C.; White, R.; Fang, X.Y.; Weaver, M.; Guo, Y.B. Microstructure evolution characteristics of Inconel 625 alloy from selective laser melting to heat treatment. *Mater. Sci. Eng. A* **2017**, *705*, 20–31. [CrossRef]
10. Son, K.; Kassner, M.E.; Lee, K.A. The Creep Behavior of Additively Manufactured Inconel 625. *Adv. Eng. Mater.* **2020**, *22*, 1900543. [CrossRef]
11. Shiojiri, M.; Saijo, H. Imaging of high-angle annular dark-field scanning transmission electron microscopy and observations of GaN-based violet laser diodes. *J. Microsc.* **2006**, *223*, 172–178. [CrossRef]
12. Parish, C.M.; Brewer, L.N. Key parameters affecting quantitative analysis of STEM-EDS spectrum images. *Microsc. Microanal.* **2010**, *16*, 259–272. [CrossRef] [PubMed]
13. Nellist, P.D.; Pennycook, S.J. Accurate structure determination from image reconstruction in ADF STEM. *J. Microsc.* **1998**, *190*, 159–170. [CrossRef]
14. Rosenauer, A.; Mehrtens, T.; Müller, K.; Gries, K.; Schowalter, M.; Satyam, P.V.; Bley, S.; Tessarek, C.; Hommel, D.; Sebald, K.; et al. Composition mapping in InGaN by scanning transmission electron microscopy. *Ultramicroscopy* **2011**, *111*, 1316–1327. [CrossRef]
15. Kim, Y.-M.; He, J.; Biegalski, M.D.; Ambaye, H.; Lauter, V.; Christen, H.M.; Pantelides, S.T.; Pennycook, S.J.; Kalinin, S.V.; Borisevich, A.Y. Probing oxygen vacancy concentration and homogeneity in solid-oxide fuel-cell cathode materials on the subunit-cell level. *Nat. Mater.* **2012**, *11*, 888–894. [CrossRef]
16. Morgan, D.G.; Ramasse, Q.M.; Browning, N.D. Application of two-dimensional crystallography and image processing to atomic resolution Z-contrast images. *J. Electron Microsc.* **2009**, *58*, 223–244. [CrossRef]
17. Buades, A.; Buades, A.; Coll, B.; Morel, J.M. A review of image denoising algorithms, with a new one. *Multiscale Model. Simul.* **2005**, *4*, 490–530. [CrossRef]
18. Binev, P.; Blanco-Silva, F.; Blom, D.; Dahmen, W.; Lamby, P.; Sharpley, R.; Vogt, T. High-Quality Image Formation by Nonlocal Means Applied to High-Angle Annular Dark-Field Scanning Transmission Electron Microscopy (HAADF–STEM). In *Modeling Nanoscale Imaging in Electron Microscopy*; Springer: Boston, MA, USA, 2012; pp. 127–145.
19. Mevenkamp, N.; Binev, P.; Dahmen, W.; Voyles, P.M.; Yankovich, A.B.; Berkels, B. Poisson noise removal from high-resolution STEM images based on periodic block matching. *Adv. Struct. Chem. Imaging* **2015**, *1*, 3. [CrossRef]
20. Mevenkamp, N.; MacArthur, K.; Tileli, V.; Ebert, P.; Allen, L.; Berkels, B.; Duchamp, M. Multi-modal and multi-scale non-local means method to analyze spectroscopic datasets. *Ultramicroscopy* **2020**, *209*, 112877. [CrossRef] [PubMed]
21. Ede, J.M.; Beanland, R. Improving Electron Micrograph Signal-to-Noise with an Atrous Convolutional Encoder-Decoder. *Ultramicroscopy* **2019**, *202*, 18–25. [CrossRef]
22. Yang, X.; De Andrade, V.; Scullin, W.; Dyer, E.L.; Kasthuri, N.; De Carlo, F.; Gürsoy, D. Low-dose x-ray tomography through a deep convolutional neural network. *Sci. Rep.* **2018**, *8*, 2575. [CrossRef] [PubMed]
23. Giannatou, E.; Papavieros, G.; Constantoudis, V.; Papageorgiou, H.; Gogolides, E. Deep learning denoising of SEM images towards noise-reduced LER measurements. *Microelectron. Eng.* **2019**, *216*, 111051. [CrossRef]
24. Potapov, P.; Lubk, A. Optimal principal component analysis of stem xeds spectrum images. *Adv. Struct. Chem. Imaging* **2019**, *5*, 4. [CrossRef]
25. Potapov, P.; Longo, P.; Okunishi, E. Enhancement of noisy EDX HRSTEM spectrum-images by combination of filtering and PCA. *Micron* **2017**, *96*, 29–37. [CrossRef]
26. Thanh, D.N.H.; Prasath, V.B.S.; Le Minh, H. A Review on CT and X-Ray Images Denoising Methods. *Informatics* **2019**, *43*, 151–159. [CrossRef]
27. Voyles, P.M. Informatics and data science in materials microscopy. *Curr. Opin. Solid State Mater. Sci.* **2017**, *21*, 141–158. [CrossRef]
28. Ghosh, S.; Das, N.; Das, I.; Maulik, U. Understanding deep learning techniques for image segmentation. *ACM Comput. Surv.* **2019**, *52*, 1–35. [CrossRef]
29. Yang, R.; Buenfeld, N.R. Binary segmentation of aggregate in SEM image analysis of concrete. *Cem. Concr. Res.* **2001**, *31*, 437–441. [CrossRef]

30. Banerjee, S.; Chakraborti, P.C.; Saha, S.K. An automated methodology for grain segmentation and grain size measurement from optical micrographs. *Measurement* **2019**, *140*, 142–150. [CrossRef]

31. Peregrina-Barreto, H.; Terol-Villalobos, I.R.; Rangel-Magdaleno, J.J.; Herrera-Navarro, A.M.; Morales-Hernández, L.A.; Manríquez-Guerrero, F. Automatic grain size determination in microstructures using image processing. *Measurement* **2013**, *46*, 249–258. [CrossRef]

32. Iskakov, A.; Kalidindi, S.R. A Framework for the Systematic Design of Segmentation Workflows. *Integrating Mater. Manuf. Innov.* **2020**, *9*, 70–88. [CrossRef]

33. Uusimaeki, T.; Wagner, T.; Lipinski, H.G.; Kaegi, R. AutoEM: A software for automated acquisition and analysis of nanoparticles. *J. Nanopart. Res.* **2019**, *21*, 122. [CrossRef]

34. Jany, B.R.; Janas, A.; Krok, F. Automatic microscopic image analysis by moving window local Fourier Transform and Machine Learning. *Micron* **2020**, *130*, 102800. [CrossRef]

35. Münch, B.; Martin, L.H.J.; Leemann, A. Segmentation of elemental EDS maps by means of multiple clustering combined with phase identification. *J. Microsc.* **2015**, *260*, 411–426. [CrossRef] [PubMed]

36. Mirzaei, M.; Rafsanjani, H.K. An automatic algorithm for determination of the nanoparticles from TEM images using circular hough transform. *Micron* **2017**, *96*, 86–95. [CrossRef]

37. Krakowiak, K.J.; Wilson, W.; James, S.; Musso, S.; Ulm, F.J. Inference of the phase-to-mechanical property link via coupled X-ray spectrometry and indentation analysis: Application to cement-based materials. *Cem. Concr. Res.* **2015**, *67*, 271–285. [CrossRef]

38. Drumetz, L.; Henrot, S.; Veganzones, M.A.; Chanussot, J.; Jutten, C. Blind hyperspectral unmixing using an extended linear mixing model to address spectral variability. In Proceedings of the 2015 7th Workshop on Hyperspectral Image and Signal Processing: Evolution in Remote Sensing (WHISPERS), Tokyo, Japan, 2–5 June 2015; pp. 1–4.

39. Ding, Q.; Colpan, M. Decision Tree Induction on HyperSpectral Cement Images. *Int. J. Comput. Intell.* **2006**, *2*, 169–175.

40. Georget, F.; Wilson, W.; Scrivener, K.L. edxia: Microstructure characterisation from quantified SEM-EDS hypermaps. *Cem. Concr. Res.* **2021**, *141*, 106327. [CrossRef]

41. Schulz, B.; Merker, G.; Gutzmer, J. Automated SEM mineral liberation analysis (MLA) with generically labelled EDX spectra in the mineral processing of rare earth element ores. *Minerals* **2019**, *9*, 527. [CrossRef]

42. Iglesias, J.C.Á.; Augusto, K.S.; Da Fonseca Martins Gomes, O.; Domingues, A.L.A.; Vieira, M.B.; Casagrande, C.; Paciornik, S. Automatic characterization of iron ore by digital microscopy and image analysis. *J. Mater. Res. Technol.* **2018**, *7*, 376–380. [CrossRef]

43. Snow, Z.; Diehl, B.; Reutzel, E.W.; Nassar, A. Toward in-situ flaw detection in laser powder bed fusion additive manufacturing through layerwise imagery and machine learning. *J. Manuf. Syst.* **2021**, *59*, 12–26. [CrossRef]

44. Wang, C.; Tan, X.P.; Tor, S.B.; Lim, C.S. Machine learning in additive manufacturing: State-of-the-art and perspectives. *Addit. Manuf.* **2020**, *36*, 101538. [CrossRef]

45. Johnson, N.S.; Vulimiri, P.S.; To, A.C.; Zhang, X.; Brice, C.A.; Kappes, B.B.; Stebner, A.P. Invited review: Machine learning for materials developments in metals additive manufacturing. *Addit. Manuf.* **2020**, *36*, 101641. [CrossRef]

46. Miyazaki, S.; Kusano, M.; Bulgarevich, D.S.; Kishimoto, S.; Yumoto, A.; Watanabe, M. Image Segmentation and Analysis for Microstructure and Property Evaluations on Ti–6Al–4V Fabricated by Selective Laser Melting. *Mater. Trans.* **2019**, *60*, 561–568. [CrossRef]

47. Gupta, S.; Sarkar, J.; Kundu, M.; Bandyopadhyay, N.R.; Ganguly, S. Automatic recognition of SEM microstructure and phases of steel using LBP and random decision forest operator. *Measurement* **2020**, *151*, 107224. [CrossRef]

48. Stadelmann, P. Image analysis and simulation software in transmission electron microscopy. *Microsc. Microanal.* **2003**, *9*, 60–61. [CrossRef]

49. Xu, Y.; Yamazaki, M.; Villars, P. Inorganic Materials Database for Exploring the Nature of Material. *Jpn. J. Appl. Phys.* **2011**, *50*, 11RH02. [CrossRef]

50. De la Peña, F.; Prestat, E.; Fauske, V.T.; Burdet, P.; Jokubauskas, P.; Nord, M.; Ostasevicius, T.; MacArthur, K.E.; Sarahan, M.; Johnstone, D.N.; et al. Hyperspy/Hyperspy: HyperSpy v1.5.2. 2019. Available online: https://hyperspy.org/ (accessed on 15 May 2021).

51. Van der Walt, S.; Schönberger, J.L.; Nunez-Iglesias, J.; Boulogne, F.; Warner, J.D.; Yager, N.; Gouillart, E.; Yu, T. scikit-image: Image processing in Python. *PeerJ* **2014**, *2*, e453. [CrossRef] [PubMed]

52. Reback, J.; Jbrockmendel; McKinney, W.; Van den Bossche, J.; Augspurger, T.; Cloud, P.; Hawkins, S.; Gfyoung; Sinhrks; Roeschke, M.; et al. Pandas-Dev/Pandas: Pandas 1.3.0. 2021. Available online: https://pandas.pydata.org/ (accessed on 5 July 2021).

53. Zack, G.W.; Rogers, W.E.; Latt, S.A. Automatic measurement of sister chromatid exchange frequency. *J. Histochem. Cytochem.* **1977**, *25*, 741–753. [CrossRef]

*materials*

MDPI

Article

Design and Additive Manufacturing of Porous Sound Absorbers—A Machine-Learning Approach

Sebastian Kuschmitz [1,*,†] , Tobias P. Ring [2,*,†] , Hagen Watschke [1] , Sabine C. Langer [2] and Thomas Vietor [1]

[1] TU Braunschweig, Institute for Engineering Design, 38106 Braunschweig, Germany; h.watschke@tu-braunschweig.de (H.W.); t.vietor@tu-braunschweig.de (T.V.)
[2] TU Braunschweig, Institute for Acoustics, 38106 Braunschweig, Germany; s.langer@tu-braunschweig.de
* Correspondence: s.kuschmitz@tu-braunschweig.de (S.K.); t.ring@tu-braunschweig.de (T.P.R.); Tel.: +49-531-391-3346 (S.K.); +49-531-391-8773 (T.P.R.)
† These authors contributed equally to this work.

Abstract: Additive manufacturing (AM), widely known as 3D-printing, builds parts by adding material in a layer-by-layer process. This tool-less procedure enables the manufacturing of porous sound absorbers with defined geometric features, however, the connection of the acoustic behavior and the material's micro-scale structure is only known for special cases. To bridge this gap, the work presented here employs machine-learning techniques that compute acoustic material parameters (Biot parameters) from the material's micro-scale geometry. For this purpose, a set of test specimens is used that have been developed in earlier studies. The test specimens resemble generic absorbers by a regular lattice structure based on a bar design and allow a variety of parameter variations, such as bar width, or bar height. A set of 50 test specimens is manufactured by material extrusion (MEX) with a nozzle diameter of 0.20 mm and a targeted under extrusion to represent finer structures. For the training of the machine learning models, the Biot parameters are inversely identified from the manufactured specimen. Therefore, laboratory measurements of the flow resistivity and absorption coefficient are used. The resulting data is used for training two different machine learning models, an artificial neural network and a k-nearest neighbor approach. It can be shown that both models are able to predict the Biot parameters from the specimen's micro-scale with reasonable accuracy. Moreover, the detour via the Biot parameters allows the application of the process for application cases that lie beyond the scope of the initial database, for example, the material behavior for other sound fields or frequency ranges can be predicted. This makes the process particularly useful for material design and takes a step forward in the direction of tailoring materials specific to their application.

Keywords: acoustics; porous materials; material extrusion; design for additive manufacturing; sound absorption; artificial neural network; machine learning

Citation: Kuschmitz, S.; Ring, T.P.; Watschke, H.; Langer, S.C.; Vietor, T. Design and Additive Manufacturing of Porous Sound Absorbers—A Machine-Learning Approach. *Materials* **2021**, *14*, 1747. https://doi.org/10.3390/ma14071747

Academic Editor: Mika Salmi

Received: 1 March 2021
Accepted: 26 March 2021
Published: 1 April 2021

1. Introduction

Porous materials are widely used for sound absorption and noise mitigation and find applications in various technical disciplines. Therein, three mechanisms can be distinguished: absorption of airborne sound, decoupling of vibrating systems and reduction of flow-induced sound. First, the most common and well-known application for porous materials is sound absorption, for example in room acoustics [1] or porous liners in aircraft engine inlet and exhaust pipe [2,3]. The scope of this work as well lies within this field. For this mechanism, the dominating effect is the dissipation of the acoustic energy when the material is placed in front of an acoustically rigid boundary. A second important aspect is the application of porous materials with the effect of decoupling vibrating structures. This effect is commonly employed for improving sound insulation of double panel walls, for example in aircraft fuselage sidewall panels [4,5] or in building acoustics [6,7]. Thereby, the two wall panels are separated by the porous material. The damped propagation of the acoustic wave inside the porous material results in a decrease of the mechanic coupling of the two panels. The effect of decoupling can be exploited as well with regard to aircraft structures. Here, the application of porous materials on the outer aircraft skin can reduce the excitation of structure-borne sound that results from an external pressure field [8]. The third application of porous materials is the reduction of flow-induced sound emitted from moving objects [9–12]. All mentioned applications make use of sheets of porous materials, for example, fibrous mineral wool, synthetic foam, or porous metals. These porous materials show favorable acoustic properties, specifically a high absorption coefficient, especially in the higher frequency regime [13]. Nevertheless, these materials typically consist of an irregular microstructure and a geometric description of the microstructure is available as homogenized values only. This makes an investigation of the material structure on the micro-scale (the micro-scale is hereafter understood as the size scale of the pores, see [14], whereas the macro-scale are the overall specimens dimensions) and its relation to the acoustic behavior rather difficult. Therefore, the basic idea of the presented paper is to bridge the gap between the absorber's micro-scale geometry and its acoustic behavior by means of machine-learning techniques. The acoustic behavior thereby is described material-specific parameters in combination with mechanical porous media models. This "detour" allows the necessary input data set to be rather small and enables the prediction of the acoustic behavior for arbitrary application cases of the material that go beyond the specific test cases with which the data has been retrieved.

1.1. Description of Porous Media by Mechanical Models

A common engineering task in acoustics is the adjustment of the appropriate amount of damping of a fluid cavity, for example to adjust reverberation time. Acoustic damping can be effectively reached by means of porous materials that are mounted onto the surface of a rigid wall and thus serve as acoustic absorbers. The general procedure for the application of porous material commonly is a first laboratory-scale characterization of the material followed by an analysis of the target system, resulting in information where and how the material at hand can effectively be employed. The laboratory tests mostly incorporate acoustic measurements of the absorption coefficient, flow resistivity and further parameters. This straightforward procedure allows for a reasonable application of available materials. Nevertheless, the possibility to directly design materials, in the sense of tailoring them with special regard to the application often is of interest. An approach to determine the necessary acoustic parameters that fulfill a given task is shown in [15] by tailoring porous media to mitigate aeroacoustic trailing edge noise. Nevertheless, a core challenge during the material design task of porous media is the difference between the geometric and acoustics properties of the material. On one side, the geometric description is essential for a subsequent manufacturing process. For porous media in acoustic applications, the geometry on both the macro-scale and the micro-scale is important. Geometric properties are, for example, the sample thickness on the macro-scale and the pore diameter (if the pores are assumed to be rather cylindrical) or even more complex descriptions of the pore geometry on the micro-scale. On the other side, the desired material properties are derived using mechanical models that resemble the acoustic behavior of the material. Thereby, the Biot model, [16,17], is the most complex one since it models the wave propagation of elastic waves within the material of both, the pressure wave in the fluid phase and the shear and pressure wave in the elastic (skeleton) phase of the material. Due to this full description, the Biot model is accepted as a reference for modeling porous materials [18]. Other, more simple models neglect the elasticity of the skeleton phase and model only the pressure wave within the fluid. They incorporate the influence of the skeleton by definition of equivalent fluid properties. These models are referred to as the class of equivalent fluid models. Prominent examples are the model introduced by Johnson et al. [19] and the model by Champoux and Allard [20]. All aforementioned models employ material parameters that represent the material in a homogenized way, the pore structure of the material is not resolved. These acoustic material parameters are hereafter referred to as Biot parameters. The Biot parameters can be classified into the parameters that can be attributed to the (equivalent) fluid phase and those attributed to the skeleton. The relevant parameters of the fluid phase are the porosity ϕ, the flow resistivity Ξ, the tortuosity α_∞ and the viscous and thermal characteristic lengths Λ, Λ'. For the special case of the Johnson–Champoux–Allard–Lafarge (JCAL) model [19–21] that is used within this work, the static thermal permeability k_0' is additionally introduced. For the full description of the material using Biot's theory, the mechanical parameters of the skeleton material are required as well. From the set of Biot parameters the flow resistivity, porosity and tortuosity can be easily explained: the porosity refers to the share of fluid volume on the overall sample volume, the flow resistivity measures the possibility of a fluid flow to pass through the material and the tortuosity indicates the degree of the geometric complexity of the pore structure. However, one of the major drawbacks of the porous material models is that at least some of the required parameters are hard to determine [18]. Indeed, some parameters can be measured, such as flow resistivity, porosity and tortuosity. Moreover, for a given material it is possible to inversely estimate these parameters, as shown for example in [22,23]. For the case of an available geometric description of the material on the micro-scale, in [24] the computation of the flow resistivity of a porous sample is computed using the Lattice–Boltzmann method. Nevertheless, in general, no connection between the Biot parameters and the micro-scale geometry is known. Therefore the application of porous material models is limited to materials that already exist and for which the required parameters can be measured or otherwise determined.

1.2. Additive Manufacturing of Porous Materials for Sound Absorption

One promising possibility to generate acoustically effective materials is the rapidly growing field of additive manufacturing (AM), also known as 3D printing. The layer-by-layer working principle of additive manufacturing processes provides new kinds of design freedom during product development. In this way, for example, undercuts, mesoscopic lattice structures, or free-form surfaces can be realized, making additive manufacturing particularly suitable for the production of acoustically effective structures [25]. The specimens investigated in this contribution consist of a micro-lattice of parallel bars as described in [26]. Hence, AM is employed within this contribution for manufacturing the required amount of specimens.

Due to the aforementioned reasons, AM provides the opportunity to significantly improve both geometric and functional product properties [27,28]. Due to the aforementioned potentials, AM processes can greatly improve the production of acoustically effective structures. Especially the material extrusion (MEX) is predestined for this application due to its process principles, for example, no need for support structures, no requirement for removing powder or resin in undercuts and channels, or the use of fine nozzle diameters. The high resolution of the microstructure in the x-y direction is only limited by the nozzle diameter, which in MEX processes can be down to 0.10 mm on the lower end, and the printer itself. Other processes such as the powder bed fusion of polymers (PBF-P) or vat photopolymerization (VAT) theoretically also offer the possibility of producing even finer microstructures. However, since the liquid material of the VAT process (capillary effects, etc.) leads to unwanted hardening of the material and the powder bed in the PBF-P process results in sintering as well as powder residues in the geometry, the MEX process is particularly well suited for this use [27].

In the past, AM techniques have been used to produce a variety of sound-absorbing materials, including Helmholtz resonators [29], straight [30], inclined [31], angled tubes [29], sound crystals [32]. Moreover, microperforated plates [33] and sounds absorbers made of micro-grids [34–36] have been manufactured (see also [26,34,37–40]). The additive manufacturing of porous absorbers has also been addressed in a rudimentary way in the literature. Ring et al. have demonstrated the additive manufacturing of the porous and acoustically effective absorber structures using MEX processes and were able to show the potential of these structures. Additionally, the authors were able to investigate relationships between the geometry and acoustic properties, the Biot parameters (tortuosity, porosity, etc.) [26,41].

Zielinski et al. have performed a 3D printing process comparison with different materials using porous absorbers, revealing that the process parameters have a crucial influence on the reproducibility of acoustically effective structures [37]. Boulvert et al. have presented studies on geometrical factors influencing the acoustic properties, in which mainly the gradation of porosity is realized via the infill of the porous absorbers [38]. This approach is similar to the one shown in [26], where grading the porosity via the infill alone does not fully exploit the design potential of additive manufacturing.

A common feature of all previously presented research results is the focus on the technical feasibility of additive manufacturing of acoustically effective porous structures and research into factors influencing the acoustic properties. However, the focus so far has been only on the manufacturing of porous geometry without defining the acoustic properties in advance. Nevertheless, the use of AM technologies for the creation of porous structures offers the possibility to efficiently introduce acoustic properties in a product-specific way, taking into account the potentials and limitations of the manufacturing process. A method for targeted adjustment of acoustic properties through defined geometries, taking into account process-related influencing factors, thus is desirable.

1.3. Aims and Scope of the Presented Work

As mentioned above, several experimental capabilities exist to investigate the acoustic behavior of porous materials. Moreover, the acoustic behavior of porous materials can be described using various mechanical models of different complexity and accuracy. Recently, AM capabilities have proven successful in manufacturing generic porous materials as well. Nevertheless, for manufacturing a porous specimen, a suitable geometric description of the specimen is required on both, the micro- and macro-scale. The acoustic description of the specimen with both, experiments and/or mechanical models, involves the description with the Biot parameters. These two "worlds" (macro-/micro-scale of the geometry and Biot parameters) are rather detached, as no general connection of geometric dimensions and the Biot parameters of porous materials is known. Indeed, for some special cases models exist to estimate the Biot parameters from the specimen geometry, for example, the tortuosity can be estimated from the porosity as shown in [14,42–44]. In order to bridge this gap, the work presented here employs Machine-Learning (ML) techniques to build a model that computes the Biot parameters for a given geometric description of a specimen.

A similar idea, the computation of the acoustic behavior of porous material using ML techniques, is presented, for example, in [45–47]. In contrast to the approach proposed within this contribution, the authors directly compute the absorption coefficient of the porous material using a neural network model. This approach is reasonable as long as the macroscopic sample dimensions are kept constant or if data from samples with different dimensions are available for the training, whereas the latter is the case in the mentioned publications. Another approach is presented in [48]. Here the authors use more material-specific parameters for prediction of the absorption coefficient of a porous absorber. However, the parameter set chosen is rather general and accounts for the type of material only by referring to the porosity. For the work presented here, a "detour" via the Biot parameters is used. It is a detour in the sense that the whole process becomes more complex, as the Biot parameters have to be derived which is a challenging process involving ill-posed mathematical problems. However, this detour has some favorable implications: the Biot parameters are material-specific parameters. This way, no mixing of the description of the material and its acoustic behavior (for example: absorption coefficient in front of a rigid wall) occurs. Having the Biot parameters available, it becomes possible to compute the acoustic behavior of the investigated material for conditions apparent in the available data (sound field, macro-scale dimensions of the specimen), but other conditions can be covered as well. For example, the behavior of the material for other frequency ranges than the measured ones can be computed. Another application example for the Biot parameters is the case of more complex simulations, such as finite element computations. Using such wave-resolving methods, the behavior of the material can be evaluated in other sound fields, such as diffuse fields. Moreover, the dimension of the necessary training data is reduced as samples with a constant thickness can be used rather than a population comprising different sample thicknesses. Due to these reasons, the "detour" via the Biot parameters is assumed to justify the higher effort and adds value to this field of research.

The resulting ML models are then used for an inverse material design approach. Therefore, materials are designed that are supposed to show a predefined acoustic behavior. Thereby, the advantage of using the Biot parameters as a basis becomes clear, as this procedure can be done by adjusting the specimen geometry on the micro-scale only but as well on the macro-scale.

The procedure presented within this work allows to adjust the acoustic properties of the absorbers before production and thus enables an efficient application of acoustically effective materials that are best suited for the requirements of the application. In turn, this procedure might even reduce secondary factors such as costs, weight, or carbon footprint.

1.4. Outline of the Paper

The paper is organized as follows: in Section 2 the methods applied within this work are presented and a description of the developed specimen is given. A general overview of the applied procedure is presented in Section 2.1, the design methodology and manufacturing process for the generic specimens are shown in Sections 2.2 and 2.3, respectively. The methods to acoustically investigate and derive Biot parameters are described in Sections 2.4 and 2.5, the applied machine learning process including the applied models is described in Section 2.6. In Section 3, the results of the acoustical inspection, parameter identification and machine learning process are shown in Sections 2.4 and 3.2. The results of the application of the process to generate new absorbers with a predefined behavior are shown in Sections 3.3 and 3.4. A conclusion and an outlook is given in Section 4.

2. Materials and Methods

In this section, the study design and used methods are presented. The goal of the work is to build a model using ML techniques that computes the Biot parameters for a given specimen micro-scale geometry. In a subsequent step, the obtained ML models are used to design new sound-absorbing structures that exhibit a predefined acoustic behavior. Within this section, first, the overall concept of the process is described, followed by a description of the generic specimen investigated here. Furthermore, the used experimental setups for manufacturing and characterization of the specimens are described and the process to identify the Biot parameters is shown. Finally, the used machine learning models and the material design approach are described.

2.1. General Description of the Applied Process

The goal to build ML models that allow the computation of Biot parameters from a geometric description of the specimens and to utilize these for material design employs a 6-step procedure. The process is sketched in Figure 1.

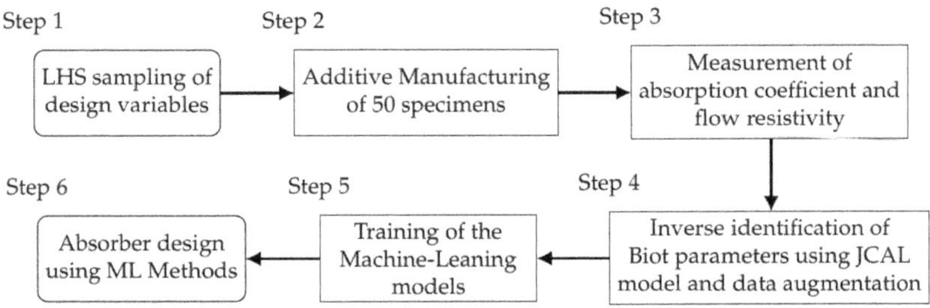

Figure 1. General procedure of the work to obtain ML models that predict Biot parameters from specimen geometry and their application to absorber design.

The generic specimens investigated within this work are presented in Section 2.2 and are parametrically defined by four so-called design variables. The values of the design variables are referred to as the design parameters and form the geometric description of the specimen on the micro-scale.

The procedure starts in Step 1 (see Section 2.2) with a Latin Hypercube Sampling (LHS) of the design variables to generate a population of 50 specimens that explore the chosen data range of the design variables. All 50 designs are shown as well in Table A1 in the Appendix A. In Step 2, the 50 designed specimens are manufactured using AM technology, this is described in detail in Section 2.3. With the design parameters, the required inputs for the ML models are available. Hence, the Biot parameters that form the outputs of the ML models have to be determined. It would be desirable to directly measure these quantities. However, only a few of the Biot parameters can be measured directly without tremendous effort. To the author's knowledge, these are the flow resistivity, the porosity and the tortuosity. Since the authors only have access to a measurement setup for the flow resistivity, an inverse parameter identification procedure is employed for the remaining Biot parameters. Therefore, in Step 3 the flow resistivity and the absorption coefficient are measured, the measurement principles are described in Section 2.4. Using these quantities, the remaining Biot parameters are inversely identified in Step 4. Therefore, the specimen is modeled using the Johnson–Champoux–Allard–Lafarge (JCAL) model, this procedure is described in detail in Section 2.5. Within this step, a data augmentation procedure is employed as well in order to enrich the database. Now both, the micro-scale geometry (the design parameters) and the Biot parameters are available. These data sets form the training data for the ML models (inputs/features: design parameters, outputs/labels: Biot parameters) that are set up and trained in Step 5. This is described in Section 2.6. Thereby, two different models, a K-Nearest Neighbor model and an Artificial Neural network, are employed. Having these models available, the design of porous materials (Step 6) becomes possible. Therefore, an inverse procedure is employed that is described in Section 3.3.

2.2. Specimen Design (Step 1)

For the additive manufacturing of porous absorber structures, a test specimen was developed within the scope of the research work, taking design for additive manufacturing regarding design potentials and limitations into account. Detailed descriptions of the specimens baseline design can be found in [26]. For the development of AM parts, a systematic development with the aid of different design tools is important to encourage the generation of ideas and to ensure a goal-oriented design [28,49,50]. However, the application of AM potentials must be considered in the early stages of product development, otherwise, the freedoms of AM cannot be fully utilized [51]. For this reason, the limitations and potentials of additive manufacturing with a focus on the porosity of the test specimens were considered during development to ensure additive manufacturability. In addition to the manufacturability of the test specimens by material extrusion, the focus is also on the possibility to vary the test specimens to be able to manufacture different variants that show only a small variation of the absorption properties. For this reason, a regular lattice structure based on a bar design was selected that functions as a porous structure.

This test specimen, as shown in Figure 2, is designed in layers and contains a defined number of bars per layer. The specimen design can generally be adjusted on the micro- and macro-scale. The micro-scale can be varied by different parameters, hereafter referred to as design variables. These are the bar width (d), the bar spacing (s), the bar height (h) and the plane angle (φ) and can be adjusted in a certain interval, see Table 1. Therein, the values are already chosen accordingly to the chosen AM process, here MEX. Table 1 shows that, for example, the bar spacing or the bar width can be varied in an interval of 0.10–0.50 mm, resulting in an increase or decrease of the bar number in the test specimen. All test specimens are printed with four outline shells resulting in a boundary of 0.80 mm with an extrusion width of 0.20 mm. The specimen diameter of 30 mm and height l = 15 mm describe the specimen on the macro-scale. These dimensions are kept constant throughout this work. The specimen height is chosen in order to balance out the required production time (proportional to the height) and the ability to generate a reasonable amount of absorption in the frequency range that is chosen for this work. The frequency range and specimen diameter are prescribed by the available impedance tube with 900–6600 Hz (see Section 2.4 for more details).

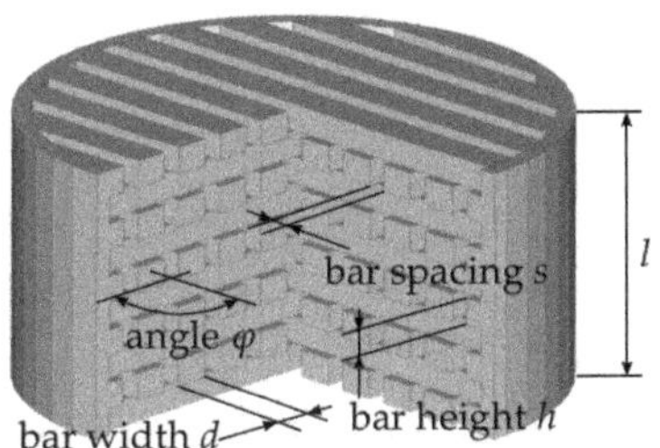

(**a**) Design variables of the investigated specimens

(**b**) Image of the investigated specimen 2

Figure 2. Overview of investigated specimen. (**a**) shows the test specimen including the design variables (from [26]) which are used to vary the test specimen. (**b**) shows the additive manufactured specimen 2.

Table 1. Overview of the design variables (see Figure 2a) and their variation ranges.

Bar Width (d)	Bar Spacing (s)	Bar Height (h)	Plane Angle (φ)
0.10–0.50 mm	0.10–1.00 mm	0.05–0.20 mm	$0°$–$90°$,

In the scope of this contribution, a large number of different test specimens are manufactured. They are expected to exhibit a broad range of varying acoustic properties to link or identify the relationship of the geometry parameters with the acoustic parameters using a machine learning approach. This way, acoustic parameters should be able to be determined during the design phase.

The parameter selection of the design variables is carried out within the parameter limits mentioned in Table 1 and employs a Design of Experiment (DOE) methodology to ensure the meaningful selection of the design parameters. The DOE-process employs a Latin Hypercube Sampling (LHS) [52–58] strategy for this purpose. Within the scope of the contribution, the sample size amounts to 50 porous absorber structures, whose parameter combinations can be found in Table A1 in the Appendix A. The data generation was done by using the software OpenSCAD, see also [59].

2.3. Additive Manufacturing of the Specimens (Step 2)

The acoustically effective porous structures were manufactured through material extrusion (MEX). Material extrusion processes offer good possibilities for the production of porous absorber structures since extremely filigree structures can be produced through changeable nozzle diameters. It is assumed that this procedure offers the highest potential that the best possible acoustic properties can be realized. However, the production of porous absorber structures also places special demands on the manufacturing process, since the use of fine nozzle diameters (down to 0.10 mm) in combination with the realization of thin walls, gaps and overhangs is at the limit of what is practically possible. For this reason, the selection and controllability of the process parameters are essential for the successful production of acoustically effective structures. If the ambient conditions (ambient temperature, humidity) or the process parameters (flow, cooling, retraction distance) are incorrect, manufacturing errors will occur, leading to dimensional deviation of the part or termination of production due to clogging of the nozzle. The material used is polylactic acid (PLA) from DasFilament (Emskirchen, Germany) to create porous absorbers [26,41]. PLA is particularly suitable for the production of microstructures by material extrusion because it has comparatively low shrinkage and thus low thermally induced stresses. In addition, PLA's lower printing temperature allows a fast cooling down of the extruded strands and makes it more suitable for parts with fine details such as overhangs and gaps, respectively, which is required for the bar design [60].

In addition, compared to other processes, such as PBF-P or VAT, there is the advantage that there is no material buildup due to sintering or unwanted polymerization due to capillary effects [27]. A total of 50 different test specimens were created as described in Section 2.2 which is similar to the amount of specimen in [47]. The resulting building times are approx. 120–240 min per specimen, depending on the number of bars and the layer height. The specimens were sliced using Simplify3D® (4.1.2, Simplify3D, LLC, Cincinnati, OH, USA, 2020) and manufactured on the X400, a pro-consumer additive manufacturing machine from German RepRap GmbH (Feldkirchen, Germany), whose original extruders were replaced by E3D Hemera (Direct Extruder, 24 V). A 0.20 mm Micro-Swiss nozzle (Ramsey, MN, USA) was used as the extrusion nozzle. The process parameters were used as listed in Table 2.

Table 2. Process parameters used in additive manufacturing.

	Nozzle Diameter 0.20 mm
Nozzle temperature (°C)	210
Bed temperature (°C)	60
Layer height [1] (mm)	0.05–0.11
Flow (%)	85
Extrusion Speed (mm/s)	36
Cooling (%)	40
Outline direction	Inside-Out
Extrusion width (mm)	0.15; 0.20; 0.25

[1] Increment of 0.005 mm.

Figure 3 shows the printing process of three specimens. The first grid layer was applied horizontally to the printer because this orientation showed less detachment from the build platform. The smaller bar widths (<0.20 mm) were achieved by targeted under extrusion because this ensured a more continuous printing process. With smaller nozzle diameters, the nozzle got clogged or the flow was not constant. Since all samples were to be produced with as far as possible the same process parameters, the 0.20 mm nozzle was chosen. The environmental conditions (temperature and humidity) were constant during the manufacturing of the test specimens and in addition, the material is dried at 45 °C for about 6 h before processing. To assess the manufacturing accuracy of the additively manufactured test specimens, an actual-target comparison of the initial geometry and the manufactured test specimen was carried out (see Section 3.1.1).

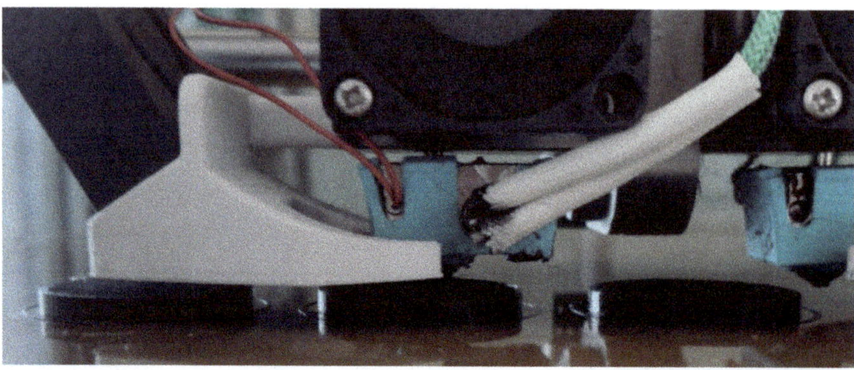

Figure 3. Additive manufacturing of three test specimens using material extrusion. Used 3D printer: X400, ppro-consumer additive manufacturing machine from German RepRap GmbH.

2.4. Acoustical Investigation of Specimen Population (Step 3)

The method presented within this work aims to train ML models to map between the geometry and the Biot parameters of porous samples. The geometry of all manufactured specimens is known from the manufacturing process. Indeed, this manufacturing process is somewhat uncertain, hence the manufactured geometry probably differs from the design parameters. Nevertheless, previous studies have shown that the manufacturing process applied here at least allows a sufficient precision for the purpose of this work [26]. Apart from the specimen geometry, the Biot parameters are of interest and have to be determined. However, only few of the Biot parameters can be measured directly and only the measurement of the flow resistivity is available to the authors, the remaining Biot parameters are inversely identified, a procedure already successfully applied in the past. Therefore, the absorption coefficient and the flow resistivity of all specimens are measured and serve as inputs to the inverse procedure. In the following sections, the measurements of the flow resistivity (Section 2.4.1) and absorption coefficient (Section 2.4.2) are described. The inverse procedure to estimate the Biot parameters from the measurements in described in Section 2.5.

2.4.1. Measurement of the Flow Resistivity

All 50 manufactured specimens, as described in Section 2.2, are measured in the laboratory regarding their flow resistivity and their absorption coefficient. For the measurement of the flow resistivity, the method with the alternating flow is employed according to ISO 9053-2:2020 [61]. The measurement setup Nor1517A of Norsonic (Oelde-Stromberg, Germany) is used (see Figure A1a). It comprises a vessel that is equipped with a sinusoidally moving piston for generating an alternating pressure in the vessel and a measurement microphone to measure the pressure. The vessel is closed by the specimen, hence a pressure drop over the specimen can be measured. As the movement of the piston generates a known air flow q and the pressure drop over the specimen Δp is measured by the microphone, the airflow resistance R can be directly evaluated by $R = \frac{\Delta p}{q}$. The flow resistivity Ξ is computed from the airflow resistance with the specimen's cross-sectional area A and the specimen thickness l by $\Xi = \frac{R A}{l}$. It should be noted that the specimen diameter used here is only 30 mm, whereas the measurement standard requires a diameter of 100 mm. To circumvent this difference, an adapter is manufactured and the measured values are corrected accordingly.

2.4.2. Measurement of the Absorption Coefficient

The absorption coefficient is measured in an impedance tube with a diameter of 30 mm using the two-microphone method according to ISO 10534-2:1998 [62]. The measurement device is an AED 1000 AcoustiTube (Dresden, Germany) (see Figure A1b). Thereby, the acoustic transfer function between the two microphones is evaluated while the system is excited by a broadband white noise signal. From the transfer function, the complex reflection factor of the specimen surface can be evaluated and the absorption coefficient can be computed from the reflection factor. Due to the measurement principle, below the so-called cut-on frequency (here: 6600 Hz) that itself relates to the tube's diameter, only plane waves travel through the tube. Thus, the presented absorption coefficients within this work are valid for normal incident plane waves only. As the cut-on frequency gives the upper limit for the possible frequency range, the lower limit is determined by the microphone distance (here: 20 mm) and results for the setting used here to 900–6600 Hz. Measurements in the frequency range below 900 Hz are assumed here to be unnecessary, as the absorption coefficient of a specimen with a height of $l = 15$ mm is expected to be negligibly low.

2.5. Inverse Parameter Identification Using the JCAL-Model (Step 4)

Based on the measured flow resistivity and absorption coefficient data described in Section 2.4, the remaining Biot parameters of the porous specimens are inversely identified.

This inverse procedure is common in the literature and has been applied often in the past, for example [23,26]. Therefore, the absorbers are modeled using the JCAL model [19–21]. The mathematical form of the model is used as given in [63] and shown in the appendix, see Appendix D. The JCAL model belongs to the class of equivalent fluid models, as the skeleton phase of the two-phase (fluid-filled pores and solid skeleton) system is assumed to be rigid. In former studies, the Biot model had been employed directly within this parameter identification schema [26], using an in-house finite-element code capable of using Biot's model for numerical computations [64]. Nevertheless, numerical experiments have shown that the skeleton's elasticity of the specimens used here can be neglected and the skeleton can be modeled as rigid. This allows the application of models belonging to the class of equivalent fluids, here the JCAL model. The JCAL model is a six parameter model, featuring the following parameters: the flow resistivity Ξ, the porosity ϕ, the high-frequency limit of the tortuosity α_∞, the viscous and thermal characteristic lengths Λ, Λ' and the static thermal permeability k'_0.

The JCAL model allows the computation of the equivalent density $\tilde{\rho}$ and the equivalent bulk modulus $\tilde{K}$ of the equivalent fluid that represents the porous material, see Equations (A1) and (A2), respectively. Using these quantities, the characteristic impedance $\tilde{Z}_c$ and the complex wavenumber $\tilde{k}$ can be computed using the Equations (1) and (2) from [23], respectively.

$$\tilde{Z}_c = \sqrt{\tilde{\rho}\,\tilde{K}} \tag{1}$$

$$\tilde{k} = \omega\sqrt{\frac{\tilde{\rho}}{\tilde{K}}} \tag{2}$$

With the quantities, $\tilde{Z}_c$ and $\tilde{k}$, the surface impedance Z_s of the porous material in front of a rigid impervious wall can be computed for the case of normal incident plane waves using Equation (3) from [65]. Thereby, the thickness of the specimen l is used, $j = \sqrt{-1}$.

$$Z_s = -j\frac{\tilde{Z}_c}{\phi\,tan(\tilde{k}\,l)} \tag{3}$$

Here, it becomes clear why the prediction of the Biot parameters is preferred over the direct prediction of the absorption coefficient. The Biot parameters (or the subset used here for the JCAL model) are material-specific parameters. The acoustic behavior of the material for the case of a material mounted in front of a rigid wall is accounted for in subsequent computation. Hence, the process proposed here finally allows the description of the material at hand, independent of the mounting conditions or the sample thickness. Using the surface impedance Z_s and the characteristic impedance of the surrounding fluid Z_0, the absorption coefficient α of the porous material can be computed:

$$\alpha = \frac{4\,\mathrm{Re}(Z_s)\,Z_0}{(\mathrm{Re}(Z_s) + Z_0)^2 + \mathrm{Im}(Z_s)^2} \tag{4}$$

Using the Equations (1)–(4), (A1) and (A2), the absorption coefficient of a material can be computed for a given set of the six Biot parameters used for the JCAL model. Based on these formulations, an inverse parameter identification procedure is set up using an evolutional algorithm (settings: see Table A2). The updating process is sketched in Figure 4.

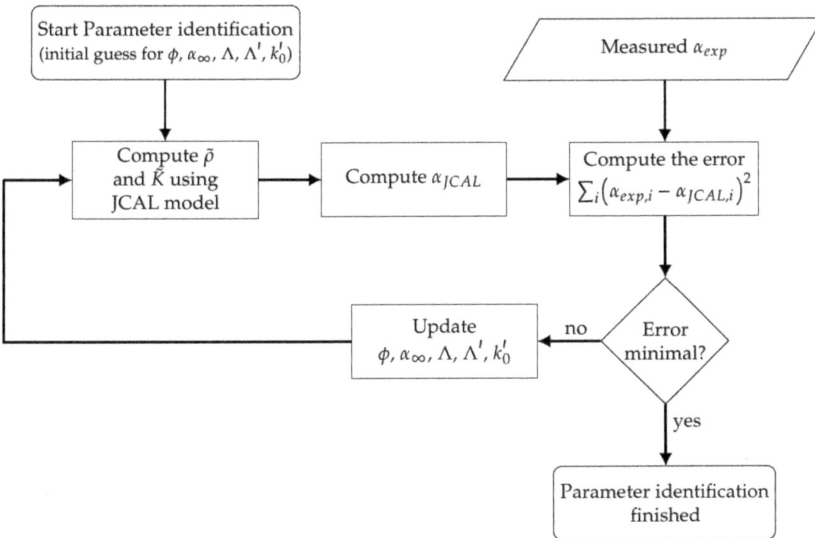

Figure 4. Flowchart of the inverse parameter identification process. Starting with an initial guess for the Biot parameters, the absorption coefficient is computed using the JCAL model and the results is compared to the measured data. Based on the error measure, the Biot parameters are updated until convergence is reached.

As described before, the flow resistivity is determined experimentally, the remaining five parameters (ϕ, α_∞, Λ, Λ', k_0') are inversely identified. Therefore, the absorption coefficient is computed (α_{JCAL}) for a first initial guess of the parameter set. The error of the computed absorption coefficient α_{JCAL} and measured absorption coefficient α_{exp} is determined using the sum of the squared differences for all frequencies i:

$$err = \sum_i \left(\alpha_{exp,i} - \alpha_{JCAL,i}\right)^2 \quad . \tag{5}$$

Based on the error measure given in Equation (5), the parameter set is updated using an evolutional algorithm as described in [66]. The objective function to be minimized yields the error measure given in Equation (5). The implementation of the evolutional algorithm is taken from the python-library scipy [67], the specific configuration is shown in Table A2 in the Appendix C.

An evolutional algorithm is a stochastic approach, hence several runs of the algorithm will yield somewhat different results, even though the initial parameter set remains unchanged. This can be due to different minimal solutions that are found. However, even if the same minimum is found in different runs, the stochastic nature will produce different results within small limits. This behavior is used here to enlarge the database (data augmentation) of the manufactured 50 specimens. The inverse parameter identification procedure is run ten times for each specimen, hence the database that can be used for training the ML models comprises 500 data sets with slightly different geometry-Biot parameter combinations. This behavior is due to the fact that the inverse parameter identification problem is ill-posed. Therefore, without additional information, it cannot be distinguished between the different results and whether one or the other outcome is the physically correct result and hence, all available results are used. In Figure A2, in the Appendix C, a correlation plot can be found. It shows the inversely estimated porosity of the samples over an analytical estimate that is computed with

$$\phi_{analyt} = \frac{s}{s+d} \quad . \tag{6}$$

It can be seen, that a general correlation (correlation coefficient $\rho = 0.63$) of the inversely identified parameter and the analytical estimate can be found. The occurring differences thereby could be attributed to several reasons, the rough analytical estimate, manufacturing imperfections as well as the inverse parameter identification procedure. Therefore, in general, the procedure is trusted. In order to further investigate the data augmentation procedure, the standard deviation of the ten parameter identification runs is computed for each specimen and Biot parameter. The results can be seen in Figure A3 in the appendix (Appendix C). The standard deviation is normalized with the mean of the identified parameters. It can be seen that the variations of most parameters have a magnitude of approx. 1×10^{-1}. Therefore, it is assumed that indeed some differences occur for each run but since the deviations are rather small, the general behavior is kept. Thus, it is assumed that the data augmentation yields meaningful results to enrich the database and the choice of input parameters is reasonable.

2.6. Machine-Learning for the Geometry-Biot Parameter Relations (Step 5)

Since Machine-Learning is a vast and still rapidly growing field, hereafter only a very short sketch of the idea and its application for engineering tasks is given. For further information, it can be referred to various textbooks, such as [68,69]. To properly explain the ML approach, it might be handy to first recall the classical model building process, as sketched in Figure 5 and described in, for example [70]:

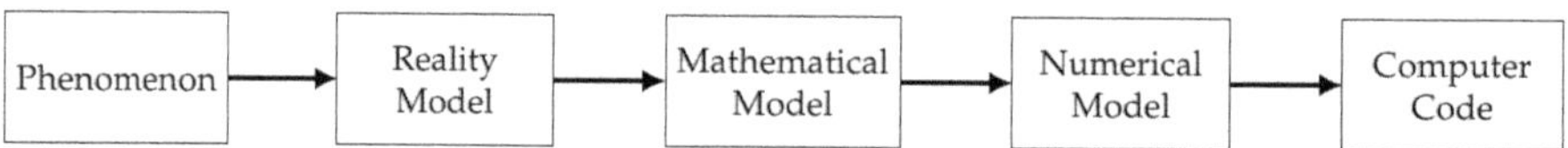

Figure 5. Classical model building process (from [70]).

This classical approach starts with an event or phenomenon happening in reality that is observed and, on a first level, is transformed into a so-called reality model. This reality-model is then described using mathematical equations, yielding the mathematical model which, in turn, is solved—often using numerical methods that lead to the numerical model—by means of a computer. Within this classical approach, the basic relation, which is "what the model should do" is coded within the mathematical equations of the mathematical model. These mathematical equations are themselves based on the laws of physics (or any other discipline). The basic idea of ML is, in contrast to this classical model building approach, that the relations between the model parameters are not described based on physical laws and mathematical equations, but that the relations are derived from existing data. This process of deriving the relations from data often is referred to as learning or training the ML model. Thereby, it is of tremendous importance that the desired relations are already "coded" within the existing data. The field of ML models can generally be divided into the sub-classes of supervised and unsupervised learning models. Thereby, supervised learning employs both, the input data for the model and so-called labels, which is the desired output of the model. On the other side, unsupervised learning models require the input data only and attempt to find some sort of structure within this data, for example, clustering.

Within the scope of this work, two different types of ML models are employed: (a) a K-Nearest Neighbor (KNN) model, see Section 2.6.2, and (b) an Artificial Neural Network (ANN) model, see Section 2.6.3. Both models belong to the class of supervised learning models and are applied here as such, whereas both models can be used for unsupervised learning as well. The reason for choosing these two different model types here is to allow for a comparison of different approaches regarding the necessary effort for the model training and the ability to predict a suitable set of geometry parameters for the AM process. The models differ significantly regarding the required computational effort for the training (few seconds for the KNN model, $\approx$25 min for the ANN model). Moreover, it is assumed that the models show a different behavior regarding the ability to inter-/extrapolates from the learned data, which is assumed to be crucial for a reasonable prediction of the Biot parameters for new specimen geometries. Such differences were already observed in the past on other problem types, therefore these models were applied here as well. Other ML model types have not been tested yet but are subject to future studies. The models are implemented using the python-library scikit-learn [71].

2.6.1. Scaling of the Input and Output Data

The input data for the ML models are geometry parameters and, for the bar width, the bar spacing and the bar height measured in mm. The plane angle in measured in $^\circ$. Except from the plane angle, the magnitude of the input data is approximately 1×10^{-1}, the magnitude of the plane angle is in the order of 1×10^{1}. The input data is scaled using mean and standard deviation of the data with:

$$x^\star = \frac{x - \overline{x}}{y} \ . \tag{7}$$

In Equation (7), the quantity $x^\star$ represents the scaled input data, $\overline{x}$ is the mean and y represents the standard deviation of the input data.

The Biot parameters, which is the output data of the ML models, have very different dimensions and scales, for example, the viscous and thermal characteristic lengths are geometric lengths, the static thermal permeability has the dimension of a surface, the porosity and tortuosity are dimensionless and the flow resistivity has the dimension $\mathrm{Pa\,s/m}^2$. Furthermore, the magnitude of these parameters is rather different and comprises a large range of values; for example, the magnitude to the flow resistivity is $1 \times 10^{3} - 1 \times 10^{5}$, the magnitude of the static thermal permeability is approx. 1×10^{-8}. It could be found that the training process becomes rather unsuccessful when the input and output data is used directly. This is expected to be a result of the large range of values within the data. Therefore, variable scaling is introduced and applied to the Biot parameters in order to result in training data comprising a smaller value range. The resulting scaled Biot parameters are shown in Table 3:

Table 3. Scaling of the Biot parameters to balance out the different magnitudes.

Biot Parameter	Scaling	Biot Parameter	Scaling
Ξ^{1}	$\Xi^\star = \Xi$	ϕ^{2}	$\phi^\star = 1 \times 10^{3}\, \phi$
$\alpha_\infty{}^{3}$	$\alpha_\infty^\star = 1 \times 10^{3}\, \alpha_\infty$	Λ^{4}	$\Lambda^\star = 1 \times 10^{6}\, \Lambda$
Λ'^{5}	$\Lambda'^\star = 1 \times 10^{6}\, \Lambda'$	$k_0'^{\,6}$	$k_0'^{\,\star} = 1 \times 10^{12}\, k_0'$

[1] flow resistivity , [2] porosity, [3] tortuosity, [4] thermal characteristic length, [5] viscous characteristic length, [6] static thermal permeability

The flow resistivity remains unscaled. This scaling is assumed to only improved the training performance of the ML models but does not affect the general procedure.

2.6.2. K-Nearest Neighbors

K-Nearest Neighbor (KNN) ML models are a very popular approach and have been successfully applied to classification problems. For further information on the topic, the reader is referred to the the vast amount of textbooks and papers on this topic, such as [69,72–74]. For the classification, the labels of the existing data serve as the classes into which new data instances shall be classified during the application of the model. As inputs, KNN models use a feature vector for each data instance. Now a class is formed by all input data instances, whose feature vectors are similar to each other. During the application of the model, new data instances with their individual feature vectors are compared to the feature vectors of the classes the model initially had learned and the new instance is classified into that class to which its own feature vector is most similar. The similarity thereby is measured using certain distance measures that are specified by the user. Another parameter that determines the behavior of the model is the number of neighbors (k) that are accounted for during the classification process. Due to expected noise within the data, it is common to compare new data instances to more than one instance. The parameter k determines the number of data sets to which the new data instance is compared. The distance measure and the number of neighbors are the so-called hyperparameters of the ML model. The appropriate choice of these hyperparameters is crucial for a successful application of ML models. KNN models can be applied to regression problems as well. In such cases, the output quantity is not a single and discrete class but is computed as a continuous quantity from the k neighbors [69].

The KNN model employed within this work uses a k-parameter of $k = 5$ and weights the different neighbors of the new data instance based on the inverse of its distance to the learned data instance. The choice of the k-parameter results from the hyperparameter tuning that was done on the validation data set. In Figure A4, the results of the cross-validation with different k-parameters are shown. It can be seen, that the cross-validation score is improved from $k = 2$ to $k = 5$ but even higher values for k do not further improve the score. Therefore, $k = 5$ is chosen. The validation and test scores for the KNN model can be found in Table 4. The model is used as a multi-output regression model.

Table 4. Training and test scores of the KNN and ANN model; uncertainty measure is two times the standard deviation of the cross-validation process (equiv. to approx. 95% confidence interval).

R^2-Score	KNN [1]	ANN [2]
cross-validation	0.80 ± 0.10	0.73 ± 0.13
test on unseen data	0.76	0.80

[1] K-Nearest Neighbor, [2] Artificial Neural Network

2.6.3. Artificial Neural Network

The approach of the Artificial Neural Networks is based on the idea of the so-called perceptron [75,76]. A perceptron can be understood as a unit that has input channels taking a data vector and yielding a certain scalar output based on the input data vector. Therefore, a weighted sum of the input vector elements is computed and the result of this sum serves as the argument to the so-called activation function. This activation function then computes the perceptron's output and can yield both, discrete or continuous values. The most simple category of ANNs are multi-layer perceptrons, for which a multitude of single perceptrons (or neurons) are organized in several layers and interconnected to each other, so the output of one neuron is the input to one or more subsequent neurons in the following layer. This connection between the layers is further associated with a certain weight, the weights of each connection within the ANN is adjusted during the training process. Therefore, several optimization algorithms can be used. This is necessary since the weight adjustment problem is ill-posed. The main hyperparameters of the ANN thus are the topology (number of neurons/layer and number of layers), the activation function(s) and the optimization algorithm. One of the main advantages of multi-layer perceptron neural networks over single perceptrons is the ability to learn and reproduce non-linear functional relations. Indeed, much more complex versions of this multi-layer perceptron prototype exist today, for example, recurrent neural networks, convolutional neural networks and many more [69,77].

The ANN employed within this work is set up as a four-layer feed-forward ANN with two hidden layers. The input and output layers have 4 and 6 neurons, respectively, according to four geometry parameters that serve as inputs and the six Biot parameters Ξ, ϕ, α_∞, Λ, Λ' and k'_0 as outputs. The hidden layers employ 4000 and 100 neurons, respectively. The activation function used for all neurons is the so-called Rectified Linear Unit (ReLU), the optimization algorithm for adjusting the weights is the Adam algorithm. The resulting ANN model is sketched in Figure 6.

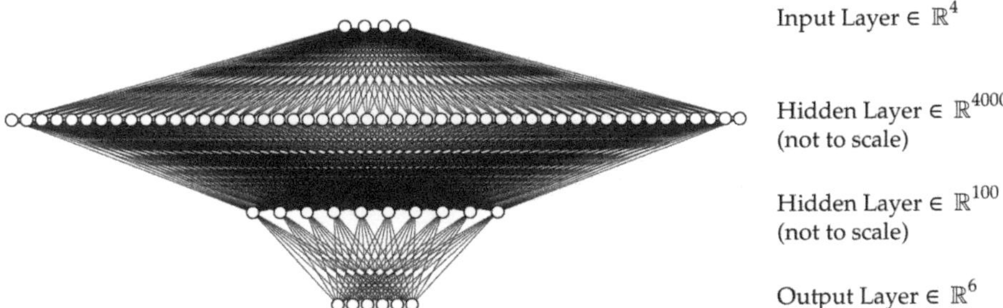

Figure 6. Sketch of the ANN used within this work, drawn using [78]. The input layer has four neurons and takes the geometry parameters w, s, h, φ of the specimen. The output layer has six neurons for the Biot parameters Ξ, ϕ, α_∞, Λ, Λ' and k_0'. The two hidden layers have 4000 and 100 neurons (not to scale), respectively. The activation function for all neurons is the ReLU function.

The hyperparameters (the number of layers, neurons, activation function and optimization algorithm) are derived during a 3-fold cross-validation procedure. Thereby, the topology mentioned above (4-4000-100-6) reached the best validation score, as shown in Figure A4. The ReLU activation function and the Adam optimization algorithm were the only choices with which a converging training process could be reached. The validation and test scores for the ANN model can be found in Table 4.

3. Results and Discussion

Within this section, the obtained results of the work are presented. In the first part (Section 3.1), the manufactured specimen are investigated regarding their actual geometry and their acoustic parameters (Steps 1–3 of Figure 1). Therefore, in Section 3.1.1, an optical inspection of two specimens is presented. These specimens are found to show the highest and lowest mean absorption coefficient over the entire frequency range and thus represent some kind of extreme values. The acoustic inspection is presented in Section 3.1.2. Here, the measured flow resistivity and absorption coefficient are presented. The results of the inverse parameter identification procedure are shown in Section 3.1.3 (Step 4). The second part (Section 3.2) of this section is devoted to the results of building the ML models (Step 5). Thereby, the results of the training processes are shown and the accuracy of the predicted Biot parameters is assessed for two specific samples. The obtained ML models are applied in Section 3.3 for the design of new absorber specimens that are intended to show a specific prescribed absorption coefficient over frequency (Step 6).

The overall idea of the procedure within this work is to generate models that connect the specimen geometry to the Biot parameters. The reason for choosing the Biot parameters (which are material-specific parameters) rather than directly predicting the behavior of the material, for example, the absorption coefficient, is that the Biot parameters can be used for subsequent analyses and other contexts. This is addressed in Section 3.4 by manufacturing a new specimen with a varied specimen height l and prediction of the absorption coefficient for a broader frequency range. Height and frequency range were kept constant for the initial specimen population with which the Biot parameters are identified. Now the Biot parameters that were obtained by the ML models are used to predict the absorption coefficient of a specimen with a larger specimen height and for a broader frequency range. This shows, that the obtained models are suitable as well to design porous absorbers with a different geometry.

3.1. Inspection of the Specimen Population

Within this section, the results of the inspection of the initially manufactured specimen population are shown. Thereby, in Section 3.1.1 the results of a first optical inspection are presented and the generally correct AM process is verified. These results (the specimen micro-scale geometry) serve as inputs (features) to the ML models. In Section 3.1.2, the results of the acoustic measurements are presented, which are the measurements of the flow resistivity and the absorption coefficient. In Section 3.1.3, the results of the inverse identification of the Biot parameters are shown. These Biot parameters serve as outputs (labels) for the ML models, whose prediction performance is presented and assessed in Section 3.2.

3.1.1. Results of the Optical Investigation of the Additively Manufactured Specimens

To assess the manufacturing accuracy of the additively manufactured test specimens, an actual-target comparison of the initial geometry and the manufactured test specimen was carried out. Specimens 2 and 39 are selected as examples since these specimens show the highest and lowest mean absorption coefficient over the frequency range of interest (see Figure 7b) and thus provide some kind of extreme values.

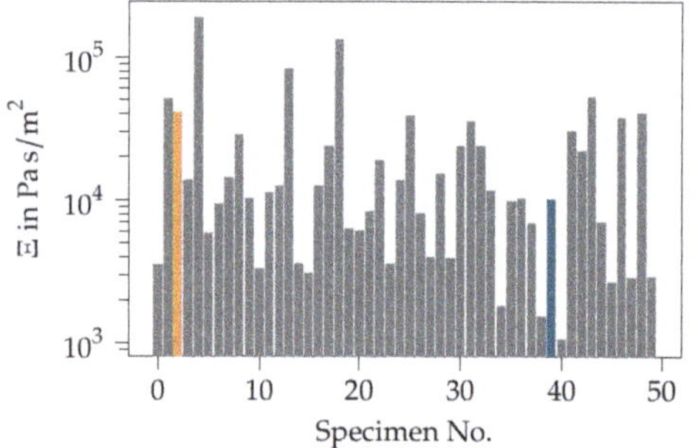

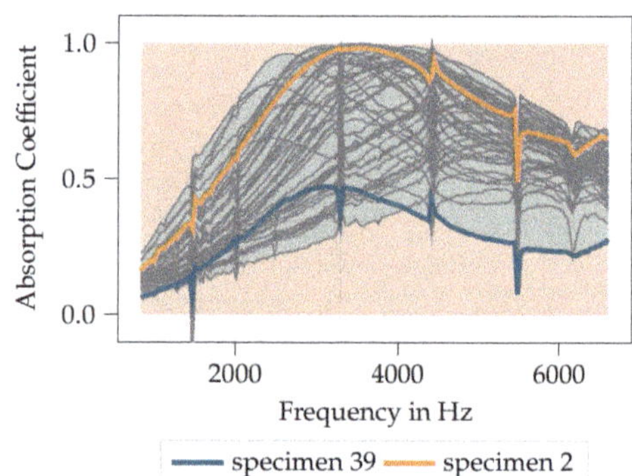

(**a**) Overview of all flow resistivity measurements for the specimen population of 50 specimens. It can be seen that specimen 2 and 39, which represent the highest and lowest mean absorption coefficient do not as well represent extreme values for the flow resistivity.

(**b**) Overview of all conducted absorption coefficient measurements for the specimen population of 50 specimens. The green and red areas indicate data ranges for which specimens are/are not already available.

Figure 7. Measured data (flow resistivity and absorption coefficient) of all 50 investigated specimens. The specimens 2 and 39 are marked in orange and blue, respectively, and represent those specimens with the highest and lowest mean absorption coefficient in the frequency range.

Test specimens 2 and 39 are generated using the variation parameters given in Table 5 using OpenSCAD [59] and should have the specified values after additive manufacturing. Manufacturing accuracy is assessed through the use of optical metrology. Specifically, a Keyence VR5200 (Neu-Isenburg, Germany) 3D surface profilometer with microlens was used. Figure 8 shows the test specimens 2 (Figure 8a) and 39 (Figure 8b) at 80× magnification.

Table 5. Nominal geometry parameters of the specimens.

	Bar Width (d)	**Bar Spacing (s)**	**Bar Height (h)**	**Angle (ϕ)**
Specimen 2	0.40 mm	0.40 mm	0.08 mm	60.00°
Specimen 39	0.40 mm	0.80 mm	0.15 mm	70.00°

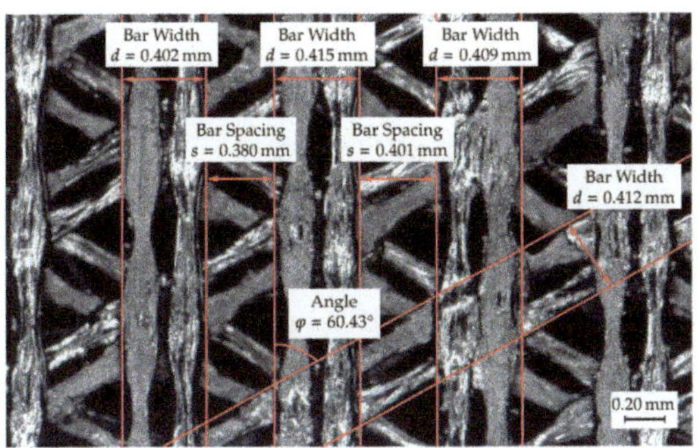

(**a**) Optical investigation of Specimen 2

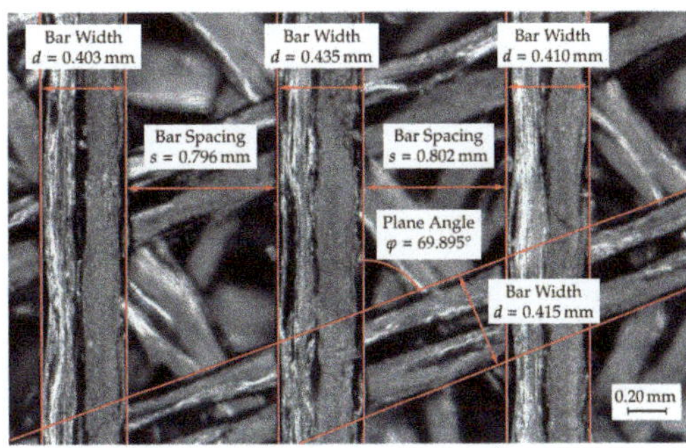

(**b**) Optical investigation of Specimen 39

Figure 8. Actual-target comparison of specimens 2 and 39 using Keyence VR5200 3D surface profilometer. (**a**) shows specimen 2 with only slight deviations in the actual-target comparison. However, some small tapers can be seen in the filament strands. (**b**) shows specimen 39, which also shows only minimal deviations in the actual-target comparison.

Test specimen 2 shows a relatively uniform production pattern, with a partial tapering of the filament strands in the spaces between the underlying bars. Two applied filament strands each form the individual webs of the acoustically effective structures, the width of which is 0.40 mm for test specimen 2. Evaluation through optical measurement technology showed that the resulting width of the bars lies within a range of 0.402–0.415 mm, which corresponds to a maximum deviation of roughly 4%. In addition to the bar width, the bar spacing was also examined, which should also be 0.40 mm. In contrast to the bar width, a small variation of the values can also be measured here. The resulting bar spacing have a value between 0.38 and 0.401 mm, which corresponds to a maximum deviation of roughly 5%. The resulting plane angle between the individual bar structures should have a value of 60°, which was achieved with a value of 60.43° with very small deviations (<1%).

A similar manufacturing analysis was also carried out for test specimen 39, see Figure 8b, by also determining the bar width, bar spacing and plane angle. In contrast, to test specimen 2, the applied filament strands are much more homogeneous in their width and only minor tapers in the bar width can be measured. At maximum, a bar width of 0.403–0.435 mm could be determined, which corresponds to a maximum deviation of roughly 9%. However, the vast majority of the bars show a much smaller deviation from the required 0.40 mm. The bar spacing of the test specimen 39 has a value between 0.796 and 0.802 mm, which corresponds to a maximum deviation of <1% of the required 0.8 mm. The defined plane angle of 70° was also satisfactorily achieved with a value of 69.895°.

All in all, the manufacturing accuracy can be classified as acceptable and only slightly fails to meet the desired dimensions. The highest measured deviation is at a bar width of 0.435 mm and thus corresponds to a deviation of 0.035 mm. However, the vast majority of the tested bars show significantly smaller deviations. Test specimen 2 also exhibits only minor deviations from the desired geometry. Clarification is needed for the partial tapering of the filament strands in Specimen 2, which may be due to insufficient thermal energy or slow cooling of the strands, resulting in elongation of the filament strands and partial tapering.

3.1.2. Results of Acoustic Measurements of the Specimen Population

The results of the measurement of the flow resistivity are shown for all 50 specimens in Figure 7a. It can be seen, that flow resistivity values in the range of 1535.90–51,061.20 Pa s/m^2 are observed. Specimen 2 and specimen 39 are marked in orange and blue, respectively.

These specimen exhibits the largest (specimen 2) and lowest (specimen 39) mean absorption coefficient over the frequency range, as shown in Figure 7b. Nevertheless, the flow resistivity alone does not seem to explain the acoustic behavior since other specimen with even higher (for example specimen 5) and lower (specimen 41) flow resistivities exist whose mean absorption coefficient lies between the ones of specimen 2 and specimen 39.

In Figure 7b, all absorption coefficient curves of all specimens are plotted over frequency. In this figure, no distinction between the different specimens is made, only the curves of specimen 2 and specimen 39 are highlighted. These curves are the curves with the highest/lowest mean absorption over the entire frequency range. All other absorption curves show a mean absorption coefficient between these two highlighted curves, however, the spectral distribution might change. Furthermore, a red and a green area is marked. These areas indicate where no data is obtained from the entire specimen population. They become relevant for the discussion of the performance of the ML-models for the absorber design in Section 3.3. For the majority of the specimens, an absorption coefficient curve with a distinct maximum can be seen. This maximum can be found in a frequency range between approx. 2800 and 4200 Hz. Classical porous materials like foams are expected to show a rather broad and flat course of the absorption coefficient for high frequencies. Nevertheless, the absorption maximum for the specimen used here could already be flattened in comparison to former studies, in the sense that the frequency range with high absorption characteristics gets broader. These measured absorption curves are used for the inverse parameter identification procedure described in Section 2.5 for defining the target function.

The absorption coefficient curves shown in Figure 7b show some "spikes" or "kinks" at specific frequency points. The reason for this behavior could not be revealed by now. Since all "spikes" are located at the same frequencies and occur for all measured specimens, it is assumed that the "spikes" result from a systematic error, probably in the transfer function computation of the applied two microphone method. Within this procedure, the measured transfer function is corrected for the microphone phase mismatch using a transfer function measured prior to the measurement campaign. This correction procedure might be ill-conditioned for the distinct frequency points. Therefore, the results away from the spikes are supposed to be trustworthy.

3.1.3. Results of the Inverse Parameter Identification

Based on the measured flow resistivity and absorption coefficient, the Biot parameters are inversely derived using the procedure described in Appendix C. In order to give a first insight into these data, a correlation analysis between the flow resistivity Ξ and the design variables is conducted. The variation of the design parameters of the specimen are expected to change the flow resistivity, therefore the plot in Figure 9 shows the flow resistivity of all specimens over the geometry parameters, these are the bar width (top left), the bar height (top right), the bar spacing (bottom left) and the plane angle (bottom right). Moreover, the correlation coefficient ρ (throughtout the manuscript, Pearson's correlation coefficient $\rho = \frac{cov(X,Y)}{\sigma_X \sigma_Y}$ is used. Here, both the flow resistivity and the geometric quantities are assumed to be the random variables X and Y, respectively) for the two quantities are computed and given in the plots.

For the bar width, (see Figure 9, top left) the stages of the LHS sampling strategy can be seen, values from 0.20 to 0.50 mm with a step width of 0.10 mm are used. Thereby, the flow resistivity generally increases with the bar width, resulting in an intermediate positive correlation of $\rho_d = 0.52$. A correlation of similar magnitude, but in the opposite direction can be detected for the bar spacing, see Figure 9, bottom left, with $\rho_s = -0.56$. Here the flow resistivity decreases with increasing bar spacing. Both correlations of the bar width and the bar spacing to the flow resistivity seem reasonable—increasing the bar width decreases the open channel area, increasing the bar spacing enlarges the open area. On the other side, the bar height (see Figure 9, top right) and the plane angle (see Figure 9, bottom right) show only a very weak correlation to the flow resistivity. In total, the plots show that no trivial dependency can be found to adjust the specimen dimensions properly in order to tailor their absorption characteristics. This finding further motivates the application of ML-models to design absorbing structures.

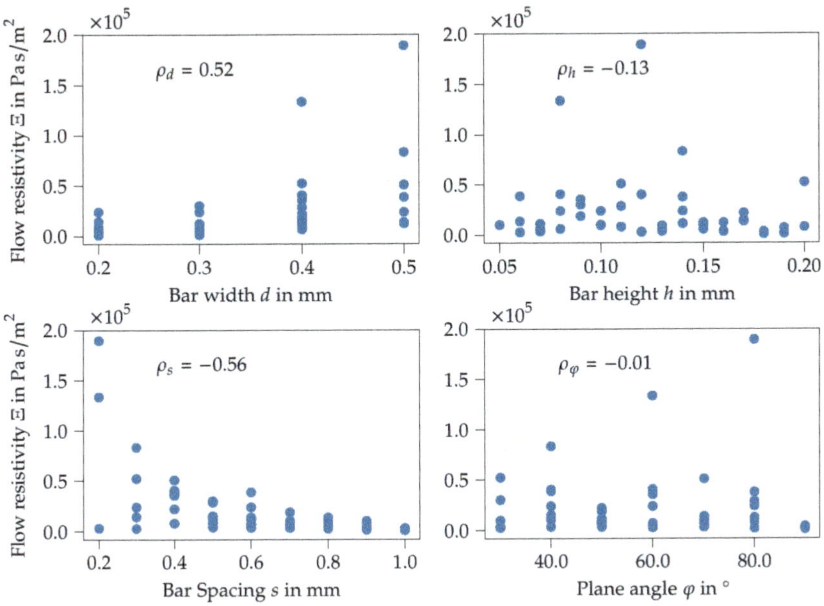

Figure 9. Correlation of the flow resistivity and the design variables of all manufactured specimens. It can be seen that only the bar width and the bar spacing show a relevant correlation with the flow resistivity, whereas the flow resistivity does not (linearly) depend on the bar height and plane angle.

3.2. Setup of Machine-Learning Models for Predicting Biot Parameters from the Specimen Geometry

The training process for both ML models employed within this work is composed as follows. The 500 available data sets generated from the genetic algorithm during the inverse parameter identification are split into two sub-sets, a training and validation data set (A) and a test data set (B). The test data set comprises 20% (100 data instances) of the available data and is used for the performance test at the end of the training process, only. The test data set is chosen randomly from the available 500 data instances. The hyperparameters, namely the number of nearest neighbors and the distance measure of the KNN model and the number of layers/neurons and activation function of the ANN model, were found using a three-fold cross-validation schema [69]. Therefore, the remaining 80% (data set A, comprising 400 data instances) of the available data is split into three parts of equal size (133 data instances), from which two (266 data instances) are used for training the model (training data set) and the third for evaluating its performance (validation data set). This procedure is run three times, so all elements of the data set A have been used at least once for training and testing, respectively. The performance is measured in terms of the cross-validation score, this is the mean of all scores during the cross-validation. The cross-validation score, as well as the test score, are measured with the so-called coefficient of determination R^2 [79]. This quantity measures the amount of variance explained by the model and thus is often used for measuring the prediction accuracy of dependent variables from independent variables [80]. The R^2 measure normally varies between 0 and 1, whereas $R^2 = 1$ refers to a perfect prediction by the model, $R^2 = 0$ refers to a model always predicting the mean of the data. Under specific conditions, values of R^2 below 0 or above 1 can be obtained as well.

The cross-validation procedure is run several times with varying hyperparameter in order to find parameters that perform well (highest R^2-score) for the given problem. Figure A4 shows the results of this hyperparameter tuning for both, the KNN and the ANN model, respectively. It can be seen, that for the KNN model a k-parameter of larger than $k = 5$ does not lead to a further improvement of the results, therefore $k = 5$ is chosen. For the ANN model, different topologies were investigated with at least two and a maximum of four hidden layers. From Figure A4 it can be seen, that the model with the 4-4000-100-6 topology performed best, therefore this model is used for all investigations. The hyperparameters activation function (ReLU) and optimization algorithm (Adam) were the only ones with which a converging solution could be obtained. Therefore, these are used as well.

After fixing the ML model design using the chosen hyperparameters mentioned above, the models are tested using the remaining share of the data for estimating their performance with regard to yet unseen data. This performance is measured using the test score. The resulting scores are summarized for both models in Table 4.

From Table 4 it can be seen, that both models reach training and test scores above 0.70 whereas the KNN model performs better during the cross-validation with a higher score and lower uncertainty. Nevertheless, the ANN model performs better on the unseen test data. This may be due to the fact that before the test procedure, both models have been trained with all data sets from the cross-validation, thus more information was available to train the model. Another effect might be that it is suspected that ANN models have a better ability to extrapolate from know to unseen data than other models. Furthermore, it should be noted that the KNN model requires only a few seconds of computing time on an intermediate laptop for the training process, whereas the ANN model requires a computation time of approx. 25 min. To summarizing, the models both show a reasonable performance in terms of prediction accuracy whereas the required computational effort for the training is very different. Nevertheless, since for the purpose of this work the focus is on the prediction capability rather than computational effort, both models seem to be reasonably applied to the absorber design procedure.

3.2.1. Performance of the KNN Model for Predicting Biot Parameters

In Figure 10, the performance of the KNN model is illustrated. Therefore, the blue line shows the absorption coefficient that is computed using the Equations (1)–(4), (A1) and (A2) with the inversely identified Biot parameters. The orange line indicates the computed absorption coefficient using the mentioned set of equations if the Biot parameters computed using the KNN model are used. For comparison, the green line shows the measured absorption coefficient. The plot Figure 10a shows the results for specimen 39, the specimen with the lowest mean absorption coefficient. It can be seen that the Biot parameters, computed using the ML model (orange), practically perfectly approximate the inversely identified parameters (solid line) with regard to the subsequently computed absorption coefficient. Both approaches match the measured data to a reasonable extend, nevertheless some deviations can be observed around the maximum of the absorption coefficient at approx. 3600 Hz.

Similar findings can be observed for the specimen with the highest mean absorption coefficient, specimen 2, see Figure 10b. For this specimen, all shown curves match nearly perfectly. Hence, the inversely identified Biot parameters resemble a good approximation of the measured absorber and the trained ML model is able to reproduce the Biot parameters from the geometry pretty well.

An overview of the performance of the KNN model over the entire training data is given in Figure 11. The six shown plots refer to the six parameters of the JCAL model, for each plot the test data is plotted in the horizontal axis, the prediction of the model is plotted in the vertical direction. A perfect training result would produce only datum points on the solid black line, the prediction would fit the test data perfectly. For each quantity, Pearson's correlation coefficient ρ is given in the plot's title. It can be seen that for the flow resistivity, see Figure 11 top left, a practically perfect fit between the test data and the ML model can be obtained. For the porosity (top middle), the tortuosity (top right) and the viscous characteristic length (bottom left) correlation coefficients above 0.90 can be obtained. Nevertheless, some outliers can be observed as well, for example for the thermal characteristic length (bottom middle), two data points show a value of approximately 0.00 m for the test data, but are predicted with values of 0.55×10^{-3} m and 0.70×10^{-3} m, respectively. It is expected, that such outliers occur when for these specific specimens the combination of the different geometry parameters has a negligible influence on the acoustic parameters. This rather low influence results in a low potential to extract information about the relation between the parameters and thus results in wrong training results. In total, the predictions of the KNN model are in good agreement with the input data, hence the training process is assumed to be successful.

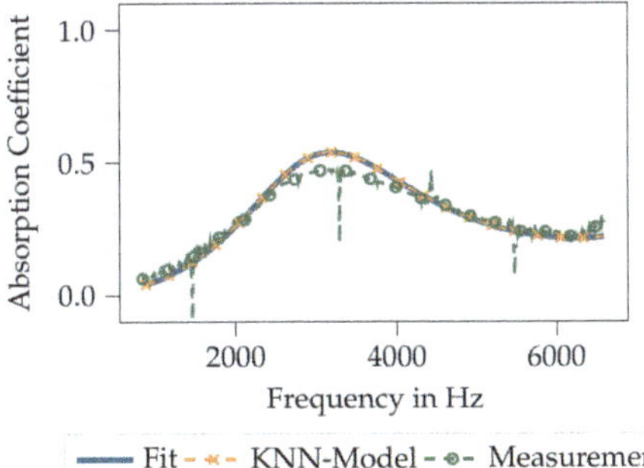

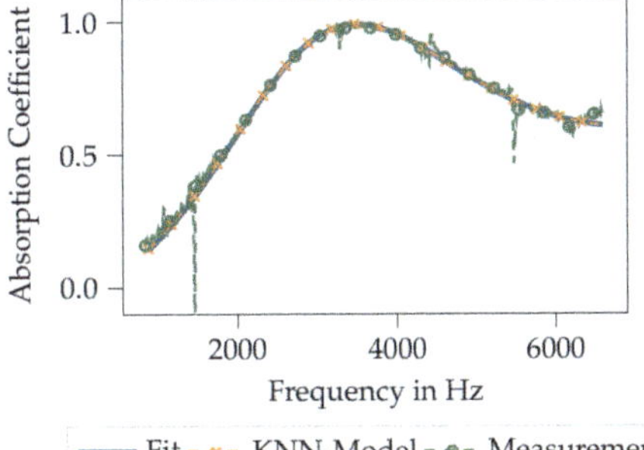

(a) Comparison for specimen 39 with lowest mean absorption coefficient. It can be seen that the ML model resembles the inversely identified parameters nearly perfectly, and both fit the measurement to a high extend.

(b) Comparison for specimen 2 with highest mean absorption coefficient. It can be seen that the ML model resembles the inversely identified parameters nearly perfectly, and both fit the measurement to a high extend.

Figure 10. Comparison of the measured absorption coefficient, the absorption coefficient computed from the inversely identified Biot parameters ('Fit') and absorption coefficient computed using the Biot parameters outputted by the KNN model.

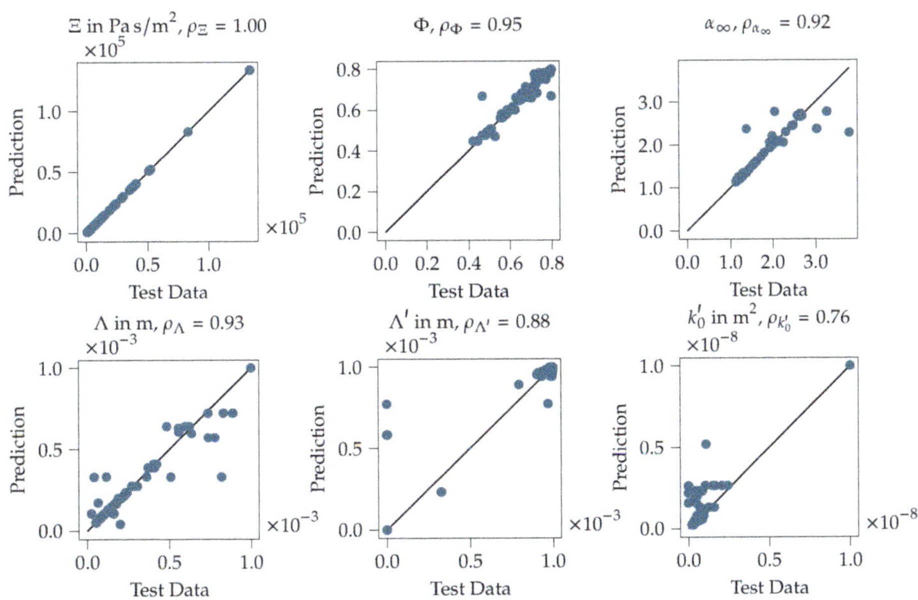

Figure 11. Correlation graphs for inversely identified Biot parameters and their corresponding prediction by the KNN model. It can be seen that all quantities are predicted with reasonable accuracy, thus the model is qualified for further application.

3.2.2. Performance of the ANN Model for Predicting Biot Parameters

Very similar observations can be made for the ANN model, as shown in Figure 12. For specimen 39 with the lowest mean absorption coefficient (see Figure 12a), the computed absorption coefficient with the Biot parameters of the ML model output fits nearly perfect the absorption coefficient computed using the inversely identified Biot parameters. Moreover, both approaches match the measured data with reasonable agreement. Here as well, the largest deviations can be found around the maximum of the absorption coefficient. For the specimen no. 2 (see Figure 12b), the findings are pretty similar. Only in the very high-frequency regime, above 6000 Hz very small deviations between the used models can be observed, still, both models match the measured data very well. Summarizing, the inverse parameter identification process as well as the subsequent training of the ML models seems to result in data models that are capable of resembling the original data and thus the printed porous absorbers reasonably well.

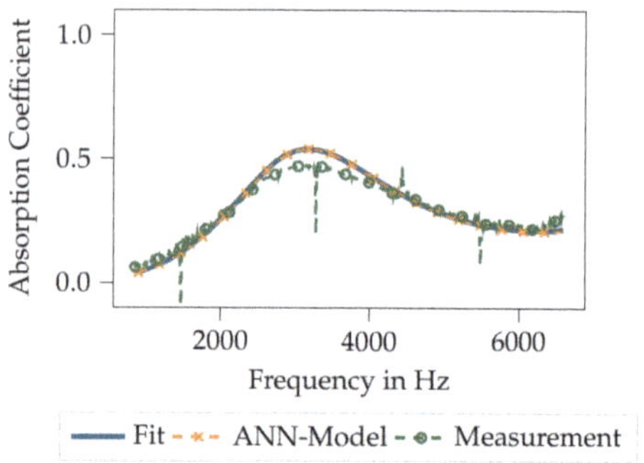

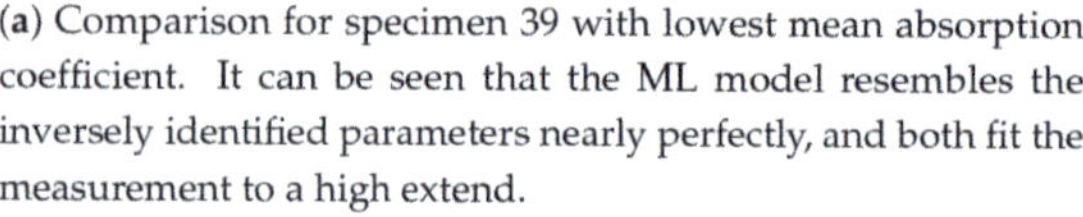

(a) Comparison for specimen 39 with lowest mean absorption coefficient. It can be seen that the ML model resembles the inversely identified parameters nearly perfectly, and both fit the measurement to a high extend.

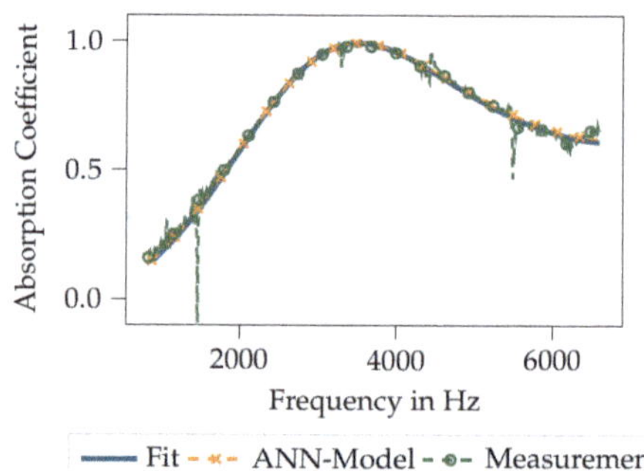

(b) Comparison for specimen 2 with highest mean absorption coefficient. It can be seen that the ML model resembles the inversely identified parameters nearly perfectly, and both fit the measurement to a high extend.

Figure 12. Comparison of the measured absorption coefficient, the absorption coefficient computed from the inversely identified Biot parameters ('Fit') and absorption coefficient computed using the Biot parameters outputted by the ANN model.

For the ANN model as well the correlation of the input data and the predictions are shown for all Biot parameters in Figure 13. The findings are similar to the KNN model; the flow resistivity shows a practically perfect fit of the test data and the model's prediction, porosity, tortuosity and the thermal characteristic length show values of the correlation above 0.90. As well, some outliers can be observed, for example for the static thermal permeability (bottom right). Here, some test data points with rather large values of $0.80 \times 10^{-8} - 1.00 \times 10^{-8}\,\mathrm{m}^2$ can be seen that are predicted by the ANN model with rather low values of $0.01 \times 10^{-8} - 0.45 \times 10^{-8}\,\mathrm{m}^2$. Here as well, it is assumed that the geometry parameter combination has only a low effect on this acoustic parameter which leads to rather inaccurate predictions. To summarizing, the training of the ANN model can be assumed to be successful here as well.

3.3. Application of the Machine-Learning Models for Absorber Design (Step 6)

To now generate porous absorbers that show predefined absorption characteristics, three target curves are defined. These three curves are referred to as low, medium and high and are shown in Figure 14. The course of the different target curves is chosen arbitrarily, nevertheless, they are intended to resemble common application cases. The idea is to generate printable absorber designs using the ML models described in the previous section and to evaluate to which extend the printed absorbers show the desired characteristics.

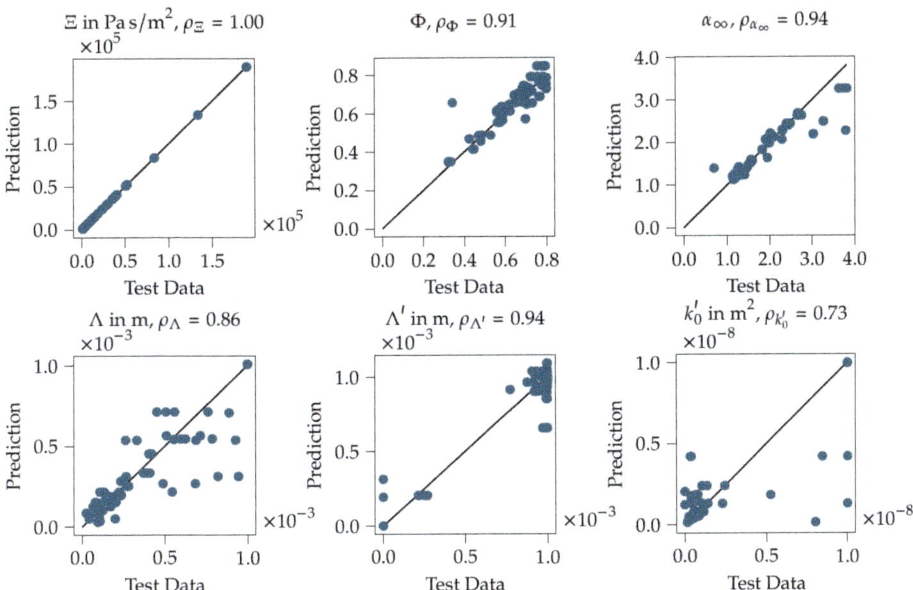

Figure 13. Correlation graphs for inversely identified Biot parameters and their corresponding prediction by the ANN model. It can be seen that all quantities are predicted with reasonable accuracy, thus the model is qualified for further application.

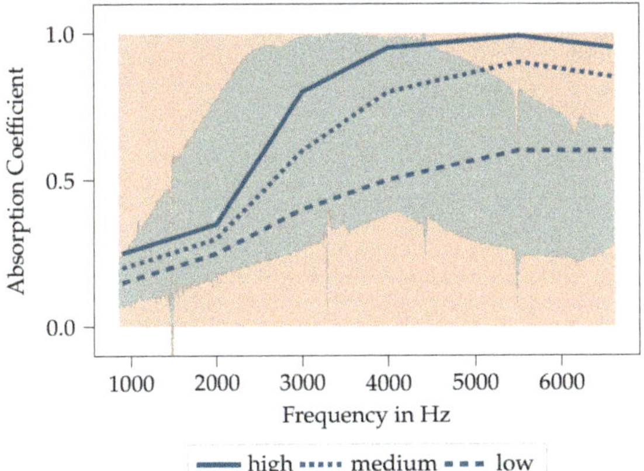

Figure 14. Target curves for the porous absorber design. Red and green areas mark known/unknown regions that are/are not covered by the existing specimen yet. The target curve "low" lies completely within the known data range, it is assumed that here only interpolation within the available data is required. The target curves "medium" and "high" leave the known range above approx. 5000 Hz, here the ability to extrapolate from the known data is required.

In Figure 14, the target curves for the absorption coefficient are plotted over frequency, thereby the frequency range is kept identical to the specimen already available. All target curves start at rather low absorption coefficients at low frequencies and show an increase of the absorption coefficient with frequency. The curves thus resemble a general characteristic of porous absorbers. Nevertheless, when compared to the absorption characteristics shown in Figure 7b, it should be noted that the frequency range with a high absorption coefficient is generally broader than it is with the available specimen. Furthermore, the red and green areas from Figure 7b are shown again here, indicating the areas for which suitable specimen designs are already available. It should be noted here that the curve low remains within the green area for all frequencies, whereas the curves medium and high leave the green area for frequencies above approx. 5200 Hz and 4300 Hz, respectively. This will be important for the result discussion in Sections 3.3.1 and 3.3.2, respectively. Hence, to generate specimens whose absorption coefficient follows these curves, an extrapolation capability of the ML models is required.

The approach to design porous absorbers using the trained ML models is as follows. As the target value of the absorption coefficient is given by the target curves and the required Biot parameters are unknown, an inverse procedure is established. The procedure is sketched in Figure 15. An optimization strategy similar to the inverse parameter identification procedure described in Appendix C is set up. The used evolutional algorithm (for parameters, see Table A5) is the same as in the parameter identification procedure, only a slight difference is made for the error measure. Here, the mean of the squared difference of the target curve α_{target} and the absorption coefficient computed using the Biot parameters from the ML model in combination with the JCAL model $a\alpha_{JCAL}$ is used:

$$err = \frac{1}{i} \sum_i \left(\alpha_{target,i} - \alpha_{JCAL,i} \right)^2 \ . \tag{8}$$

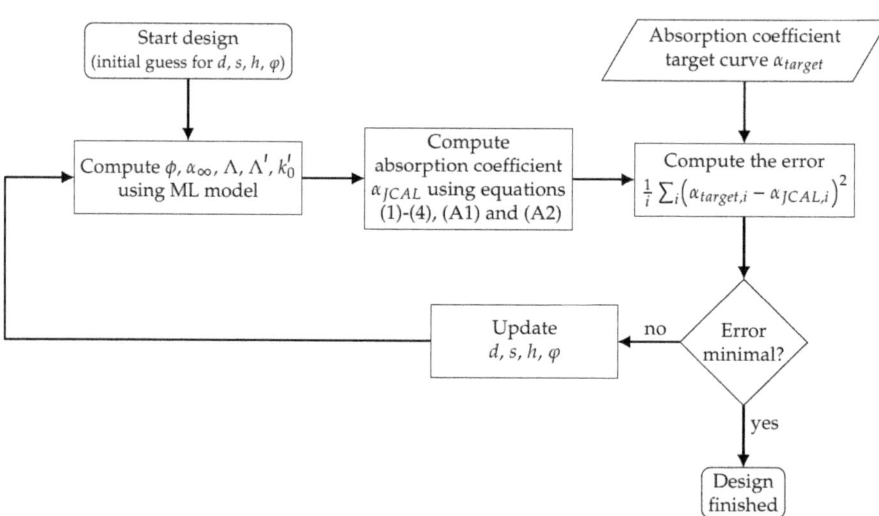

Figure 15. Flow chart of the inverse absorber design process (Step 6 from the flowchart in Figure 1). Based on an initial guess of the design parameters, the corresponding Biot parameters and the resulting absorption coefficient is computed. Based on the given error measure with respect to the target, the design parameters are updated until convergence is reached.

Starting with an initial guess for the geometry parameters, the Biot parameters are computed using the ML model and the resulting absorption coefficient is computed using Equations (1)–(4), (A1) and (A2). The results are compared to the chosen absorption coefficient target curve and the error is computed. Based on this error measure, the optimization algorithm updates the geometry parameter set. These new geometry parameters are again fed into the ML model to compute the resulting Biot parameters with whom the loop starts again until the requested error bound is reached. The optimization based on the evolutional algorithm is set up as a bounded optimization. The bounds are chosen based on the applied MEX process and given in Table 6.

Table 6. Bounds for the evolutional optimization algorithm during absorber design.

Quantity	Lower Bound	Upper Bound
bar width (mm)	0.15	0.50
bar height (mm)	0.10	0.30
bar spacing (mm)	0.10	1.00
plane angle ($^\circ$)	30	90

After completing the material design using the process described here, the resulting designed absorbers are printed equivalently to the specimen used for the data generation. The resulting specimens are measured in the impedance tube regarding their absorption coefficient and using a 3D surface profilometer regarding their geometry. The results are presented in the following sections.

3.3.1. Specimen Design with K-Nearest Neighbor Approach

The first attempt for generating printable porous absorbers that show a prescribed absorption characteristic is done using the KNN model. The design parameters for all three target curves and the corresponding measured values are shown in Table 7. In Table 7, it can be seen that the actually manufactured geometry is rather close to the design parameters. The maximum evaluated deviation is 0.04 mm.

Table 7. Design parameters and measured values of the specimens designs using the KNN model. The printed specimens are measured using a 3D surface profilometer, quantities marked with a dash ('-') could not be measured due to the measurement principle.

Quantity	Target "low"		Target "medium"		Target "high"	
	Design	Measured	Design	Measured	Design	Measured
bar width (mm)	0.17	0.21	0.26	0.24	0.20	0.19
bar height (mm)	0.12	–	0.11	–	0.15	–
bar spacing (mm)	0.93	0.89	0.44	0.41	0.33	0.325
plane angle ($^\circ$)	63.40	63.399	63.000	63.256	33.40	33.466

In Figure A5, in the Appendix G, the KNN designs are shown in orange together with the geometry parameters of the initial 50 specimen population. It can be seen that all designs found by the KNN model lie within the learned data range. Indeed, the designs are new in the sense that most chosen values were not investigated and learned before. However, the data range given from the initial specimen population is not left by the KNN model.

The results of the predicted and measured absorption coefficient are shown in Figure 16, whereas Figure 16a shows the results with the target curve "low", Figure 16b,c show the results for the target curves "medium" and "high", respectively. For all plots, the blue curves show the relevant target curve, the orange line shows the expected absorption coefficient of the designed material that is based on the JCAL model and the green line shows the measured data of the printed specimen. In the figure caption, the optimization error after the material design is given, this is the resulting error between the target curve and the expected absorption coefficient of the designed material as given by Equation (8). It should be noted that for computing the orange design curves, the actual manufactured parameter values from Table 7 are used.

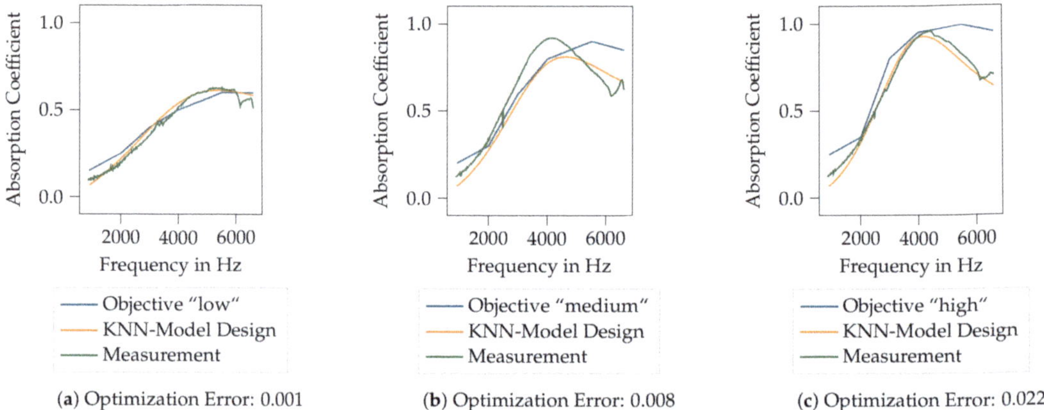

(a) Optimization Error: 0.001   **(b)** Optimization Error: 0.008   **(c)** Optimization Error: 0.022

Figure 16. Design of porous media using an KNN machine-learning model for three different target curves. It can be seen that, especially for the targets "low" and "high" the predictions are met with high accuracy. However, the KNN model is not able to extrapolate from the know data range, as the prediction does not follow the target curves for "medium" and "high" above approx. 5000 Hz.

For the target curve "low", see Figure 16a, it can be seen that the material design using the KNN model can produce reasonable results. The optimization error after the material design is very low, the predicted absorption coefficient matches the target curve fairly well. Furthermore, the measured data shows that the absorption coefficient of the printed specimen follows the predicted absorption coefficient very well. For the case of a low absorption coefficient, the proposed procedure seems to produce reasonable results. The results for the target function "medium" as shown in Figure 16b, the results are rather different. For instance, it can be stated that the predicted absorption coefficient is able to follow the target curve only in the frequency regime below approx. 4000 Hz. Before commenting on the course of the predicted absorption coefficient, it should be noted that the measured data fit the designed material rather well. Indeed, the deviations between the measured and the predicted data are higher than for the target curve "low" but the general course is met rather well. Interesting here is the course of the predicted absorption coefficient in the higher frequency regime above 4000 Hz. Here, the predicted absorption coefficient is not able to follow the target curve anymore. This behavior is crucial and special to the KNN model. It is assumed here that the reason for this behavior is the inability of the KNN model to extrapolate from already learned information. Recalling the introduction of the target curves in Figure 14, it can be seen that the target curves "medium" and "high" leave the area of the learned data between 5200 Hz and 4300 Hz, respectively. Hence, to generate designs that exhibit an absorption coefficient beyond the learned data, the applied models must have the ability to extrapolate from the learned data to some extend. The results of the design process for the target curve "high" are shown in Figure 16c. Regarding the predicted absorption coefficient, a similar behavior as for the target curve "medium" can be observed. The predicted absorption coefficient follows the target rather well below 4000 Hz but is not able to follow for higher frequencies. Accordingly, the optimization error with a value of 0.022 is rather high. Nevertheless, the measured data show a very good agreement with the predicted absorption coefficient, hence the computed geometry parameters prove to work for practical applications.

It should be noted here that all measured absorption coefficient curves show a rather strange behavior above 6200 Hz, as the absorption coefficient curves show some sort of "kink" and increases for larger frequencies, whereas it monotonically decreases for lower frequencies. This behavior is not common for porous absorbers since such materials are expected to show a rather smooth course of the absorption coefficient over frequency. It is expected here that this "kink" is caused by an imperfect fit of the specimen within the impedance tube that leads to enclosed air-filled voids between the specimen and the tube's wall. Such voids might serve as a high-frequency resonator, resulting in a corrupted absorption coefficient measurement. The imperfect fit of the specimen again might be due to manufacturing inaccuracies.

To summarize, the KNN model can generate printable geometry descriptions for the design of porous media. Nevertheless, it is required that the requested designs lie somehow within the previously learned data range, the ability to extrapolate outside the learned data range seems rather limited.

3.3.2. Specimen Design with Artificial Neural Network Approach

The material design task using the ANN model leads to rather different designs compared to the KNN model. The predicted parameter values for all geometry parameters and target curves are shown in Figure A5 in the Appendix G. It can be seen that the ANN model generates both, geometry values inside and outside the data range provided by the specimen population. Hence, some ability to extrapolate from the learned data range can be confirmed. A comparison of the manufactured and designed geometry parameters is shown in Table 8.

Table 8. Design parameters and measured values of the specimens designs using the ANN model. The printed specimens are measured using 3D surface profilometer, quantities marked with a dash ('-') could not be measured due to the measurement principle.

Quantity	Target "low"		Target "medium"		Target "high"	
	Design	Measured	Design	Measured	Design	Measured
bar width (mm)	0.15	0.16	0.17	0.162	0.15	0.16
bar height (mm)	0.12	-	0.20	-	0.30	-
bar spacing (mm)	0.89	0.843	0.66	0.665	0.10	0.11
plane angle ($^\circ$)	78.00	78.107	30.00	30.296	62.40	61.927

It can be seen that generally similar parameter ranges as obtained with the KNN model are reached. The manufacturing accuracy as well is similar to the KNN model, the maximum deviation between the manufactured and predicted geometry parameters is 0.05 mm for the bar spacing, target curve "low". All other (measurable) parameters are manufactured with a deviation of 0.01 mm.

In Figure 17, the predicted and measured absorption coefficient data is shown for all three target curves. Similar to the corresponding plot for the KNN model, the blue curve shows the target curve, the orange curve shows the predicted absorption coefficient and the green curve shows the measured absorption coefficient of the manufactured material. The optimization error is given in the plot's captions. For the target curve "low", see Figure 17a, a good agreement of all shown curves can be seen. The optimization error with a value of 0.001 is very low, this means that the ANN model can compute suitable geometry parameters for this design. Furthermore, the measured data are in very good agreement with the predicted absorption coefficient curve as well, thus the predicted geometry parameters prove appropriate for the practical application. For this target curve as well as for the others, the "kink" in the course of the absorption coefficient curve, already seen for the KNN model, can be observed as well and is expected to be a result of resonator effects caused by air-filled voids between the specimen and the impedance tube's walls.

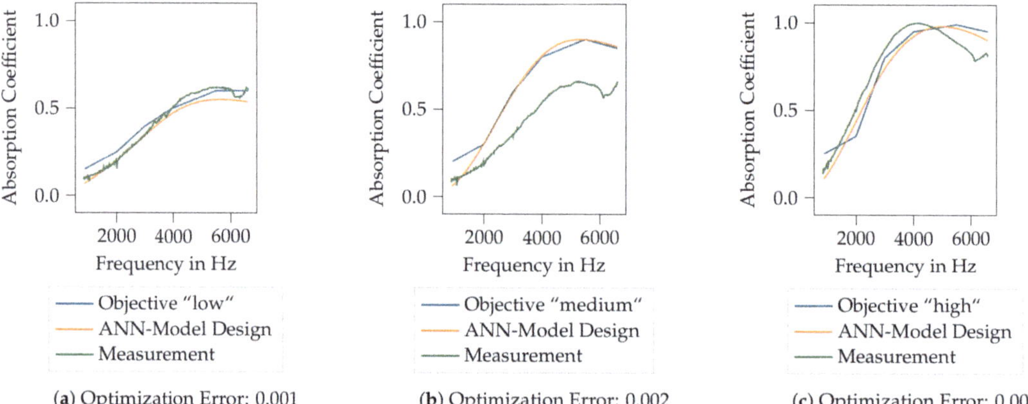

(**a**) Optimization Error: 0.001   (**b**) Optimization Error: 0.002   (**c**) Optimization Error: 0.002

Figure 17. Design of porous media using an ANN machine-learning model for three different target curves. It can be seen that, especially for the targets "low" and "high" the predictions are met with reasonable accuracy. The ANN model is able to extrapolate from the know data range, as the prediction follows the target curves for "medium" and "high" above approx. 5000 Hz. The large deviations of the measurement for the target "medium" are expected to result from manufacturing inaccuracies.

A rather undesirable result is shown in Figure 17b for the target curve "medium". For instance, the predicted absorption coefficient curve (orange) is in good agreement with the target curve (blue), this is confirmed by the low optimization error of 0.002. Nevertheless, the manufactured specimen does not follow the designed material, the absorption coefficient is rather low with a maximum of approx. 0.66 at 5200 Hz, whereas the predicted absorption coefficient should have been 0.89 at this frequency. A specific reason for this deviation yet remains unknown. Compared to the KNN model design, it can be stated that the value of the bar spacing predicted using the ANN model is rather high (ANN model: 0.66 mm; KNN model 0.44 mm). Nevertheless, it is expected that the relations between the geometry and the resulting Biot parameters/the absorption coefficient are highly nonlinear, hence simple value comparisons for single values might not be feasible here. This issue could be thought of as a negative drawback from using ML models. Since ML models rely on learned relations only rather than implementing physical models, the interpretability of the model outputs is rather limited.

For the target curve "high", the results are shown in Figure 17c. Here again, a rather good agreement of all three curves is observed. The optimization error between the target curve and the predicted absorption coefficient with a value of 0.002 is equivalently low as for the other target curves. Furthermore, the measurement data follow the predicted curve rather well, nevertheless, some deviations can be observed. For example, the absorption maximum is shifted approx. 200 Hz to lower frequencies and the declining of the absorption coefficient for frequencies above the absorption maximum is more strongly present as it is with the design data. It might be the case as well that here a physical limit is reached for which the applied lattice-style absorber design is not able to generate such high absorption coefficients for high frequencies. Nevertheless, the observed deviations are accepted to be rather small and the procedure in total seems to prove successful.

Regardless of the deviations in the high-frequency range, Figure 17c shows a very interesting behavior that should be hereafter discussed. Recalling the definition of the target design curves in Figure 14, it can be observed that the target curves "medium" and "high" leave the learned data range for high frequencies. Hence, to generate absorber designs that show an absorption coefficient comparable to these target curves, the applied ML models need some ability to extrapolate beyond the learned data range. For the KNN model, it could be shown that this model type does not have this capability. The ANN model, on the other side, shows this capability to a reasonable extent—the designs (orange curves) for the target curve "medium" and "high" follow the target curves and thus leave the learned data range. Moreover, at least the design for the target curve "high" shows that the design parameters are rather reasonable as the measured data as well leaves the learned data range.

3.4. Design of a Material with Different Height and Prediction of Another Frequency Range

The ML models are built to predict the Biot parameters from the specimen's micro-scale geometry. Thus, material parameters of the porous material are obtained. The "detour" via the Biot parameters opens the opportunity to design other specimens with varying macro-scale geometries, such as the specimen height. Moreover, the possibility to predict the acoustic behavior for other application cases, for example other frequency ranges becomes possible. This is due to the fact that Biot parameters are material-specific parameters and thus no mixing of the material description and the acoustic behavior of the material in specific conditions (sound field, mounting of the specimen) occurs. Therefore is assumed that, if the Biot parameters identified on specimens with a small height can be used as well to predict the behavior of specimen with a larger height, the Biot parameters can be trusted as they give a reasonable description of the material.

In order to verify this assumption, the design parameters of the specimen with target curve "high", computed with the KNN model are used, see Table 7. Again, the ML model is used to compute the Biot parameters and the resulting absorption coefficient is computed using the Equations (1)–(4), (A1) and (A2) for a new specimen height l = 30 mm. The specimen is manufactured using the design parameters and the resulting absorption coefficient is measured. Unlike for the initial specimen geometry, here the tested frequency range is enlarged to 150–6600 Hz. The results are shown in Figure 18.

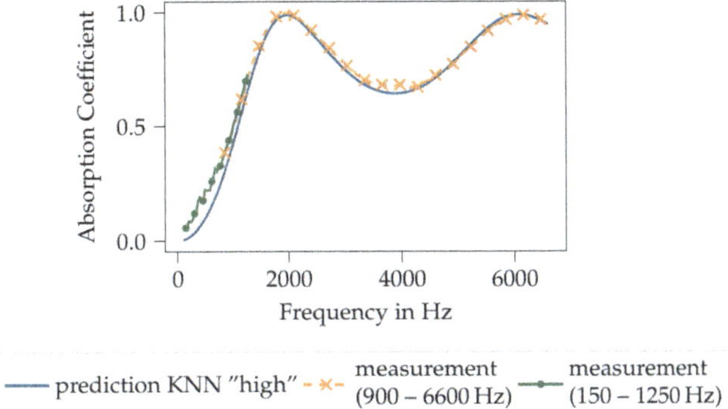

Figure 18. Computed and measured absorption coefficient of a specimen with a height l = 30 mm and enlarged frequency range 150–6600 Hz. The computation employs the Biot parameters obtained from the KNN model for traget curve "high" and computes the absorption coefficient using the Equations (1)–(4), (A1) and (A2) for the new specimen height.

In Figure 18 it can be seen that the computed and measured absorption coefficient coincide very well. This holds for both, the now strongly changed course of the absorption coefficient as well as the prediction in the low frequency regime. It is assumed that such a result could not have been obtained by means of ML alone. Therefore, the specimen height and the enlarged frequency range would have needed to be included in the input data, which here is not the case. Accordingly, many more samples would have been necessary to be produced in order to cover the larger dimension of the design space properly. This has been circumvented by computing the Biot parameters, which are material-specific parameters. Thus, the indeed larger effort using the Biot parameters on the other side can be viewed as a means of dimensionality reduction for the input data as well. Moreover, the obtained Biot parameters can be used in subsequent analyses: the Biot parameters are invariant to the type of sound field and mounting conditions of the material. Thus, this material-based description can be used for computations in other field types (diffuse field, varying angle of the incident waves, other frequency range as shown here, etc.) or even more complex analyses such as wave-resolving finite element computations as well. This could not be covered with the direct computation of the absorption coefficient from the ML models since those data are limited to the conditions with which the training data are obtained, here a planar normal incident sound field with a specific frequency range.

4. Conclusions and Outlook

The presented work aims to bridge the gap between the micro-scale geometry and acoustic material parameters of porous absorbers to allow a reasonable absorber design. The core challenge is that porous absorbers can be characterized from an acoustics point of view utilizing the Biot parameters and by geometric dimensions on the micro-scale. These two "worlds" are separated, the estimation of the Biot parameters from the geometry is possible only for simple special cases. The work presented here shows a way of connecting these two "worlds" through machine learning. In the following, the work is briefly summarized and an outlook on future works is given.

4.1. Conclusions

The presented work investigates a generic test specimen constructed by a micro-lattice of parallel bars. The specimens are manufactured using additive manufacturing, a specifically appropriate production procedure for porous materials. In order to connect the micro-scale geometry and the acoustic material parameters, namely the Biot parameters, the range of reasonable and manufacturable geometry values is sampled using a Latin Hypercube Strategy and an amount of 50 specimens is manufactured. The resulting specimens are optically and acoustically investigated, the latter comprising measurements of the absorption coefficient and flow resistivity. The Biot parameters of the specimens are derived using an inverse parameter identification scheme. Using the geometry of the specimens as inputs (features) and the Biot parameters as outputs (labels) two Machine-Learning regression models (K-Nearest Neighbors (KNN) and Artificial Neural Network (ANN)) are built and trained by means of supervised learning. The resulting models are then applied to a material design procedure with the goal of generating printable absorber structures that exhibit predefined absorption characteristics. This procedure finally is capable of computing the micro-scale geometry of the specimens that is required to obtain a porous absorber that shows the desired acoustic behavior, which is measured here in terms of the absorption coefficient.

The approach of using the Biot parameters here to design porous absorbers is chosen, as it exhibits some favorable implications. The Biot parameters can be viewed as material-specific parameters of the porous material and thus are invariant to the macro-geometry (for example the thickness) of the material, the mounting conditions or the sound field. If a quantity describing the behavior of the material was considered, for example, the absorption coefficient, a mixing of material description and material behavior would occur. This mixing is circumvented by predicting the Biot parameters directly and thus, these parameters can be used for other subsequent analyses as well without the limitation on how the parameters were obtained before.

A first aspect of the work presented here is the additive manufacturing of the specimen geometry, especially on the micro-scale. Here, a material extrusion process is employed, thereby specific means as targeted under-extrusion and the use of a 0.20 mm nozzle prove capable of reliably producing the required fine microstructures. This is verified using optical inspection using a profilometer. It can be shown that only small deviations of the desired and actually manufactured geometry exist. The acoustic inspection using the flow resistivity and absorption coefficient deliver the inputs for the inverse parameter identification procedure to obtain the Biot parameters. A first correlation analysis of the flow resistivity and the micro-scale geometry parameters of the specimen reveals that no trivial dependency between these parameters can be found. This finding motivates the use of machine-learning to exploit underlying dependencies that are coded in the data. Another important ingredient for the work presented here is the inverse parameter identification procedure that is used for obtaining the Biot parameters. Therefore, the specimens are modeled using a mechanical model for porous media, here the Johnson–Champoux–Allard–Lafarge (JCAL) model. This model takes the Biot parameters of the specimen as inputs. With this model, the absorption coefficient of the specimen can be computed for the case of mounting the specimen in front of an impervious rigid wall, which is the case for the measurements in the impedance tube that has been used here. Using the JCAL model, the inverse parameter identification procedure is set up using an evolutional algorithm. The results of the inverse identification of the Biot parameters are compared to analytical estimates and found to be trustworthy.

The machine-learning models are set up using the micro-scale geometry of the specimens as input and the Biot parameters as outputs. A correlation analysis shows that the models investigated here, namely the K-Nearest Neighbor model and the Artificial Neural Network are able to reasonably compute the Biot parameters from the specimen geometry. The obtained models are then applied to design porous absorbers that show a predefined acoustic behavior, here characterized by the absorption coefficient. Therefore, target curves of the absorption coefficient are defined and the specimen' micro-scale geometry is inversely computed. It can be shown, that both models are able to generate printable absorber structures. However, some differences can be observed. A major difference between the models is their ability to extrapolate from the trained data—the KNN model is not able to produce specimens beyond the initial population, whereas the ANN model is able to compute designs that lie outside the initially learned data range. However, the measured absorption coefficient of the designed specimens matches the prediction in all three cases for the KNN model, the ANN model is able to predict the absorption coefficient only for two of three cases correctly. Finally, the procedure is applied to predict and generate another specimen with a different macro-geometry (here, the specimen height) as well. It can be shown that this application case can be handled by the proposed procedure fairly well.

Summarizing, the procedure to generate machine-learning based models that predict Biot parameters from porous absorbers micro-scale geometry can be viewed as successful. It was possible to generate models that are capable of the intended task and that these models can be successfully applied to design new absorbing structures. The higher effort for obtaining Biot parameters and thus focusing on the material description rather than on the acoustic behavior allow the application of the procedure for a multitude of practical applications, such as the generation of input parameters for finite element analyses or the idea of tailoring materials specifically for their intended application case.

4.2. Outlook

Although the presented process already delivers good results, some open questions remain. These will be hereafter discussed. For instance, the design of the presented work is rather complex, which is mainly due to the necessary computation of the Biot parameters. Therefore, it would be feasible to re-conduct the study and to measure more of the Biot parameters directly. This would result in a lower dimension of the optimization problem and probably more accurate results of the inverse parameter identification of the remaining Biot parameters. Another aspect for future works is the fact that not all prediction results of the machine-learning models could be verified by measurements. For example, it should be clarified why the result obtained with the ANN for the target curve "medium" shows high deviations compared to the other results. Moreover, it is assumed that the specimen design induces some physical limit that prevents from obtaining an even better acoustic behavior. The exploration of these limits is also subject to future work.

Another aspect regarding the application of the proposed method is the exploration of the physical relation of the Biot parameters and the specimen's geometry. As already stated, for some special cases the Biot parameters can be estimated from the material micro-scale.

This will be further investigated using the models obtained within this work and indeed is planned as the next step.

There is also a need for further research in the field of additive manufacturing processes, for example in the field of microstructures. An extended understanding of the dependencies between process parameters and the resulting microstructure is important. This knowledge would also establish the connection between deviations in the shape and form of the microstructure and resulting acoustic properties. The derivation of design rules for the manufacturing of microstructures utilizing material extrusion is necessary to ensure the realization of structures that are as small as possible. The topic of robust additive manufacturing of microstructures also has great research potential.

In addition, there is also a need for further research in the area of the resulting dimensional accuracy and also the shape of the additively manufactured microstructures. Figure 19 shows test specimens 26, whose bars have a bar width of 0.20 mm, which means that only one strand is needed to realize the bar width. In contrast to Figure 8, the test specimens 2 and 39 consists of bars with a bar width of 0.40 mm, which means that two parallel strands are required to realize the bar width. Figure 19 shows that the test specimens with only one strand for the realization of the bar width result in more homogeneous results in the geometric expression of the strands. Specifically, the one-strand design allows a much more constant application of the strands without fluctuating cross-sections. Furthermore, the actual-target comparison of specimen 26 shows desired dimensional results after the manufacturing process with only minor deviations. The reason for the poorer geometric expressions in the two-strand design may lie in the individual process parameters such as the cooling of the specimen during layer application or the nozzle temperature, but this needs to be investigated in further studies.

(a) Optical investigation of Specimen 26

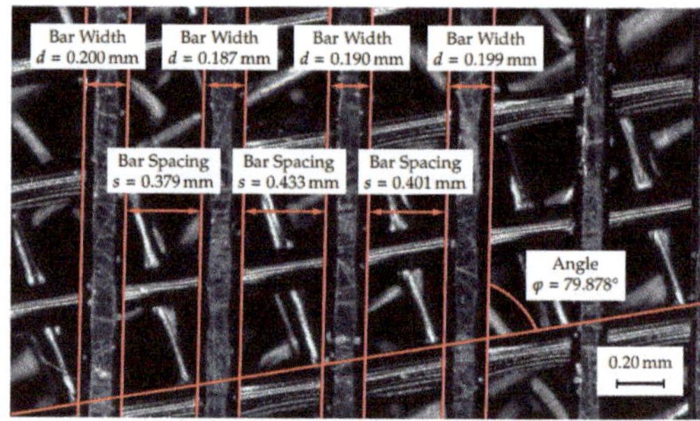

(b) Specimen 26—measurement

Figure 19. Actual-target comparison of specimen 26 with the Keyence VR5200 3D surface profilometer. (**a**) shows specimen 26 without measurement lines at 80× magnification. (**b**) shows specimen 26 with measurement lines and only slight deviations in the actual-target comparison

Author Contributions: Conceptualization: S.K., T.P.R., H.W., S.C.L. and T.V.; methodology & validation: S.K. and T.P.R.; additive manufacturing and characterization: S.K. and H.W.; specimen characterization by acoustic measurements, implementation and application of all acoustic models and the inverse parameter identification procedure, implementation and application of the machine-learning process including the models: T.P.R.; formal analysis & software: T.P.R.; project administration: S.C.L. and T.V.; writing—original draft: S.K. and T.P.R.; writing—review & editing: S.K., T.P.R., H.W., S.C.L. and T.V.; All authors have read and agreed to the published version of the manuscript.

Funding: This research received no external funding.

Data Availability Statement: Data sharing is not applicable to this article.

Acknowledgments: We acknowledge the support by the German Research Foundation, the "12plus6 funding initiative" by the Faculty of Mechanical Engineering of the Technische Universität Braunschweig and the Open Access Publication Funds of Technische Universität Braunschweig.

Conflicts of Interest: The authors declare no conflict of interest.

Abbreviations

The following symbols and abbreviations are used in this manuscript:

A	specimen cross sectional area (flow resistivity measurement)
d	bar width
h	bar height
k	number of neighbors in KNN model
k_0'	static thermal permeability
l	specimen height
q	volume flow through the specimen (flow resistivity measurement)
R	airflow resistance
R	coefficient of determination (performance measure of ML model training)
s	bar spacing
α	absorption coefficient
α_∞	tortuosity (high frequency limit)
Δp	pressure drop over the specimen (flow resistivity measurement)
Λ	viscous characteristic length
Λ'	thermal characteristic length
Ξ	flow resistivity
ρ	Pearson's Correlation Coefficient
φ	plane angle
ϕ	porosity
AM	additive manufacturing
ANN	artificial neural network
DOE	design of experiment
JCAL	Johnson–Champoux–Allard–Lafarge
KNN	k-nearest neighbor
LHS	latin hypercube sampling
MEX	material extrusion
ML	machine learning
PBF-P	powder bed fusion of polymers
PET-G	glycol modified polyethylene terephthalate
PLA	polylactide
VAT	vat photopolymerization

Appendix A. Overview of the Design Parameter Space LHS Sampling

Table A1. Overview of the generated test Spec.s. The design parameters (d, s, h, φ) were varied using a Latin hypercube sampling. The process parameters (layer height, the extrusion width) are used during the manufacturing process and are given here for information only.

	Variation Parameters (Spec.)				Process Parameters (AM)	
	Bar Width (d)	Bar Spacing (s)	Bar Height (h)	Angle (φ)	Layer Height	Extrusion Width
Spec. 0	0.20 mm	0.70 mm	0.16 mm	40.00°	0.08 mm	0.20 mm
Spec. 1	0.50 mm	0.40 mm	0.11 mm	70.00°	0.11 mm	0.25 mm
Spec. 2	0.40 mm	0.40 mm	0.08 mm	60.00°	0.08 mm	0.20 mm
Spec. 3	0.40 mm	0.80 mm	0.06 mm	70.00°	0.06 mm	0.20 mm
Spec. 4	0.50 mm	0.20 mm	0.12 mm	80.00°	0.06 mm	0.25 mm
Spec. 5	0.30 mm	0.90 mm	0.15 mm	50.00°	0.075 mm	0.15 mm
Spec. 6	0.30 mm	0.70 mm	0.13 mm	50.00°	0.065 mm	0.15 mm

Table A1. *Cont.*

| | Variation Parameters (Spec.) | | | | Process Parameters (AM) | |
	Bar Width (d)	Bar Spacing (s)	Bar Height (h)	Angle (φ)	Layer Height	Extrusion Width
Spec. 7	0.20 mm	0.30 mm	0.17 mm	40.00°	0.085 mm	0.20 mm
Spec. 8	0.40 mm	0.50 mm	0.11 mm	80.00°	0.11 mm	0.20 mm
Spec. 9	0.30 mm	0.90 mm	0.10 mm	40.00°	0.10 mm	0.15 mm
Spec. 10	0.20 mm	0.90 mm	0.12 mm	50.00°	0.06 mm	0.20 mm
Spec. 11	0.30 mm	0.80 mm	0.07 mm	80.00°	0.07 mm	0.15 mm
Spec. 12	0.30 mm	0.50 mm	0.16 mm	80.00°	0.08 mm	0.15 mm
Spec. 13	0.50 mm	0.30 mm	0.14 mm	40.00°	0.07 mm	0.25 mm
Spec. 14	0.20 mm	0.90 mm	0.07 mm	30.00°	0.07 mm	0.20 mm
Spec. 15	0.30 mm	0.20 mm	0.06 mm	70.00°	0.06 mm	0.15 mm
Spec. 16	0.50 mm	0.80 mm	0.15 mm	80.00°	0.075 mm	0.25 mm
Spec. 17	0.20 mm	0.30 mm	0.08 mm	80.00°	0.08 mm	0.20 mm
Spec. 18	0.40 mm	0.20 mm	0.08 mm	60.00°	0.08 mm	0.20 mm
Spec. 19	0.20 mm	0.70 mm	0.07 mm	50.00°	0.07 mm	0.20 mm
Spec. 20	0.20 mm	0.70 mm	0.08 mm	70.00°	0.08 mm	0.20 mm
Spec. 21	0.20 mm	0.50 mm	0.11 mm	50.00°	0.11 mm	0.20 mm
Spec. 22	0.40 mm	0.70 mm	0.09 mm	50.00°	0.09 mm	0.20 mm
Spec. 23	0.30 mm	0.60 mm	0.18 mm	90.00°	0.09 mm	0.15 mm
Spec. 24	0.50 mm	0.60 mm	0.17 mm	70.00°	0.085 mm	0.25 mm
Spec. 25	0.50 mm	0.60 mm	0.06 mm	40.00°	0.06 mm	0.25 mm
Spec. 26	0.20 mm	0.40 mm	0.20 mm	80.00°	0.10 mm	0.20 mm
Spec. 27	0.20 mm	0.50 mm	0.16 mm	90.00°	0.08 mm	0.20 mm
Spec. 28	0.40 mm	0.50 mm	0.17 mm	40.00°	0.085 mm	0.20 mm
Spec. 29	0.30 mm	0.90 mm	0.13 mm	60.00°	0.065 mm	0.15 mm
Spec. 30	0.30 mm	0.30 mm	0.10 mm	40.00°	0.10 mm	0.15 mm
Spec. 31	0.40 mm	0.40 mm	0.09 mm	60.00°	0.09 mm	0.20 mm
Spec. 32	0.50 mm	0.60 mm	0.14 mm	60.00°	0.07 mm	0.25 mm
Spec. 33	0.30 mm	0.60 mm	0.14 mm	50.00°	0.07 mm	0.15 mm
Spec. 34	0.20 mm	1.00 mm	0.19 mm	80.00°	0.095 mm	0.20 mm
Spec. 35	0.40 mm	0.90 mm	0.10 mm	30.00°	0.10 mm	0.20 mm
Spec. 36	0.30 mm	0.70 mm	0.05 mm	70.00°	0.05 mm	0.15 mm
Spec. 37	0.40 mm	0.80 mm	0.13 mm	60.00°	0.065 mm	0.20 mm
Spec. 38	0.30 mm	0.90 mm	0.19 mm	90.00°	0.095 mm	0.15 mm
Spec. 39	0.40 mm	0.80 mm	0.15 mm	70.00°	0.075 mm	0.20 mm
Spec. 40	0.20 mm	1.00 mm	0.18 mm	90.00°	0.09 mm	0.20 mm
Spec. 41	0.30 mm	0.50 mm	0.09 mm	30.00°	0.09 mm	0.15 mm
Spec. 42	0.40 mm	0.40 mm	0.17 mm	50.00°	0.085 mm	0.20 mm
Spec. 43	0.40 mm	0.30 mm	0.20 mm	30.00°	0.10 mm	0.20 mm
Spec. 44	0.30 mm	0.60 mm	0.19 mm	70.00°	0.095 mm	0.15 mm
Spec. 45	0.20 mm	0.80 mm	0.18 mm	30.00°	0.09 mm	0.20 mm
Spec. 46	0.40 mm	0.40 mm	0.14 mm	80.00°	0.07 mm	0.20 mm
Spec. 47	0.20 mm	0.30 mm	0.06 mm	60.00°	0.06 mm	0.20 mm
Spec. 48	0.40 mm	0.40 mm	0.12 mm	40.00°	0.06 mm	0.20 mm
Spec. 49	0.30 mm	1.00 mm	0.12 mm	60.00°	0.06 mm	0.15 mm

Appendix B. Photographs of the Used Measurement Setups

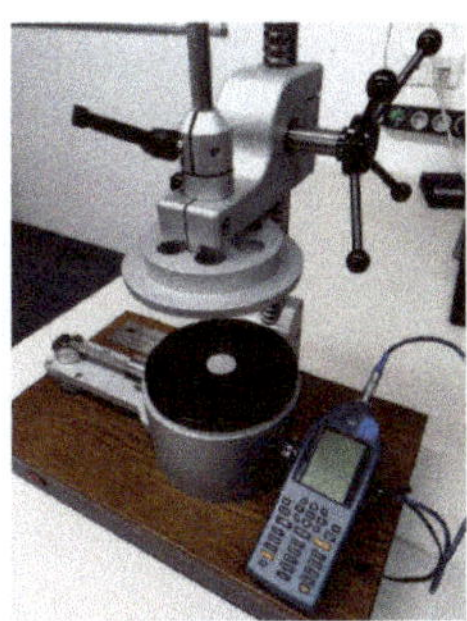

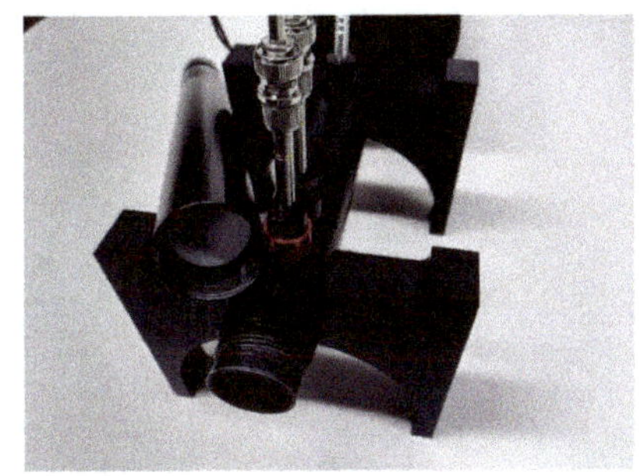

(a) Flow resistivity test stand NOR 1517 with specimen and adapter

(b) Impedance tube AED 1000 with specimen and 1/4″ micro-phones

Figure A1. Measurement test stands used within this work.

Appendix C. Inverse Parameter Identification for the Biot Parameters

Table A2. Settings of the evolutional algorithm used for the inverse parameter identification procedure. The implementation of the algorithm is used from the python library scipy [67] an follows [66].

Property/Parameter	Setting/Value
objective function	takes the Biot parameters, applies Equations (1)–(A2) to compute the absorption coefficient, computes the error using Equation (5) with regard to the measured results and returns the error
bounds	$\phi \in [0.20, 0.80] \quad \alpha_\infty \in [0, 10]$ $\Lambda \in \left[1 \times 10^{-7}, 1 \times 10^{-3}\right]$ $\Lambda' \in \left[1 \times 10^{-7}, 1 \times 10^{-3}\right]$ $k_0' \in \left[1 \times 10^{-12}, 1 \times 10^{-8}\right]$
convergence tolerance (rel.)	1×10^{-2}
mutation parameter	0.50
recombination parameter	0.70
max. iterations	1000
population size	75

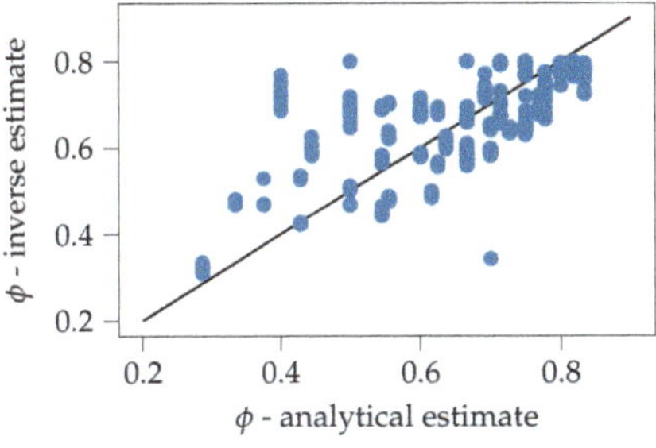

Figure A2. A correlation plot for the inversely estimated porosity of the specimen (vertical axis) and an analytical porosity estimate, computed using $s/(s + d)$. A general correlation (correlation coefficient ρ = 0.63) can be found. Therefore, it is assumed that the general procedure can be trusted.

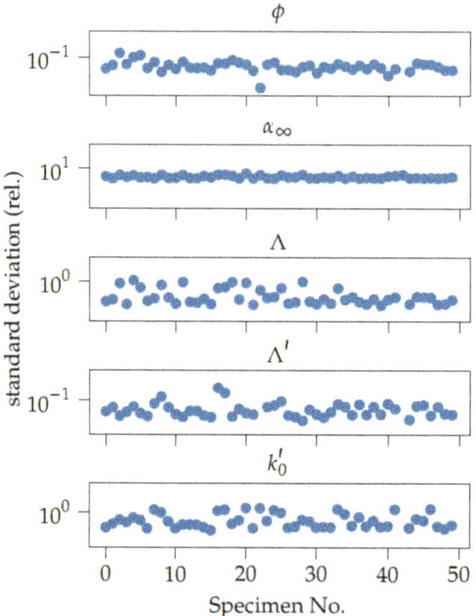

Figure A3. Relative standard deviation for the data augmentation procedure. The standard deviation is computed for all ten runs of the inverse parameter estimates for each specimen and normalized by the mean of the ten runs. It can be seen, that the data varies about approx. 10% for most Biot parameters.

Appendix D. Mathematical Formulation o the JCAL Model Used Here

The mathematical formulation of the JCAL model used within this work is shown in Equations (A1) and (A2), respectively. It is taken from [63]. Table A3 lists the used symbols and the described physical quantity.

$$\tilde{\rho} = \frac{\alpha_\infty \rho_0}{\phi}\left[1 + \frac{\Xi\phi}{j\omega\rho_0\alpha_\infty}\sqrt{1 + j\frac{4\alpha_\infty^2\eta\rho_0\omega}{\Xi^2\Lambda^2\phi^2}}\right] \tag{A1}$$

$$\tilde{K} = \frac{\frac{\gamma P_0}{\phi}}{\gamma - (\gamma - 1)\left[1 - j\frac{\phi\kappa}{k_0'C_p\rho_0\omega}\sqrt{1 + j\frac{4k_0'^2 C_p\rho_0\omega}{\kappa\Lambda'^2\phi^2}}\right]^{-1}} \tag{A2}$$

Table A3. Symbols used in the JCAL model.

Symbol	Quantity	Symbol	Quantity
Ξ	flow resistivity	η	dynamic viscosity of the fluid
α_∞	tortuosity	ρ_0	ambient density of the fluid
ϕ	porosity	ω	angular frequency
Λ	viscous characteristic length	C_p	heat capacity at constant pressure
Λ'	thermal characteristic length	γ	heat capacity ratio
k_0'	static thermal permeability	κ	thermal conductivity

Appendix E. Settings of the Machine Learning Models

Table A4. Settings of the KNN and ANN model (implementation see python-library scikit-learn [71]).

(a) Settings of the KNN model	
parameter	**setting**
n neighbors	5
weights	distance (weighting by inverse of the distance)
algorithm	auto (default, chooses according input data)
leaf size	30 (default)
p	2 (default)
metric	minkowski (default)
metric params	None (default)
n jobs	Nonde (default)
parameter	**setting**
hidden layers	4000–100
activation	ReLU
solver	Adam
alpha	1×10^{-4} (default)
batch size	auto (default)
learning rate	constant (default)
learning rate init	1×10^{-3} (default)
power t	0.50 (default)
max iter	50.00
shuffle	True (default)
random state	None (default)
tol	1×10^{-9}
verbose	False (default)
warm start	False (default)
momentum	0.90 (default)
nesterovs momentum	True (default)
early stopping	False (default)
validation fraction	0.10 (default)
beta 1	0.90 (default)
beta 2	1.00 (default)
epsilon	1×10^{-8} (default)
n iter no change	100
max fun	15.00 (default)

Appendix F. Results of the Hyperparameter-Tuning of the ML Models

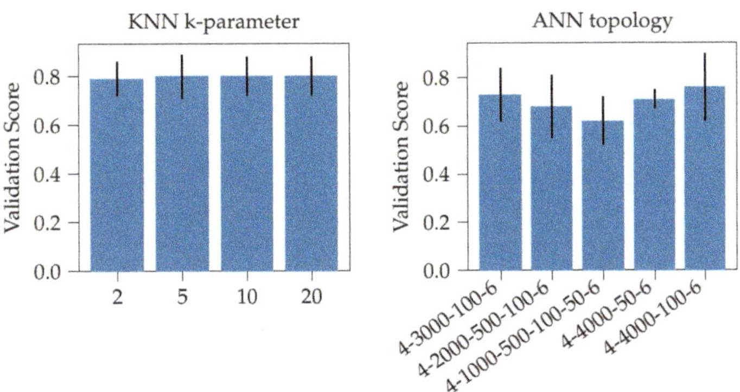

Figure A4. Hyperparameter tuning of the KNN and ANN model. The tuning is done on the validation data during the 3-fold cross-validation. The error bars indicate the 95% confidence interval of the variation during the cross-validation procedure.

Appendix G. Settings of the Evolutional Algorithm for the Inverse Absorber Design and Results

Table A5. Settings of the evolutional algorithm used for the inverse specimen design procedure. The implementation of the algorithm is used from the python library scipy [67] an follows [66].

Property/Parameter	Setting/Value
objective function	takes the design variables, inputs into ML model, computes Biot parameters, applies Equations (1)–(A2) to compute the absorption coefficient, computes and returns the error using Equation (8) with regard to target curve
bounds	see Table 6
convergence tolerance (rel.)	1×10^{-2}
mutation parameter	0.50
recombination parameter	0.70
max. iterations	1000
population size	60

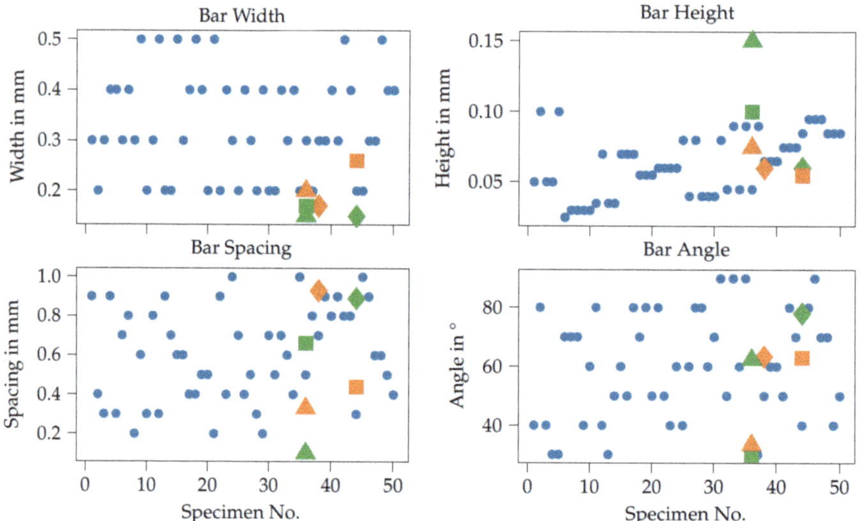

Figure A5. Overview over all designed Spec.s using the proposed ML approach. The colors refer to the ML model and the Spec. start population: Blue: basic population, Orange: KNN model, Green: ANN model. The symbols indicate the absorption coefficient target curve: ◆: target curve "low", ■: target curve: "medium", ▲: target curve: "high".

References

1. Labia, L.; Shtrepi, L.; Astolfi, A. Improved Room Acoustics Quality in Meeting Rooms: Investigation on the Optimal Configurations of Sound-Absorptive and Sound-Diffusive Panels. *Acoustics* **2020**, *2*, 451–473. [CrossRef]
2. Nayfeh, A.H.; Kaiser, J.E.; Telionis, D.P. Acoustics of Aircraft Engine-Duct Systems. *AIAA J.* **1975**, *13*, 130–153. [CrossRef]
3. Sutliff, D.L.; Elliott, D.; Jones, M.; Hartley, T.C. Attenuation of FJ44 Turbofan Engine Noise With a Foam-Metal Liner Installed Over-the-Rotor. In Proceedings of the 15th AIAA/CEAS Aeroacoustics Conference (30th AIAA Aeroacoustics Conference), Miami, FL, USA, 11–13 May 2009.
4. Wilby, J.F.; Scharton, T. *Acoustic Transmission through a Fuselage Sidewall*; National Aeronatics and Space Administration: Cambridge, MA, USA, 1973.
5. Blech, C.; Appel, C.K.; Ewert, R.; Delfs, J.W.; Langer, S.C. Numerical prediction of passenger cabin noise due to jet noise by an ultra–high–bypass ratio engine. *J. Sound Vib.* **2020**, *464*, 114960. [CrossRef]

6. Antonio, J.; Tadeu, A.; Godinho, L. Analytical evaluation of the acoustic insulation provided by double infinite walls. *J. Sound Vib.* **2003**, *263*, 113–129. [CrossRef]
7. Van den Wyngaert, J.C.; Schevenels, M.; Reynders, E.P. Predicting the sound insulation of finite double-leaf walls with a flexible frame. *Appl. Acoust.* **2018**, *141*, 93–105. [CrossRef]
8. Beck, S.; Langer, S. Modeling of flow-induced sound in porous materials. *Int. J. Numer. Methods Eng.* **2014**, *98*, 44–58. [CrossRef]
9. Geyer, T.; Sarradj, E.; Fritzsche, C. Measurement of the noise generation at the trailing edge of porous airfoils. *Exp. Fluids* **2010**, *48*, 291–308. [CrossRef]
10. Geyer, T.F.; Sarradj, E. Trailing Edge Noise of Partially Porous Airfoils. In Proceedings of the 20th AIAA/CEAS Aeroacoustics Conference, Atlanta, GA, USA, 16–20 June 2014. [CrossRef]
11. Ewert, R.; Appel, C.; Dierke, J.; Herr, M. RANS/CAA Based Prediction of NACA 0012 Broadband Trailing Edge Noise and Experimental Validation. In Proceedings of the 15th AIAA/CEAS Aeroacoustics Conference (30th AIAA Aeroacoustics Conference), Miami, FL, USA, 11–13 May 2009. [CrossRef]
12. Herr, M.; Dobrzynski, W. Experimental Investigations in Low-Noise Trailing Edge Design. *AIAA J.* **2005**, *43*, 1167–1175. [CrossRef]
13. Yang, M.; Sheng, P. Sound absorption structures: From porous media to acoustic metamaterials. *Annu. Rev. Mater. Res.* **2017**, *47*, 83–114. [CrossRef]
14. Koponen, A.; Kataja, M.; Timonen, J.v. Tortuous flow in porous media. *Phys. Rev. E* **1996**, *54*, 406. [CrossRef]
15. Herr, M.; Rossignol, K.S.; Delfs, J.; Lippitz, N.; Mößner, M. Specification of porous materials for low-noise trailing-edge applications. In Proceedings of the 20th AIAA/CEAS Aeroacoustics Conference, Atlanta, GA, USA, 16–20 June 2014; p. 3041.
16. Biot, M.A. Theory of Propagation of Elastic Waves in a Fluid-Saturated Porous Solid. I. Low-Frequency Range. *J. Acoust. Soc. Am.* **1956**, *28*, 168–178. [CrossRef]
17. Biot, M.A. Theory of Propagation of Elastic Waves in a Fluid-Saturated Porous Solid. II. Higher Frequency Range. *J. Acoust. Soc. Am.* **1956**, *28*, 179–191. [CrossRef]
18. Dazel, O.; Brouard, B.; Dauchez, N.; Geslain, A. Enhanced Biot's finite element displacement formulation for porous materials and original resolution methods based on normal modes. *Acta Acust. United Acust.* **2009**, *95*, 527–538. [CrossRef]
19. Johnson, D.L.; Koplik, J.; Dashen, R. Theory of dynamic permeability and tortuosity in fluid-saturated porous media. *J. Fluid Mech.* **1987**, *176*, 379–402. [CrossRef]
20. Champoux, Y.; Allard, J. Dynamic tortuosity and bulk modulus in air-saturated porous media. *J. Appl. Phys.* **1991**, *70*, 1975–1979. [CrossRef]
21. Lafarge, D.; Lemarinier, P.; Allard, J.F.; Tarnow, V. Dynamic compressibility of air in porous structures at audible frequencies. *J. Acoust. Soc. Am.* **1997**, *102*, 1995–2006. [CrossRef]
22. Ogam, E.; Fellah, Z.E.A.; Sebaa, N.; Groby, J.P. Non-ambiguous recovery of Biot poroelastic parameters of cellular panels using ultrasonicwaves. *J. Sound Vib.* **2011**, *330*, 1074–1090. [CrossRef]
23. Atalla, Y.; Panneton, R. Inverse acoustical characterization of open cell porous media using impedance tube measurements. *Can. Acoust.* **2005**, *33*, 11–24.
24. Kutscher, K.; Geier, M.; Krafczyk, M. Multiscale simulation of turbulent flow interacting with porous media based on a massively parallel implementation of the cumulant lattice Boltzmann method. *Comput. Fluids* **2019**, *193*, 103733. [CrossRef]
25. Rosen, D.W. Research supporting principles for design for additive manufacturing. *Virtual Phys. Prototyp.* **2014**, *9*. [CrossRef]
26. Ring, T.P.; Langer, S.C. Design, Experimental and Numerical Characterization of 3D-Printed Porous Absorbers. *Materials* **2019**, *12*, 3397. [CrossRef] [PubMed]
27. Gebhardt, A. *Additive Fertigungsverfahren: Additive Manufacturing und 3D-Drucken für Prototyping—Tooling—Produktion*; 5. neu bearbeitete und erweiterte auflage ed.; Carl Hanser: München, Germany, 2016.
28. Kumke, M.; Watschke, H.; Hartogh, P.; Bavendiek, A.K.; Vietor, T. Methods and tools for identifying and leveraging additive manufacturing design potentials. *Int. J. Interact. Des. Manuf. (IJIDeM)* **2018**, *12*, 481–493. [CrossRef]
29. Cai, X.; Guo, Q.; Hu, G.; Yang, J. Ultrathin low-frequency sound absorbing panels based on coplanar spiral tubes or coplanar Helmholtz resonators. *Appl. Phys. Lett.* **2014**, *105*, 121901. [CrossRef]
30. Jiang, C.; Moreau, D.; Doolan, C. Acoustic Absorption of Porous Materials Produced by Additive Manufacturing with Varying Geometries. In Proceedings of the ACOUSTICS 2017, Perth, Australia, 19–22 November 2017.
31. Liu, Z.; Zhan, J.; Fard, M.; Davy, J.L. Acoustic properties of a porous polycarbonate material produced by additive manufacturing. *Mater. Lett.* **2016**, *181*, 296–299. [CrossRef]
32. Guild, M.D.; García-Chocano, V.M.; Kan, W.; Sanchez-Dehesa, J. Acoustic metamaterial absorbers based on multi-scale sonic crystals. *J. Acoust. Soc. Am.* **2014**, *136*, 2076. [CrossRef]
33. Liu, Z.; Zhan, J.; Fard, M.; Davy, J.L. Acoustic properties of multilayer sound absorbers with a 3D printed micro-perforated panel. *Appl. Acoust.* **2017**, *121*, 25–32. [CrossRef]
34. Fotsing, E.R.; Dubourg, A.; Ross, A.; Mardjono, J. Acoustic properties of periodic micro-structures obtained by additive manufacturing. *Appl. Acoust.* **2019**, *148*, 322–331. [CrossRef]
35. Cai, X.; Yang, J.; Hu, G.; Lu, T. Sound absorption by acoustic microlattice with optimized pore configuration. *J. Acoust. Soc. Am.* **2018**, *144*, EL138–EL143. [CrossRef]
36. Guild, M.D.; Rothko, M.; Sieck, C.F.; Rohde, C.; Orris, G. 3D printed sound absorbers using functionally-graded sonic crystals. *J. Acoust. Soc. Am.* **2018**, *143*, 1714. [CrossRef]

37. Zieliński, T.G.; Opiela, K.C.; Pawłowski, P.; Dauchez, N.; Boutin, T.; Kennedy, J.; Trimble, D.; Rice, H.; Van Damme, B.; Hannema, G.; et al. Reproducibility of sound-absorbing periodic porous materials using additive manufacturing technologies: Round robin study. *Addit. Manuf.* **2020**, *36*, 101564. [CrossRef]

38. Boulvert, J.; Costa-Baptista, J.; Cavalieri, T.; Perna, M.; Fotsing, E.R.; Romero-García, V.; Gabard, G.; Ross, A.; Mardjono, J.; Groby, J.P. Acoustic modeling of micro-lattices obtained by additive manufacturing. *Appl. Acoust.* **2020**, *164*, 107244. [CrossRef]

39. Setaki, F.; Tenpierik, M.; Turrin, M.; van Timmeren, A. Acoustic absorbers by additive manufacturing. *Build. Environ.* **2014**, *72*, 188–200. [CrossRef]

40. Boulvert, J.; Cavalieri, T.; Costa-Baptista, J.; Schwan, L.; Romero-García, V.; Gabard, G.; Fotsing, E.R.; Ross, A.; Mardjono, J.; Groby, J.P. Optimally graded porous material for broadband perfect absorption of sound. *J. Appl. Phys.* **2019**, *126*, 175101. [CrossRef]

41. Ring, T.; Kuschmitz, S.; Watschke, H.; Vietor, T.; Langer, S. Additive Fertigung und Charakterisierung akustisch wirksamer Materialien. In Proceedings of the Tagungsband DAGA 2018-44. Jahrestagung für Akustik, Munich, Germany, 19–22 March 2018; pp. 451–455.

42. Comiti, J.; Renaud, M. A new model for determining mean structure parameters of fixed beds from pressure drop measurements: Application to beds packed with parallelepipedal particles. *Chem. Eng. Sci.* **1989**, *44*, 1539–1545. [CrossRef]

43. Yun, M.; Yu, B.; Xu, P.; Wu, J. Geometrical models for tortuosity of streamlines in three-dimensional porous media. *Can. J. Chem. Eng.* **2006**, *84*, 301–309. [CrossRef]

44. Bo-Ming, Y.; Jian-Hua, L. A geometry model for tortuosity of flow path in porous media. *Chin. Phys. Lett.* **2004**, *21*, 1569. [CrossRef]

45. Iannace, G.; Ciaburro, G.; Trematerra, A. Modelling sound absorption properties of broom fibers using artificial neural networks. *Appl. Acoust.* **2020**, *163*, 107239. [CrossRef]

46. Ciaburro, G.; Iannace, G.; Ali, M.; Alabdulkarem, A.; Nuhait, A. An artificial neural network approach to modelling absorbent asphalts acoustic properties. *J. King Saud Univ.-Eng. Sci.* **2020**. [CrossRef]

47. Gardner, G.C.; O'Leary, M.E.; Hansen, S.; Sun, J. Neural networks for prediction of acoustical properties of polyurethane foams. *Appl. Acoust.* **2003**, *64*, 229–242. [CrossRef]

48. Liu, J.; Bao, W.; Shi, L.; Zuo, B.; Gao, W. General regression neural network for prediction of sound absorption coefficients of sandwich structure nonwoven absorbers. *Appl. Acoust.* **2014**, *76*, 128–137. [CrossRef]

49. Watschke, H.; Kuschmitz, S.; Heubach, J.; Lehne, G.; Vietor, T. A Methodical Approach to Support Conceptual Design for Multi-Material Additive Manufacturing. *Proc. Des. Soc. Int. Conf. Eng. Des.* **2019**, *1*, 659–668. [CrossRef]

50. Blösch-Paidosh, A.; Shea, K. Design Heuristics for Additive Manufacturing Validated Through a User Study1. *J. Mech. Des.* **2019**, *141*. [CrossRef]

51. Pradel, P.; Zhu, Z.; Bibb, R.; Moultrie, J. Investigation of design for additive manufacturing in professional design practice. *J. Eng. Des.* **2018**, *29*, 165–200. [CrossRef]

52. Iman, R.L. Latin Hypercube Sampling. In *Encyclopedia of Quantitative Risk Analysis and Assessment*; Melnick, E.L., Everitt, B.S., Eds.; John Wiley & Sons, Ltd.: Chichester, UK, 2008. [CrossRef]

53. Stocki, R. A method to improve design reliability using optimal Latin hypercube sampling. *Comput. Assist. Mech. Eng. Sci.* **2005**, *12*, 393–411.

54. Helton, J.C.; Davis, F.J. Latin hypercube sampling and the propagation of uncertainty in analyses of complex systems. *Reliab. Eng. Syst. Saf.* **2003**, *81*, 23–69. [CrossRef]

55. Sheikholeslami, R.; Razavi, S. Progressive Latin Hypercube Sampling: An efficient approach for robust sampling-based analysis of environmental models. *Environ. Model. Softw.* **2017**, *93*, 109–126. [CrossRef]

56. Florian, A. An efficient sampling scheme: Updated Latin Hypercube Sampling. *Probabilistic Eng. Mech.* **1992**, *7*, 123–130. [CrossRef]

57. Iman, R.L.; Conover, W.J. A distribution-free approach to inducing rank correlation among input variables. *Commun. Stat.-Simul. Comput.* **1982**, *11*, 311–334. [CrossRef]

58. McKay, M.D.; Beckman, R.J.; Conover, W.J. A Comparison of Three Methods for Selecting Values of Input Variables in the Analysis of Output from a Computer Code. *Technometrics* **1979**, *21*, 239. [CrossRef]

59. OpenSCAD. Available online: https://github.com/openscad/openscad (accessed on 23 February 2021).

60. Ngo, T.D.; Kashani, A.; Imbalzano, G.; Nguyen, K.T.; Hui, D. Additive manufacturing (3D printing): A review of materials, methods, applications and challenges. *Compos. Part B Eng.* **2018**, *143*, 172–196. [CrossRef]

61. *Acoustics—Materials for Acoustical Applications—Determination of Airflow Resistance*; Technical Report ISO 9053-2:2020; International Standards Organization: Geneva, Switzerland, 2020.

62. *Acoustics—Determination of Sound Absorption Coefficient and Impedance in Impedance Tubes—Part 2: Transfer-Function Method*; Technical Report ISO 10534-2:1998; International Standards Organization: Geneva, Switzerland, 1998.

63. APMR—Acoustical Porous Material Recipes. Available online: https://apmr.matelys.com/ (accessed on 22 February 2021).

64. Beck, S.C.; Müller, L.; Langer, S.C. Numerical assessment of the vibration control effects of porous liners on an over-the-wing propeller configuration. *CEAS Aeronaut. J.* **2016**, *7*, 275–286. [CrossRef]

65. Allard, J.; Atalla, N. *Propagation of Sound in Porous Media: Modelling Sound Absorbing Materials 2e*; John Wiley & Sons: Hoboken, NJ, USA, 2009.

66. Storn, R.; Price, K. Differential evolution–a simple and efficient heuristic for global optimization over continuous spaces. *J. Glob. Optim.* **1997**, *11*, 341–359. [CrossRef]
67. Virtanen, P.; Gommers, R.; Oliphant, T.E.; Haberland, M.; Reddy, T.; Cournapeau, D.; Burovski, E.; Peterson, P.; Weckesser, W.; Bright, J.; et al. SciPy 1.0: Fundamental Algorithms for Scientific Computing in Python. *Nat. Methods* **2020**, *17*, 261–272. [CrossRef] [PubMed]
68. Mitchell, R.; Michalski, J.; Carbonell, T. *An Artificial Intelligence Approach*; Springer: Berlin/Heidelberg, Germany, 2013.
69. Kubat, M. *An Introduction to Machine Learning*; Springer: Berlin/Heidelberg, Germany, 2017.
70. Ring, T.P. Effiziente Unsicherheitsquantifizierung in der Akustik mittels eines Multi-Modell-Verfahrens. Ph.D. Thesis, TU Braunschweig, Braunschweig, Germany, 2020.
71. Pedregosa, F.; Varoquaux, G.; Gramfort, A.; Michel, V.; Thirion, B.; Grisel, O.; Blondel, M.; Prettenhofer, P.; Weiss, R.; Dubourg, V.; et al. Scikit-learn: Machine Learning in Python. *J. Mach. Learn. Res.* **2011**, *12*, 2825–2830.
72. Ho, T.K. Nearest neighbors in random subspaces. In *Joint IAPR International Workshops on Statistical Techniques in Pattern Recognition (SPR) and Structural and Syntactic Pattern Recognition (SSPR)*; Springer: Berlin/Heidelberg, Germany, 1998; pp. 640–648.
73. Kramer, O. K-nearest neighbors. In *Dimensionality Reduction with Unsupervised Nearest Neighbors*; Springer: Berlin/Heidelberg, Germany, 2013; pp. 13–23.
74. Laaksonen, J.; Oja, E. Classification with learning k-nearest neighbors. In Proceedings of the International Conference on Neural Networks (ICNN'96), Washington, DC, USA, 3–6 June 1996; Volume 3, pp. 1480–1483.
75. Rosenblatt, F. The perceptron: A probabilistic model for information storage and organization in the brain. *Psychol. Rev.* **1958**, *65*, 386–408. [CrossRef]
76. Rosenblatt, F. Perceptron simulation experiments. *Proc. IRE* **1960**, *48*, 301–309. [CrossRef]
77. LeCun, Y.; Bengio, Y.; Hinton, G. Deep learning. *Nature* **2015**, *521*, 436–444. [CrossRef]
78. LeNail, A. NN-SVG: Publication-Ready Neural Network Architecture Schematics. *J. Open Source Softw.* **2019**, *4*, 747. [CrossRef]
79. Barrett, J.P. The Coefficient of Determination—Some Limitations. *Am. Stat.* **1974**, *28*, 19–20. [CrossRef]
80. Nagelkerke, N.J. A note on a general definition of the coefficient of determination. *Biometrika* **1991**, *78*, 691–692. [CrossRef]

Review

Additive Manufacturing Processes in Medical Applications

Mika Salmi

Department of Mechanical Engineering, Aalto University, 02150 Espoo, Finland; mika.salmi@aalto.fi

Abstract: Additive manufacturing (AM, 3D printing) is used in many fields and different industries. In the medical and dental field, every patient is unique and, therefore, AM has significant potential in personalized and customized solutions. This review explores what additive manufacturing processes and materials are utilized in medical and dental applications, especially focusing on processes that are less commonly used. The processes are categorized in ISO/ASTM process classes: powder bed fusion, material extrusion, VAT photopolymerization, material jetting, binder jetting, sheet lamination and directed energy deposition combined with classification of medical applications of AM. Based on the findings, it seems that directed energy deposition is utilized rarely only in implants and sheet lamination rarely for medical models or phantoms. Powder bed fusion, material extrusion and VAT photopolymerization are utilized in all categories. Material jetting is not used for implants and biomanufacturing, and binder jetting is not utilized for tools, instruments and parts for medical devices. The most common materials are thermoplastics, photopolymers and metals such as titanium alloys. If standard terminology of AM would be followed, this would allow a more systematic review of the utilization of different AM processes. Current development in binder jetting would allow more possibilities in the future.

Keywords: additive manufacturing; rapid manufacturing; rapid prototyping; 3D printing; medical; implants; dental; processes; methods; clinical

Citation: Salmi, M. Additive Manufacturing Processes in Medical Applications. *Materials* **2021**, *14*, 191. https://doi.org/10.3390/ma14010191

Received: 26 November 2020
Accepted: 27 December 2020
Published: 3 January 2021

Publisher's Note: MDPI stays neutral with regard to jurisdictional claims in published maps and institutional affiliations.

1. Introduction

Additive manufacturing (AM), or, in a non-technical context, 3D printing, is a process where physical parts are manufactured using computer-aided design and objects are built on a layer-by-layer basis [1]. Usually, these procedures are called toolless processes. There are other processes, such as incremental sheet forming or laser forming, that build objects on a layer-by-layer basis as well but do so by adding the form, not the material [2,3]. These processes are not counted as an additive manufacturing process even though they have been similarly used in making, for example, customized medical products [4,5]. Currently, additive manufacturing is utilized and being investigated for use in areas such as the medical, automotive, aerospace and marine industries, as well as industrial spare parts [6–10]. Additive manufacturing is referred to as a manufacturing method where complexity or customization is free [11]. However, this requires marking and tracing of the different parts compared to mass production of the same kind of parts. Nevertheless, when comparing AM against conventional manufacturing, it has a much higher potential for customization and complex geometries. However, when comparing cost, additive manufacturing is usually not cheaper if the geometry is designed for mass production and only the manufacturing cost is calculated [12]. It would suffice to reiterate the whole product design and look at the economics over the entire product lifecycle [13]. AM is currently developing fast, and new players are entering the market all the time. There have been substantial investments in new companies, such as Carbon, Desktop Metal and Formlabs, as well as internal development in large companies in other areas, such as HP and GE. Even though the basic principles of the different AM processes have stayed the same, there are now more development resources to take the next step forward for these technologies, and this will also open up new possibilities in medical applications [14–17].

In the medical field, every patient is unique, and therefore, AM has a high potential to be utilized for personalized and customized medical applications. The most common medical clinical uses are personalized implants, medical models and saw guides [18]. In the dental field, AM is utilized on splints, orthodontic appliances, dental models and drill guides. However, AM has also been explored for making artificial tissues and organs [19]. In medicine, there is a background in digitalization of medical imaging, and that digitalization allows for reconstructing 3D models from patients' anatomy. A typical workflow for personalized medical devices starts with imaging or capturing the patient's geometry using computed tomography or other 3D scanning methods [20]. Then, these data are manipulated to obtain a 3D model of the patient's anatomy, and this can be an example already of additive manufacturing such as a medical model. Moreover, the geometry can be utilized to design patient-specific implants, and this design can be additively manufactured. After manufacturing, there is quite often a need for post-processing, such as polishing [21]. When the medical device is ready, the final step is the clinical application and follow-up.

The usage of AM is usually related to the question of what the benefits are compared to existing processes and technologies. Most often, the questions are related to whether it is cheaper to manufacture, but the whole lifecycle of the product and process should be investigated. The actual manufacturing prices cannot be the only performance indicator. Table 1 summarizes some of the benefits of AM in the medical and dental fields. Quite often, similar benefits can be found in other subject areas than medical and dental fields, for example, the industrial side, such as digital storage for industrial spare parts, which reflects heavily to digital storage of dental data.

Table 1. Some of the benefits of additive manufacturing (AM) in medical and dental fields.

Reference	Findings	Area
Ballard et al. [22]	cost and time savings	Orthopedic and maxillofacial surgery
Choonora et al. [23]	personalization	Transplants
Mahmoud et al. [24]	cost savings	Pathology specimens for students
Tack et al. [25]	time savings, improved medical outcome, decreased radiation exposure	Surgery
Ballard et al. [26]	incorporation of antibiotics	Implants
Lin et al. [27]	personalization, cost savings	Dental
Javaid et al. [28]	cost and time savings, personalization, digital storage	Dental
Aho et al. [29]	personalization	Pharmacy
Salmi et al. [30]	reduction of manual work	Dental appliances
Aquino et al. [31]	personalization, on-demand manufacturing	Pharmacy
Javaid et al. [32]	accuracy, cost and time savings, personalization, fully automated and digitized manufacturing	Orthopedics
Emelogu et al. [33]	supply chain possibilities	Implants
Gibson et al. [10]	surgeon as designer, innovation potential	Surgery
Haleem et al. [34]	ability to use different materials	Medical
Murr et al. [35]	ability to make complex geometries	Implants
Peltola et al. [36]	template for forming implants	Implants
Ramakrishnaiah et al. [37]	rough and porous surface texture, better stabilization and osseointegration	Dental implants
Nazir et al. [38]	design iterations, supply chain possibilities, complex geometries	Medical devices
Yang et al. [39]	improved understanding of anatomy and accuracy of surgery	Surgery

Since AM is a class of manufacturing processes, it is important to understand what the bases of these processes are, how those differ from each other and to describe how the process works. This review aims to explore which AM processes and materials are utilized in medical and dental applications. It especially focuses on which processes are less studied to determine research gaps. The limitation of this study is that the aim was not to explore all the possible materials used in the applications.

The current review was guided by the following research questions:

1. What are the basic benefits of AM in medical applications?
2. What AM processes based on ISO/ASTM process classification are utilized in medical applications?
3. What are the example materials utilized in founded process and application combinations?
4. Based on the findings, what are the process and application areas that could show future scientific potential?

2. Additive Manufacturing Processes

The ASTM and ISO standardization organization categorizes the AM process into seven different categories: powder bed fusion (PBF), material extrusion (ME), VAT photopolymerization (VP), material jetting (MJ), binder jetting (BJ), sheet lamination (SL) and directed energy deposition (DED) [1]. Each category includes many different vendors, solutions and material options. In this article, ASTM/ISO categories were followed. This was problematic, since the standard terminology is still not utilized in most studies, and often trade names are used for processes. To clarify different processes and principles, Table 2 lists the names of the process classes and a short description, common starting material form, trade names and how well the process is used to manufacture the plastic type of materials, metals or ceramics. Some of the processes for certain materials are in the development and research phase, such as directed energy deposition VAT photopolymerization and material jetting for metals, and some seem not to exist at all, such as sheet lamination of ceramics or directed energy deposition of plastics and ceramics. It is possible that there are scientific studies and trials of these, but no commercial providers exist. Commonly, new process and material combinations are developed based on demand, which highlights large industries and a substantial need. Usually, this leads to the selection of a commonly used material since that can be utilized in many areas.

Table 2. Characteristics of different AM processes.

AM Process	Short Description	Material Form	Plastics	Metals	Ceramics	Trade/Other Names
Powder bed fusion (PBF)	thermal energy fuses regions of a powder bed	powder	+++	+++	+	selective laser sintering (SLS), direct metal laser sintering (DMLS), selective laser melting (SLM)
Material extrusion (MEX)	material dispensed through a nozzle	filament, pellets, paste	+++	++	++	fused deposition modeling (FDM), (fused filament fabrication) FFF
VAT photo-polymerization (VP)	liquid photopolymer in a vat is cured by light	liquid	+++	+	++	SLA, digital light projection (DLP)
Material jetting (MJ)	droplets of material are selectively deposited	liquid	+++	+	+	PolyJet, NJP
Binder jetting (BJ)	a liquid bonding agent is selectively deposited	powder	+++	++	+	3D printing (3DP), ColorJet printing (CJP)
Sheet lamination (SL)	sheets of material are bonded	sheets	++	++	-	laminated object manufacturing (LOM), ultrasonic additive manufacturing (UAM)
Directed energy deposition (DED)	focused thermal energy used to fuse materials by melting when depositing	powder, wire	-	+++	+	laser-engineered net shaping (LENS), EBAM

Note: +++, widely available/many studies exist; ++, available/several studies exist; +, R&D phase/studies exist; -, no studies exist.

Each AM process and piece of equipment require a 3D model of the object that they will manufacture, and the most used format for that is stereolithography, standard triangle language, standard tessellation language (STL). The STL model is then sliced into layers and further processed to commands for the specific AM machine. To additively manufacture the part, a raw material is required, such as power, filament, liquid, paste sheet or pellets. The raw material can then be, for example, melted, dispensed, cured or fused to make parts on a layer-by-layer basis. Terminology in AM varies and, as an example, the powder bed fusion process can be called selective laser melting (SLM), selective laser sintering (SLS) or direct metal laser sintering (DMLS). For material extrusion, the most used terms are fused deposition modeling (FDM) or fused filament fabrication (FFF). As a first invented AM process, stereolithography (SLA) has been very commonly used for processes in the VAT photopolymerization class, but digital light projection (DLP) is also used if the light source is a DPL projector. Trade names in material jetting are PolyJet and NanoParticle Jetting. Binder jetting is often called 3D printing (3DP) or ColorJet printing (CJP). Sheet lamination processes are laminated object manufacturing (LOM) and ultrasonic additive manufacturing (UAM). Directed energy deposition processes are laser-engineered net shaping (LENS) and electron beam additive manufacturing. In addition, many others exist on the market.

3. Medical Applications of Additive Manufacturing

Medical applications of additive manufacturing can be classified in several ways [40,41], but this article follows application classes-based classification. AM applications can be classified into the following classes: "models for preoperative planning, education and training", "inert implants", "tools, instruments and parts for medical devices", "medical aids, supportive guides, splints and prostheses" and "biomanufacturing" [42]. For a more general classification, this can be modified so that implants do not need to be inert, and models for preoperative planning, education and training could also include postoperative and operative models using the term "medical models". Figure 1 shows an example of an application in each category including a (a) preoperative model of a skull and heart, (b) craniomaxillofacial implants, (c) a dental drilling guide, reduction forceps, nasal and throat swabs, (d) personalized and mobilizing external support and (e) a scaffold for zygomatic bone replacement and resorbable orbital implants.

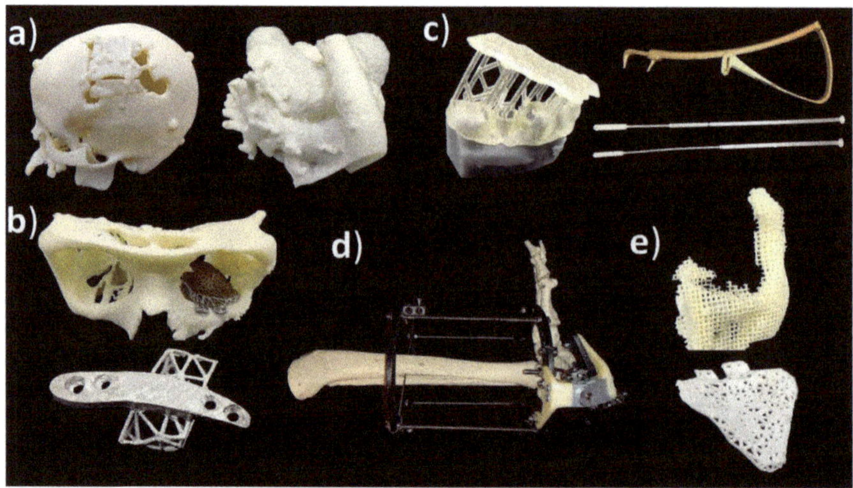

Figure 1. (**a**) Medical models; (**b**) implants; (**c**) tools, instruments and parts for medical devices; (**d**) medical aids, supportive guides, splints and prostheses; (**e**) biomanufacturing.

Classification of medical applications of additive manufacturing:

1. Medical models;
2. Implants;
3. Tools, instruments and parts for medical devices;
4. Medical aids, supportive guides, splints and prostheses;
5. Biomanufacturing.

3.1. Medical Models

Medical models are based on patient anatomy, and they can be used for pre- and postoperative operative planning and training; training medical students; and informing patients and patients' families [25,43]. The geometry can be transformed, for example, by taking only interesting sections or scaling it up or down. If models are used for training, such as bone drilling, haptic response might be desirable to be close to the bone. Medical models are widely used in the craniomaxillofacial area, but there are also cases, for example, from different limbs and other bone structures such as the spine and pelvis [25,44]. If these are utilized in the operating theater, it might be recommended that the models be sterilized, but usually, the material option can be quite freely selected which highlights also that these are one of the most common applications. Figure 2 shows a typical process workflow for manufacturing medical models starting from patient anatomy captured via medical imaging, such as computed tomography (CT), magnetic resonance imaging (MRI) or ultrasound, followed by constructing a 3D model geometry for AM using segmentation algorithms [45,46]. After AM, there is often a need for postprocessing such as removing the support structures.

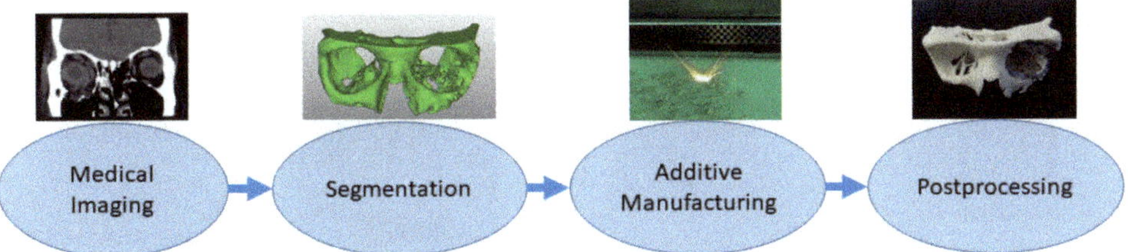

Figure 2. Typical process flow for medical models.

3.2. Implants

Implants are directly or indirectly additively manufactured to replace defective or missing tissue [47,48]. This class also includes dental applications such as crowns and bridges [49]. The material needs to be tissue-compatible and requirements are strict, and approval processes take a long time. Surface properties might affect cell adhesion. Some of the latest studies have explored how to embed materials inside implants, for example, as a type of drug delivery system [50,51]. In personalized implants, AM is a favorable solution, and a typical process requires the capture of a patient's anatomy similar to medical models. Then, this digital 3D model of the patient anatomy is used as a design reference to enable patient-specific fitting [52,53]. Most typical implants are made from metals using the powder bed fusion process, and this requires different postprocessing steps such as machining the supports, polishing and heat treatments. Before clinical operation, implants need to be sterilized. Figure 3 shows the typical process flow for implants made by additive manufacturing starting from medical imaging and segmentation followed by 3D modeling of the implant proceeding to AM, postprocessing and sterilization.

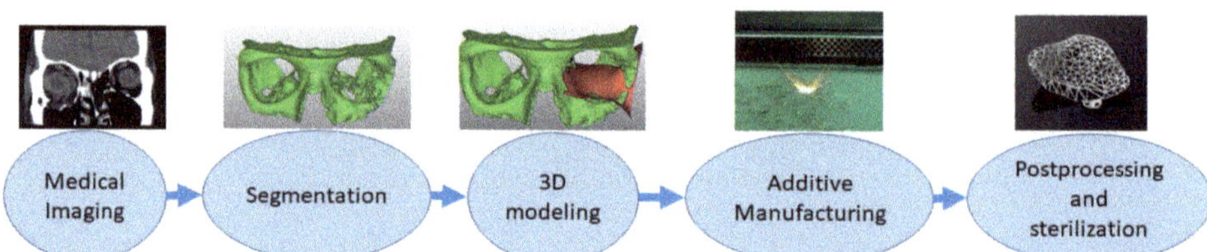

Figure 3. Typical process flow for implants.

3.3. Tools, Instruments and Parts for Medical Devices

Tools, instruments and parts for medical devices allow or enhance a clinical operation. They might utilize patient-specific dimensions and shapes, for example, in drilling guides [54], and can be invasive and need a sterilization process, since they can be in contact with body fluids, membranes, tissues and organs for a limited time. This class includes surgical instruments and orthodontic appliances [55–57]. One of the largest and most successful businesses in this class is using the VAT photopolymerization process to create molds for vacuum forming clear orthodontic aligners [58]. When patient-specific dimensions are utilized, the process is similar to that of implants and preoperative models from medical imaging or 3D scanning. 3D modeling can be conducted by referring to the 3D model of the patient's anatomy or from scratch if a patient-specific geometry or fitting is not needed. Postprocessing might include support removal, heat treatments, machining and sterilization. Tools, instruments and parts for medical devices are typically made with the process flow shown in Figure 4. For example, the process starts by taking an impression of the patient's teeth, 3D scanning it, followed by 3D modeling, VAT photopolymerization AM, postprocessing and using the part made as a mold for soft orthodontic aligners.

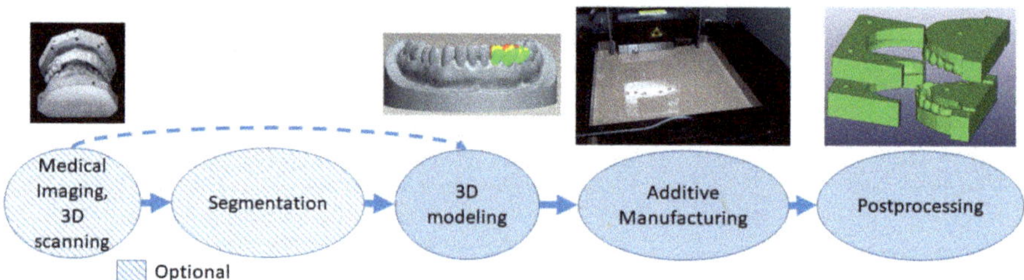

Figure 4. Typical process flow for tools, instruments and parts for medical devices.

3.4. Medical Aids, Supportive Guides, Splints and Prostheses

In this class, parts made with additive manufacturing are external to the body, and these can be combined with standard appliances to allow customization. Long-term and postoperative supports, motion guides, fixators, external prostheses, prosthesis sockets, personalized splints and orthopedic applications are examples of applications in this class [59–61]. The process can start from medical imaging followed by segmentation, 3D scanning or 3D measurements that can provide data directly for use in the 3D modeling phase. Alternative manufacturing methods for additive manufacturing are quite often computer numerical control (CNC) technologies [62]. Parts may require different kinds of postprocessing depending on the application such as support removal, heat treatments and painting or coating. The typical process flow for medical aids, supportive guides, splints and prostheses using AM is presented in Figure 5. The example case is a personalized and mobilizing external support for a pilon fracture, where 3D modeling is based on measuring the patient's ankle movement and adjusting the additive manufacturing pieces to locate the hinge so that it controls the movement under force close to the free movement of the ankle.

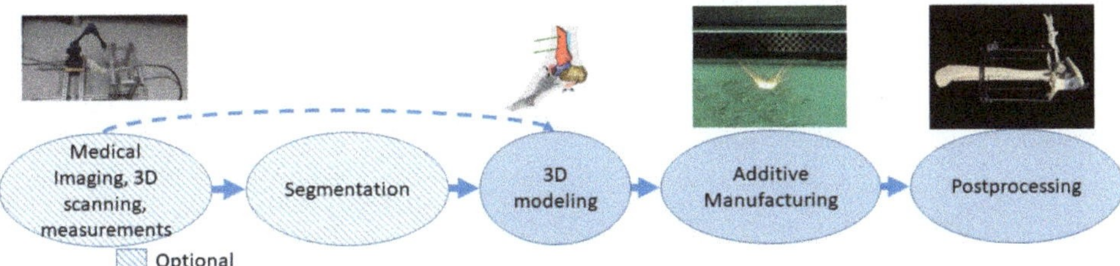

Figure 5. The typical process flow for medical aids, supportive guides, splints and prostheses.

3.5. Biomanufacturing

Biomanufacturing is a combination of additive manufacturing and tissue engineering [63]. Materials need to be biologically compatible and often active with the body so many different polymers, ceramics and composite materials are used [64]. Porous structures with cultivation and a 3D matrix can affect cell specialization. The materials can be osteoinductive, osteoconductive or resorbable [65]. Shapes can be personalized to correspond to defects [66]. For personalized shapes, the patient's geometry needs to be captured using medical imaging or 3D scanning. In the 3D modeling phase, micro- and macrostructures are modeled, and porous structures are often used for attracting cells and cell growth. The process often needs to be sterile or parts made with the ability to be sterilized after printing. Before final application, there might also be the need for cell growth in vitro or in vivo. Figure 6 shows an example of an orbital floor resorbable implant stating patient geometry with CT and segmentation followed by 3D modeling and AM of the implant. After manufacturing, the implant is sterilized.

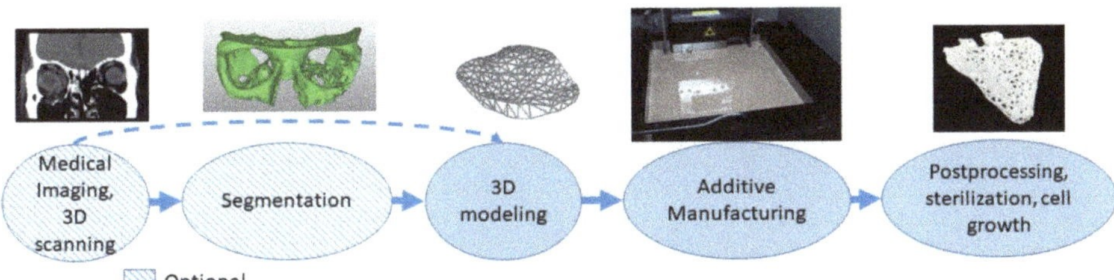

Figure 6. Typical process flow for biomanufacturing.

4. Different AM Processes in Medical Applications

Different AM processes are utilized in the applications described in Section 3. A search for the applications was first performed using ISO/ASTM terminology with a combination of AM medical application terms in a specific class and followed with a search using trade names or other commonly used names for the processes, such as the manufacturer name, for the hard-to-find application areas for certain process. The aim was to find at least some examples for each category and, moreover, determine which areas lack those applications and processes and why. The databases used for the search were Scopus, Web of Science and Google Scholar. When at least three certain application class and processes were found, the search focus was then directed to other processes and applications. More specific search terms are shown in Table 3.

Table 3. Search terms.

AM Process	Application	and Process Term	or Manufacturer
PBF		powder bed fusion or PFB or selective laser sintering or SLS or direct laser sintering or DMLS	-
MEX	medical or dental or implants or surgery or clinical	material extrusion or fused filament fabrication or FFF or fused deposition modeling of FDM	-
VP		VAT photopolymerization or stereolithography or SLA	-
MJ		material jetting or Polyjet or nano particle jetting	Objet
BJ		binder jetting or Colorjet printing	Zcorp or Zprinter
SL		sheet lamination or LOM or laminated object manufacturing	Mcor or Fabrisonic
DED		directed energy deposition or DED or laser engineered net shaping or LENS	-

There are previous studies regarding certain processes and/or application areas of medical applications of AM such as powder bed fusion of metal implants [67], additive manufacturing of medical instruments [55], biomaterials in medical additive manufacturing [68] and medical phantoms and regenerated tissue and organ applications with additive manufacturing [69]. Previous studies have not usually classified the AM processes or reviewed only a single process. Some studies focused only on utilized materials and some only on applications without any information about the AM processes or materials. Based on findings from the literature, Table 4 shows the different AM processes and materials used or explored in the medical application classes formed.

Table 4. Different AM processes in medical applications.

Application Area	PBF	MEX	VP	MJ	BJ	SL	DED
Medical models [44,50,70–79]	PA, PP	ABS+, PLA	Photocurable resin	VeroWhite, VeroClear, TangoPlus, Multi-material	ZP150, ZP151, PMMA	Paper	
Implants [37,47,49,50,67,80–88]	Ti6Al4VTi64, Co–Cr–Mo, Al2O3–ZrO2	PEEK	Clear resin V4, ATZ, NextDent C&B		ZP150, TCP, nickel-based alloy 625, Titanium		Ti6Al4V
Tools, instruments and parts for medical devices [54–56,89–96]	PA, Co–Cr, Ti	ABS, ABS+, PLA,	ProtoGen O-XT 18420, Dental SG, Dental LT, Clear resin V2, Photocurable resin WaterShed XC 11122	TangoPlus, HeartPrint Flex, MED610		Paper	
Medical aids, supportive guides, splints and prostheses [60,61,97–101]	PA	ABS, PLA, Nylon	Clear resin, Ciba–Geigy 5170, Somos 6110, Epoxy	Multi-material, Full Cure 720, ABS like, VeroWhite	ZP151, Stainless steel		
Biomanufacturing [63,102–107]	PLA, PLGA	PCL, PLA, PLGA, TCP	PDLLA, HA		VisiJet PXL, Calcium phosphate, barium titanate		

Note: polyamide (PA), polypropylene (PP), polyether ether ketone (PEEK), acrylonitrile butadiene styrene (ABS), polylactic acid (PLA), poly lactic-co-glycolic acid (PLGA), polycaprolactone (PCL), tricalcium phosphate (TCP), alumina toughened zirconia (ATZ), poly(DL-lactic acid) (PDLLA), hydroxyapatite (HA), Poly(methyl methacrylate) (PMMA).

5. Discussion and Conclusions

5.1. Limitations of Previous Reviews

There are many reviews on the medical applications of AM [28,32,40,55,108–115]. Most of them focus on certain application areas, such as surgical instruments [55], orthopedics [32], cardiology [108], or have a material focus in certain area such as metals for prosthodontic applications [115]. Some studies have looked at different AM processes, for example, material extrusion and binder jetting for producing pharmaceuticals [116] or ceramics for dental applications using material extrusion, binder jetting, VAT photopolymerization or powder bed fusion processes [111]. Furthermore, some only looked at the general applications of AM [112] or the whole process chain developments including, for example, the design phase [40]. There are reviews which focus on applications and even categorize those in some of the AM processes, such as VAT photopolymerization, powder bed fusion and material extrusion [109,110], but the other processes are missing. In dentistry, there are reviews with a material focus, categorizing different AM processes, such as material extrusion, binder jetting, VAT photopolymerization, powder bed fusion, material jetting and binder jetting [28,113], or focusing tightly, for example, on metallic implants in dentistry utilizing the powder bed fusion process [114].

However, usually, the limitations are that the utilized AM processes are not described at all, trade names, etc., are used to describe the technology or the reviews do not tie the applications to the particular AM processes. This is biggest obstacle for systematic reviews and statistics. As an example, the material extrusion process in the literature can be found with the terms: fused filament fabrication, fused deposition modeling and filament freeform fabrication. These terms are related because Stratasys registered and trademarked FDM; thus, other producers needed to invent an alternative term. Similar challenges also occur in other fields in AM. On the other hand, materials are usually well described, and the most common ones are thermoplastics, photopolymers and metals such as titanium alloys.

5.2. Processes Utilized Rarely—DED, SL

Directed energy deposition is utilized mostly only in implants, and even in those, it is quite a rare process. This might be related to the poor accuracy and surface quality. Commonly, the process is used only for metal materials. One possible application would be to explore repairing parts for medical devices, since in other industries, it is already explored for repairing applications [117]. However, strict regulations in medical devices might limit the utilization since each repair might request its own approval process if the repair process is too similar in all cases. The materials usually used in directed energy deposition are metals, such as titanium alloys, for implants and, therefore, it limits its use, for example, in medical models. Sheet lamination is only used for medical models or phantoms [72,95], which makes sense, since it can be used to make full-color realistic looking models [118]. Other areas using starting material as a sheet are challenging, since new material examples are not usually in the form of sheets. On the metal side, the ability to combine different metals in different layers would open new potential, for example, by adding functional possibility through bi-metal parts for certain implants and medical devices [119]. Sheet lamination devices are quite rare, and there are only two different vendors in the market: metal-based Fabrisonic and paper-based Mcor, which seems to be in receivership status.

5.3. Well Established Processes—PBF, MEX, VP

Powder bed fusion, material extrusion and VAT photopolymerization are utilized in all categories and are well-established processes for all the categories of medical and dental applications [67,120,121]. This might be explained as they are also the most common processes in the industry; there are affordable devices available for research, and as starting materials to explore, they are easily found in powder, resin, pellet or paste forms. In addition, mixing new materials with existing ones is achievable. For implants, the most common materials are metals, especially titanium alloys and for other applications, polymers such as polyamide, ABS, Nylon and different photocurable resins. Material extrusion of reinforced composites would be interesting material where light weight is required, such as for prostheses, and there are new developments in the material extrusion of metals which could work in many categories [122–125].

5.4. Processes Well Established in Some Application Areas—MJ, BJ

Material jetting is not used for implants and biomanufacturing. The reason for this may be that the technology is quite challenging, as the material needs to be pushed through multiple small nozzles similarly to inkjet 2D printing. Otherwise, material jetting is a quite accurate process, so it is very well suitable for medical models requiring high accuracy. Material jetting is also capable of making parts with multiple materials and colors, which is already utilized in medical models [126].

Binder jetting is not utilized for tools, instruments and parts for the medical devices category. The reason might be that parts made with binder jetting are not usually very strong compared to other methods [107,127]. For medical models, it is an excellent technique since it can use colors and does not need support structures and the process is fast and affordable [128]. The most common materials are the gypsum-based powders ZP150 and ZP151. Moreover, metals and biomaterials are used. Currently, HP and Desktop Metal are developing production systems for metal parts based on binder jetting, and this would allow the utilization of the binder jetting process very well also for tools, instruments and parts for medical devices [129].

5.5. Future Possibilities

In the future, it is recommended to utilize a standard terminology in the research conducted on medical and dental applications of additive manufacturing as well as industrial side [130]. This would allow a more systematic comparison of the utilization of different AM methods. Current and future developments in the AM processes, devices and materials would allow for increasing applications in the medical and dental area (Figure 7). It is possible that some of the processes, such as sheet lamination, will not be used at all in the future, or it might be that there will be new players in the market. However, sheet lamination of metals provides interesting possibilities for multi-metals. Some processes, such as binder jetting and material jetting, have a high potential to be more widely used in medical and dental applications with the possibility to make metal parts. Directed energy deposition has potential in repairing metallic parts and material extrusion for composite parts. Material extrusion and material jetting will have possibilities in multi-material parts. Research gaps and, thus, future research possibilities were determined by utilizing the development and possibilities of AM technologies (Table 2) and comparing those to findings from literature (Table 4).

Future research possibilities in medical AM with different processes can be seen in the following areas:

1. Directed energy deposition—repairing medical parts especially in tools, instruments and parts for medical devices;
2. Sheet lamination—multi-metal parts in medicine, especially in tools, instruments and parts for medical devices;
3. Material extrusion—composite parts and multi-material, especially in medical aids, supportive guides, splints and prostheses;
4. Material extrusion—metal parts especially in implants and tools, instruments and parts for medical devices;
5. Binder jetting—metal parts especially in implants and tools, instruments and parts for medical devices;
6. Material jetting—multi-material parts, especially in medical models and biomanufacturing.

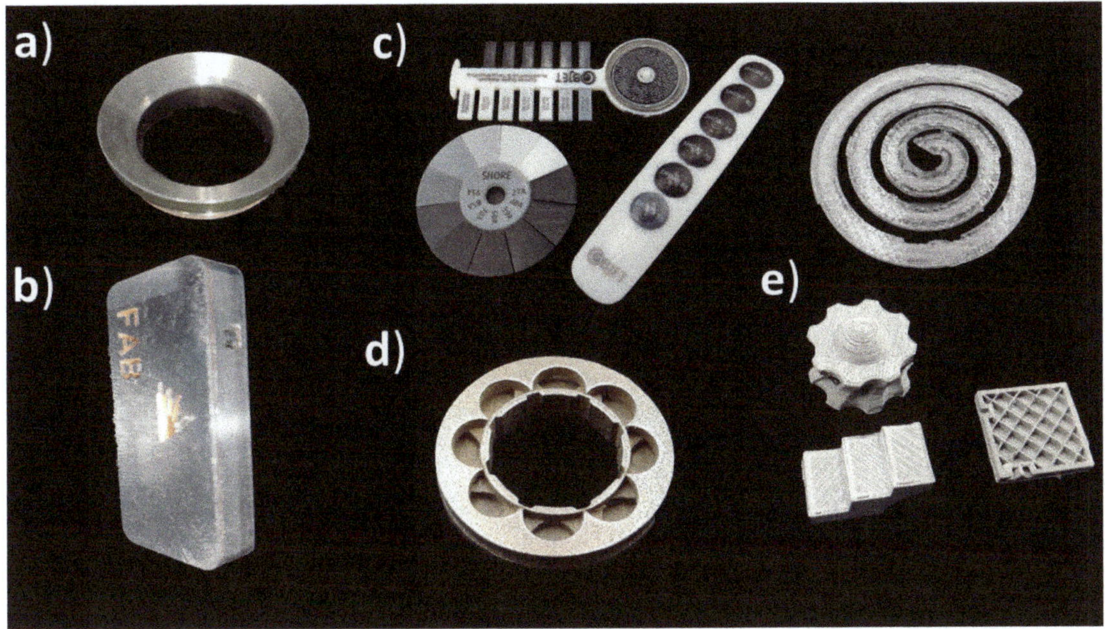

Figure 7. Future possibilities for AM in the medical and dental fields. (**a**) Directed energy deposition repairing, (**b**) multi-material metal parts with sheet lamination, (**c**) multi-material parts with material jetting, (**d**) binder jetting for metal parts, (**e**) material extrusion of metal and composite parts.

Funding: This research was funded by the Academy of Finland (grant number 325509) within the framework of the project "Direct Digital Manufacturing in Health Care Production and Operations (DiDiMinH)".

Conflicts of Interest: The author declares no conflict of interest.

References

1. ASTM International. *ISO/ASTM52900—15 Standard Terminology for Additive Manufacturing—General Principles—Terminology*; ASTM International: West Conshohocken, PA, USA, 2015.
2. Lehtinen, P.; Väisänen, T.; Salmi, M. The effect of local heating by laser irradiation for aluminum, deep drawing steel and copper sheets in incremental sheet forming. *Phys. Procedia* **2015**, *78*, 312–319. [CrossRef]
3. Casalino, G.; Ludovico, A.D.; Ancona, A.; Lugarà, P.M. Stainless Steel 3D Laser Forming for Rapid Prototyping. In *International Congress on Applications of Lasers & Electro-Optics*; Laser Institute of America: Orlando, FL, USA, 2001; Volume 2001, pp. 808–816.
4. Ambrogio, G.; De Napoli, L.; Filice, L.; Gagliardi, F.; Muzzupappa, M. Application of Incremental Forming process for high customised medical product manufacturing. *J. Mater. Process. Technol.* **2005**, *162*, 156–162. [CrossRef]
5. Eksteen, P.D.; Van der Merwe, A.F. Incremental sheet forming (ISF) in the manufacturing of titanium based plate implants in the bio-medical sector. In Proceedings of the 42nd Computers and Industrial Engineering, Cape Town, South Africa, 16–18 July 2012; pp. 15–18.
6. Calle, M.A.; Salmi, M.; Mazzariol, L.M.; Alves, M.; Kujala, P. Additive manufacturing of miniature marine structures for crashworthiness verification: Scaling technique and experimental tests. *Mar. Struct.* **2020**, *72*, 102764. [CrossRef]
7. Kestilä, A.; Nordling, K.; Miikkulainen, V.; Kaipio, M.; Tikka, T.; Salmi, M.; Auer, A.; Leskelä, M.; Ritala, M. Towards space-grade 3D-printed, ALD-coated small satellite propulsion components for fluidics. *Addit. Manuf.* **2018**, *22*, 31–37. [CrossRef]
8. Delic, M.; Eyers, D.R. The effect of additive manufacturing adoption on supply chain flexibility and performance: An empirical analysis from the automotive industry. *Int. J. Prod. Econ.* **2020**, *228*, 107689. [CrossRef]
9. Kretzschmar, N.; Chekurov, S.; Salmi, M.; Tuomi, J. Evaluating the readiness level of additively manufactured digital spare parts: An industrial perspective. *Appl. Sci.* **2018**, *8*, 1837. [CrossRef]
10. Gibson, I.; Srinath, A. Simplifying medical additive manufacturing: Making the surgeon the designer. *Procedia Technol.* **2015**, *20*, 237–242. [CrossRef]

11. Conner, B.P.; Manogharan, G.P.; Martof, A.N.; Rodomsky, L.M.; Rodomsky, C.M.; Jordan, D.C.; Limperos, J.W. Making sense of 3-D printing: Creating a map of additive manufacturing products and services. *Addit. Manuf.* **2014**, *1*, 64–76. [CrossRef]
12. Baumers, M.; Dickens, P.; Tuck, C.; Hague, R. The cost of additive manufacturing: Machine productivity, economies of scale and technology-push. *Technol. Forecast. Soc. Chang.* **2016**, *102*, 193–201. [CrossRef]
13. Thomas, D.S.; Gilbert, S.W. Costs and cost effectiveness of additive manufacturing. *Nist Spec. Publ.* **2014**, *1176*, 12.
14. Van Noort, R. The future of dental devices is digital. *Dent. Mater.* **2012**, *28*, 3–12. [CrossRef] [PubMed]
15. Jockusch, J.; Özcan, M. Additive manufacturing of dental polymers: An overview on processes, materials and applications. *Dent. Mater. J.* **2020**, *39*, 345–354. [CrossRef] [PubMed]
16. Ghomi, E.R.; Khosravi, F.; Neisiany, R.E.; Singh, S.; Ramakrishna, S. Future of Additive Manufacturing in Healthcare. *Curr. Opin. Biomed. Eng.* **2020**, *17*, 100255. [CrossRef]
17. Singh, S.; Ramakrishna, S. Biomedical applications of additive manufacturing: Present and future. *Curr. Opin. Biomed. Eng.* **2017**, *2*, 105–115. [CrossRef]
18. Pettersson, A.; Salmi, M.; Vallittu, P.; Serlo, W.; Tuomi, J.; Mäkitie, A.A. Main Clinical Use of Additive Manufacturing (Three-Dimensional Printing) in Finland Restricted to the Head and Neck Area in 2016–2017. *Scand. J. Surg.* **2020**, *109*, 166–173. [CrossRef] [PubMed]
19. Zadpoor, A.A.; Malda, J. Additive Manufacturing of Biomaterials, Tissues, and Organs. *Ann. Biomed. Eng.* **2016**, *1*, 1–11. [CrossRef]
20. Bibb, R.; Eggbeer, D.; Paterson, A. *Medical Modelling: The Application of Advanced Design and Development Techniques in Medicine*; Woodhead Publishing: Cambridge, UK, 2006.
21. Kumbhar, N.N.; Mulay, A.V. Post processing methods used to improve surface finish of products which are manufactured by additive manufacturing technologies: A review. *J. Inst. Eng. India Ser. C* **2018**, *99*, 481–487. [CrossRef]
22. Ballard, D.H.; Mills, P.; Duszak, R., Jr.; Weisman, J.A.; Rybicki, F.J.; Woodard, P.K. Medical 3D printing cost-Savings in Orthopedic and Maxillofacial Surgery: Cost analysis of operating room time saved with 3D printed anatomic models and surgical guides. *Acad. Radiol.* **2020**, *27*, 1103–1113. [CrossRef]
23. Choonara, Y.E.; du Toit, L.C.; Kumar, P.; Kondiah, P.P.; Pillay, V. 3D-printing and the effect on medical costs: A new era? *Expert Rev. Pharm. Outcomes Res.* **2016**, *16*, 23–32. [CrossRef]
24. Mahmoud, A.; Bennett, M. Introducing 3-dimensional printing of a human anatomic pathology specimen: Potential benefits for undergraduate and postgraduate education and anatomic pathology practice. *Arch. Pathol. Lab. Med.* **2015**, *139*, 1048–1051. [CrossRef]
25. Tack, P.; Victor, J.; Gemmel, P.; Annemans, L. 3D-printing techniques in a medical setting: A systematic literature review. *Biomed. Eng. Online* **2016**, *15*, 115. [CrossRef] [PubMed]
26. Ballard, D.H.; Tappa, K.; Boyer, C.J.; Jammalamadaka, U.; Hemmanur, K.; Weisman, J.A.; Alexander, J.S.; Mills, D.K.; Woodard, P.K. Antibiotics in 3D-printed implants, instruments and materials: Benefits, challenges and future directions. *J. 3d Print. Med.* **2019**, *3*, 83–93. [CrossRef] [PubMed]
27. Lin, L.; Fang, Y.; Liao, Y.; Chen, G.; Gao, C.; Zhu, P. 3D printing and digital processing techniques in dentistry: A review of literature. *Adv. Eng. Mater.* **2019**, *21*, 1801013. [CrossRef]
28. Javaid, M.; Haleem, A. Current status and applications of additive manufacturing in dentistry: A literature-based review. *J. Oral Biol. Craniofacial Res.* **2019**, *9*, 179–185. [CrossRef] [PubMed]
29. Aho, J.; Bøtker, J.P.; Genina, N.; Edinger, M.; Arnfast, L.; Rantanen, J. Roadmap to 3D-printed oral pharmaceutical dosage forms: Feedstock filament properties and characterization for fused deposition modeling. *J. Pharm. Sci.* **2019**, *108*, 26–35. [CrossRef] [PubMed]
30. Salmi, M.; Paloheimo, K.; Tuomi, J.; Ingman, T.; Mäkitie, A. A digital process for additive manufacturing of occlusal splints: A clinical pilot study. *J. R. Soc. Interface* **2013**, *10*, 20130203. [CrossRef]
31. Aquino, R.P.; Barile, S.; Grasso, A.; Saviano, M. Envisioning smart and sustainable healthcare: 3D Printing technologies for personalized medication. *Futures* **2018**, *103*, 35–50. [CrossRef]
32. Javaid, M.; Haleem, A. Current status and challenges of Additive manufacturing in orthopaedics: An overview. *J. Clin. Orthop. Trauma* **2019**, *10*, 380–386. [CrossRef]
33. Emelogu, A.; Marufuzzaman, M.; Thompson, S.M.; Shamsaei, N.; Bian, L. Additive manufacturing of biomedical implants: A feasibility assessment via supply-chain cost analysis. *Addit. Manuf.* **2016**, *11*, 97–113. [CrossRef]
34. Haleem, A.; Javaid, M. 3D printed medical parts with different materials using additive manufacturing. *Clin. Epidemiol. Glob. Health* **2020**, *8*, 215–223. [CrossRef]
35. Murr, L.E.; Gaytan, S.M.; Medina, F.; Lopez, H.; Martinez, E.; Machado, B.I.; Hernandez, D.H.; Martinez, L.; Lopez, M.I.; Wicker, R.B. Next-generation biomedical implants using additive manufacturing of complex, cellular and functional mesh arrays. *Philos. Trans. R. Soc. A Math. Phys. Eng. Sci.* **2010**, *368*, 1999–2032. [CrossRef] [PubMed]
36. Peltola, M.J.; Vallittu, P.K.; Vuorinen, V.; Aho, A.A.; Puntala, A.; Aitasalo, K.M. Novel composite implant in craniofacial bone reconstruction. *Eur. Arch. Oto Rhino Laryngol.* **2012**, *269*, 623–628. [CrossRef] [PubMed]
37. Ramakrishnaiah, R.; Mohammad, A.; Divakar, D.D.; Kotha, S.B.; Celur, S.L.; Hashem, M.I.; Vallittu, P.K.; Rehman, I.U. Preliminary fabrication and characterization of electron beam melted Ti–6Al–4V customized dental implant. *Saudi J. Biol. Sci.* **2017**, *24*, 787–796. [CrossRef] [PubMed]

38. Nazir, A.; Azhar, A.; Nazir, U.; Liu, Y.; Qureshi, W.S.; Chen, J.; Alanazi, E. The rise of 3D Printing entangled with smart computer aided design during COVID-19 era. *J. Manuf. Syst.* **2020**. [CrossRef] [PubMed]
39. Yang, T.; Lin, S.; Xie, Q.; Ouyang, W.; Tan, T.; Li, J.; Chen, Z.; Yang, J.; Wu, H.; Pan, J. Impact of 3D printing technology on the comprehension of surgical liver anatomy. *Surg. Endosc.* **2019**, *33*, 411–417. [CrossRef] [PubMed]
40. Javaid, M.; Haleem, A. Additive manufacturing applications in medical cases: A literature based review. *Alex. J. Med.* **2018**, *54*, 411–422. [CrossRef]
41. Tuomi, J.; Paloheimo, K.; Vehviläinen, J.; Björkstrand, R.; Salmi, M.; Huotilainen, E.; Kontio, R.; Rouse, S.; Gibson, I.; Mäkitie, A.A. A novel classification and online platform for planning and documentation of medical applications of additive manufacturing. *Surg. Innov.* **2014**, *21*, 553–559. [CrossRef]
42. Tuomi, J.; Paloheimo, K.; Björkstrand, R.; Salmi, M.; Paloheimo, M.; Mäkitie, A.A. Medical applications of rapid prototyping— From applications to classification. In *Innovative Developments in Design and Manufacturing—Advanced Research in Virtual and Rapid Prototyping*; CRC Press: Boca Raton, FL, USA, 2010; pp. 701–704.
43. Ventola, C.L. Medical applications for 3D printing: Current and projected uses. *Pharm. Ther.* **2014**, *39*, 704.
44. Salmi, M. Possibilities of preoperative medical models made by 3D printing or additive manufacturing. *J. Med. Eng.* **2016**, *2016*, 6191526. [CrossRef]
45. Bücking, T.M.; Hill, E.R.; Robertson, J.L.; Maneas, E.; Plumb, A.A.; Nikitichev, D.I. From medical imaging data to 3D printed anatomical models. *PLoS ONE* **2017**, *12*, e0178540. [CrossRef]
46. Marro, A.; Bandukwala, T.; Mak, W. Three-dimensional printing and medical imaging: A review of the methods and applications. *Curr. Probl. Diagn. Radiol.* **2016**, *45*, 2–9. [CrossRef] [PubMed]
47. Salmi, M.; Tuomi, J.; Paloheimo, K.; Paloheimo, M.; Björkstrand, R.; Mäkitie, A.A.; Mesimäki, K.; Kontio, R. Digital design and rapid manufacturing in orbital wall reconstruction. In *Innovative Developments in Design and Manufacturing—Advanced Research in Virtual and Rapid Prototyping*; CRC Press: Boca Raton, FL, USA, 2010; pp. 339–342.
48. Wang, X.; Xu, S.; Zhou, S.; Xu, W.; Leary, M.; Choong, P.; Qian, M.; Brandt, M.; Xie, Y.M. Topological design and additive manufacturing of porous metals for bone scaffolds and orthopaedic implants: A review. *Biomaterials* **2016**, *83*, 127–141. [CrossRef] [PubMed]
49. Tahayeri, A.; Morgan, M.; Fugolin, A.P.; Bompolaki, D.; Athirasala, A.; Pfeifer, C.S.; Ferracane, J.L.; Bertassoni, L.E. 3D printed versus conventionally cured provisional crown and bridge dental materials. *Dent. Mater.* **2018**, *34*, 192–200. [CrossRef] [PubMed]
50. Akmal, J.S.; Salmi, M.; Mäkitie, A.; Björkstrand, R.; Partanen, J. Implementation of industrial additive manufacturing: Intelligent implants and drug delivery systems. *J. Funct. Biomater.* **2018**, *9*, 41. [CrossRef]
51. Prasad, L.K.; Smyth, H. 3D Printing technologies for drug delivery: A review. *Drug Dev. Ind. Pharm.* **2016**, *42*, 1019–1031. [CrossRef]
52. Hieu, L.C.; Bohez, E.; Vander Sloten, J.; Phien, H.N.; Esichaikul, V.; Binh, P.H.; An, P.V.; To, N.C.; Oris, P. Design and manufacturing of personalized implants and standardized templates for cranioplasty applications. In Proceedings of the 2002 IEEE International Conference on Industrial Technology, IEEE ICIT'02, Bankok, Thailand, 11–14 December 2002; IEEE: New York, NY, USA, 2002; Volume 2, pp. 1025–1030.
53. Hieu, L.C.; Bohez, E.; Vander Sloten, J.; Phien, H.N.; Vatcharaporn, E.; Binh, P.H.; An, P.V.; Oris, P. Design for medical rapid prototyping of cranioplasty implants. *Rapid Prototyp. J.* **2003**, *9*, 175–186. [CrossRef]
54. Liu, K.; Zhang, Q.; Li, X.; Zhao, C.; Quan, X.; Zhao, R.; Chen, Z.; Li, Y. Preliminary application of a multi-level 3D printing drill guide template for pedicle screw placement in severe and rigid scoliosis. *Eur. Spine J.* **2017**, *26*, 1684–1689. [CrossRef]
55. Culmone, C.; Smit, G.; Breedveld, P. Additive manufacturing of medical instruments: A state-of-the-art review. *Addit. Manuf.* **2019**, *27*, 461–473. [CrossRef]
56. Kontio, R.; Björkstrand, R.; Salmi, M.; Paloheimo, M.; Paloheimo, K.S.; Tuomi, J.; Mäkitie, A. Designing and additive manufacturing a prototype for a novel instrument for mandible fracture reduction. *Surgery* **2012**, *S1*, 1–3. [CrossRef]
57. Chin, S.; Wilde, F.; Neuhaus, M.; Schramm, A.; Gellrich, N.; Rana, M. Accuracy of virtual surgical planning of orthognathic surgery with aid of CAD/CAM fabricated surgical splint—A novel 3D analyzing algorithm. *J. Cranio Maxillofac. Surg.* **2017**, *45*, 1962–1970. [CrossRef]
58. Gibson, I.; Rosen, D.; Stucker, B. Direct digital manufacturing. In *Additive Manufacturing Technologies*; Springer: Berlin/Heidelberg, Germany, 2015; pp. 375–397.
59. Herbert, N.; Simpson, D.; Spence, W.D.; Ion, W. A preliminary investigation into the development of 3-D printing of prosthetic sockets. *J. Rehabil. Res. Dev.* **2005**, *42*, 141. [CrossRef] [PubMed]
60. Ten Kate, J.; Smit, G.; Breedveld, P. 3D-printed upper limb prostheses: A review. *Disabil. Rehabil. Assist. Technol.* **2017**, *12*, 300–314. [CrossRef] [PubMed]
61. Paterson, A.M.; Donnison, E.; Bibb, R.J.; Ian Campbell, R. Computer-aided design to support fabrication of wrist splints using 3D printing: A feasibility study. *Hand Ther.* **2014**, *19*, 102–113. [CrossRef]
62. Rosicky, J.; Grygar, A.; Chapcak, P.; Bouma, T.; Rosicky, J. Application of 3D scanning in prosthetic & orthotic clinical practice. In Proceedings of the 7th International Conference on 3D Body Scanning Technologies, Lugano, Switzerland, 30 November–1 December 2016; pp. 88–97.
63. Bártolo, P.J.; Chua, C.K.; Almeida, H.A.; Chou, S.M.; Lim, A. Biomanufacturing for tissue engineering: Present and future trends. *Virtual Phys. Prototyp.* **2009**, *4*, 203–216. [CrossRef]

64. Bidanda, B.; Bártolo, P.J. *Virtual Prototyping & Bio Manufacturing in Medical Applications*; Springer: Berlin/Heidelberg, Germany, 2007.

65. Danna, N.R.; Leucht, P. Designing Resorbable Scaffolds for Bone Defects. *Bull. Nyu Hosp. Jt. Dis.* **2019**, *77*, 39–44.

66. Zadpoor, A.A. Design for additive bio-manufacturing: From patient-specific medical devices to rationally designed meta-biomaterials. *Int. J. Mol. Sci.* **2017**, *18*, 1607. [CrossRef]

67. Lowther, M.; Louth, S.; Davey, A.; Hussain, A.; Ginestra, P.; Carter, L.; Eisenstein, N.; Grover, L.; Cox, S. Clinical, industrial, and research perspectives on powder bed fusion additively manufactured metal implants. *Addit. Manuf.* **2019**, *28*, 565–584. [CrossRef]

68. Tappa, K.; Jammalamadaka, U. Novel biomaterials used in medical 3D printing techniques. *J. Funct. Biomater.* **2018**, *9*, 17. [CrossRef]

69. Wang, K.; Ho, C.; Zhang, C.; Wang, B. A review on the 3D printing of functional structures for medical phantoms and regenerated tissue and organ applications. *Engineering* **2017**, *3*, 653–662. [CrossRef]

70. Alku, P.; Murtola, T.; Malinen, J.; Kuortti, J.; Story, B.; Airaksinen, M.; Salmi, M.; Vilkman, E.; Geneid, A. OPENGLOT–An open environment for the evaluation of glottal inverse filtering. *Speech Commun.* **2019**, *107*, 38–47. [CrossRef]

71. Mäkitie, A.A.; Salmi, M.; Lindford, A.; Tuomi, J.; Lassus, P. Three-dimensional printing for restoration of the donor face: A new digital technique tested and used in the first facial allotransplantation patient in Finland. *J. Plast. Reconstr. Aesthetic Surg.* **2016**, *69*, 1648–1652. [CrossRef] [PubMed]

72. Szymor, P.; Kozakiewicz, M.; Olszewski, R. Accuracy of open-source software segmentation and paper-based printed three-dimensional models. *J. Cranio Maxillofac. Surg.* **2016**, *44*, 202–209. [CrossRef] [PubMed]

73. Mäkitie, A.; Paloheimo, K.S.; Björkstrand, R.; Salmi, M.; Kontio, R.; Salo, J.; Yan, Y.; Paloheimo, M.; Tuomi, J. Medical applications of rapid prototyping–three-dimensional bodies for planning and implementation of treatment and for tissue replacement. *Duodecim* **2010**, *126*, 143–151. [PubMed]

74. Salmi, M.; Paloheimo, K.; Tuomi, J.; Wolff, J.; Mäkitie, A. Accuracy of medical models made by additive manufacturing (rapid manufacturing). *J. Cranio Maxillofac. Surg.* **2013**, *41*, 603–609. [CrossRef] [PubMed]

75. Ionita, C.N.; Mokin, M.; Varble, N.; Bednarek, D.R.; Xiang, J.; Snyder, K.V.; Siddiqui, A.H.; Levy, E.I.; Meng, H.; Rudin, S. Challenges and limitations of patient-specific vascular phantom fabrication using 3D Polyjet printing. In *Medical Imaging 2014: Biomedical Applications in Molecular, Structural, and Functional Imaging*; International Society for Optics and Photonics: Washington, DC, USA, 2014; Volume 9038, p. 90380M.

76. Aitasalo, K.M.; Piitulainen, J.M.; Rekola, J.; Vallittu, P.K. Craniofacial bone reconstruction with bioactive fiber-reinforced composite implant. *Head Neck* **2014**, *36*, 722–728. [CrossRef]

77. Bergstroem, J.S.; Hayman, D. An overview of mechanical properties and material modeling of polylactide (PLA) for medical applications. *Ann. Biomed. Eng.* **2016**, *44*, 330–340. [CrossRef]

78. Panesar, S.S.; Magnetta, M.; Mukherjee, D.; Abhinav, K.; Branstetter, B.F.; Gardner, P.A.; Iv, M.; Fernandez-Miranda, J.C. Patient-specific 3-dimensionally printed models for neurosurgical planning and education. *Neurosurg. Focus* **2019**, *47*, E12. [CrossRef]

79. Wheat, E.; Vlasea, M.; Hinebaugh, J.; Metcalfe, C. Sinter structure analysis of titanium structures fabricated via binder jetting additive manufacturing. *Mater. Des.* **2018**, *156*, 167–183. [CrossRef]

80. Akmal, J.S.; Salmi, M.; Hemming, B.; Linus, T.; Suomalainen, A.; Kortesniemi, M.; Partanen, J.; Lassila, A. Cumulative inaccuracies in implementation of additive manufacturing through medical imaging, 3D thresholding, and 3D modeling: A case study for an end-use implant. *Appl. Sci.* **2020**, *10*, 2968. [CrossRef]

81. Tuomi, J.T.; Björkstrand, R.V.; Pernu, M.L.; Salmi, M.V.; Huotilainen, E.I.; Wolff, J.E.; Vallittu, P.K.; Mäkitie, A.A. In vitro cytotoxicity and surface topography evaluation of additive manufacturing titanium implant materials. *J. Mater. Sci. Mater. Med.* **2017**, *28*, 53. [CrossRef]

82. Balažic, M.; Recek, D.; Kramar, D.; Milfelner, M.; Kopač, J. Development process and manufactu-ring of modern medical implants with LENS technology. *J. Achiev. Mater. Manuf. Eng.* **2009**, *32*, 46–52.

83. Honigmann, P.; Sharma, N.; Okolo, B.; Popp, U.; Msallem, B.; Thieringer, F.M. Patient-specific surgical implants made of 3D printed PEEK: Material, technology, and scope of surgical application. *Biomed Res. Int.* **2018**, *2018*, 4520636. [CrossRef] [PubMed]

84. Schwarzer, E.; Holtzhausen, S.; Scheithauer, U.; Ortmann, C.; Oberbach, T.; Moritz, T.; Michaelis, A. Process development for additive manufacturing of functionally graded alumina toughened zirconia components intended for medical implant application. *J. Eur. Ceram. Soc.* **2019**, *39*, 522–530. [CrossRef]

85. Igawa, K.; Mochizuki, M.; Sugimori, O.; Shimizu, K.; Yamazawa, K.; Kawaguchi, H.; Nakamura, K.; Takato, T.; Nishimura, R.; Suzuki, S. Tailor-made tricalcium phosphate bone implant directly fabricated by a three-dimensional ink-jet printer. *J. Artif. Organs* **2006**, *9*, 234–240. [CrossRef] [PubMed]

86. Sharma, N.; Honigmann, P.; Cao, S.; Thieringer, F. Dimensional characteristics of FDM 3D printed PEEK implant for craniofacial reconstructions. *Trans. Addit. Manuf. Meets Med.* **2020**, *2*. [CrossRef]

87. Jardini, A.L.; Larosa, M.A.; Macedo, M.F.; Bernardes, L.F.; Lambert, C.S.; Zavaglia, C.; Maciel Filho, R.; Calderoni, D.R.; Ghizoni, E.; Kharmandayan, P. Improvement in cranioplasty: Advanced prosthesis biomanufacturing. *Procedia Cirp* **2016**, *49*, 203–208. [CrossRef]

88. Wilkes, J.; Hagedorn, Y.; Meiners, W.; Wissenbach, K. Additive manufacturing of ZrO2-Al2O3 ceramic components by selective laser melting. *Rapid Prototyp. J.* **2013**, *19*, 51–57. [CrossRef]

89. Kukko, K.; Akmal, J.S.; Kangas, A.; Salmi, M.; Björkstrand, R.; Viitanen, A.-K.; Partanen, J.; Joshua, M. Pearce Additively Manufactured Parametric Universal Clip-System: An Open Source Approach for Aiding Personal Exposure Measurement in the Breathing Zone. *Appl. Sci.* **2020**, *10*, 6671. [CrossRef]

90. Salmi, M.; Tuomi, J.; Sirkkanen, R.; Ingman, T.; Makitie, A. Rapid tooling method for soft customized removable oral appliances. *Open Dent. J.* **2012**, *6*, 85–89. [CrossRef]

91. Salmi, M.; Akmal, J.S.; Pei, E.; Wolff, J.; Jaribion, A.; Khajavi, S.H. 3D Printing in COVID-19: Productivity Estimation of the Most Promising Open Source Solutions in Emergency Situations. *Appl. Sci.* **2020**, *10*, 4004. [CrossRef]

92. Tiwary, V.K.; Arunkumar, P.; Deshpande, A.S.; Rangaswamy, N. Surface enhancement of FDM patterns to be used in rapid investment casting for making medical implants. *Rapid Prototyp. J.* **2019**, *25*, 904–914. [CrossRef]

93. Cloonan, A.J.; Shahmirzadi, D.; Li, R.X.; Doyle, B.J.; Konofagou, E.E.; McGloughlin, T.M. 3D-printed tissue-mimicking phantoms for medical imaging and computational validation applications. *3D Print. Addit. Manuf.* **2014**, *1*, 14–23. [CrossRef] [PubMed]

94. Sharma, N.; Cao, S.; Msallem, B.; Kunz, C.; Brantner, P.; Honigmann, P.; Thieringer, F.M. Effects of Steam Sterilization on 3D Printed Biocompatible Resin Materials for Surgical Guides—An Accuracy Assessment Study. *J. Clin. Med.* **2020**, *9*, 1506. [CrossRef] [PubMed]

95. Jahnke, P.; Schwarz, S.; Ziegert, M.; Schwarz, F.B.; Hamm, B.; Scheel, M. Paper-based 3D printing of anthropomorphic CT phantoms: Feasibility of two construction techniques. *Eur. Radiol.* **2018**, *29*, 1384–1390. [CrossRef] [PubMed]

96. Väyrynen, V.O.; Tanner, J.; Vallittu, P.K. The anisotropicity of the flexural properties of an occlusal device material processed by stereolithography. *J. Prosthet. Dent.* **2016**, *116*, 811–817. [CrossRef] [PubMed]

97. Huotilainen, E.; Salmi, M.; Lindahl, J. Three-dimensional printed surgical templates for fresh cadaveric osteochondral allograft surgery with dimension verification by multivariate computed tomography analysis. *Knee* **2019**, *26*, 923–932. [CrossRef]

98. Cascón, W.P.; Revilla-León, M. Digital workflow for the design and additively manufacture of a splinted framework and custom tray for the impression of multiple implants: A dental technique. *J. Prosthet. Dent.* **2018**, *120*, 805–811. [CrossRef]

99. Lathers, S.; La Belle, J. Advanced manufactured fused filament fabrication 3D printed osseointegrated prosthesis for a transhumeral amputation using Taulman 680 FDA. *3D Print. Addit. Manuf.* **2016**, *3*, 166–174. [CrossRef]

100. Freeman, D.; Wontorcik, L. Stereolithography and prosthetic test socket manufacture: A cost/benefit analysis. *JPO J. Prosthet. Orthot.* **1998**, *10*, 17–20. [CrossRef]

101. Sengeh, D.M. Advanced Prototyping of Variable Impedance Prosthetic Sockets For Trans-Tibial Amputees: Polyjet Matrix 3D Printing of Comfortable Prosthetic Sockets Using Digital Anatomical Data. Ph.D. Thesis, Massachusetts Institute of Technology, Cambridge, MA, USA, 2012.

102. Mohseni, M.; Bas, O.; Castro, N.J.; Schmutz, B.; Hutmacher, D.W. Additive biomanufacturing of scaffolds for breast reconstruction. *Addit. Manuf.* **2019**, *30*, 100845. [CrossRef]

103. Poh, P.S.; Chhaya, M.P.; Wunner, F.M.; De-Juan-Pardo, E.M.; Schilling, A.F.; Schantz, J.; van Griensven, M.; Hutmacher, D.W. Polylactides in additive biomanufacturing. *Adv. Drug Deliv. Rev.* **2016**, *107*, 228–246. [CrossRef] [PubMed]

104. Cao, S.; Han, J.; Sharma, N.; Msallem, B.; Jeong, W.; Son, J.; Kunz, C.; Kang, H.; Thieringer, F.M. In Vitro Mechanical and Biological Properties of 3D Printed Polymer Composite and β-Tricalcium Phosphate Scaffold on Human Dental Pulp Stem Cells. *Materials* **2020**, *13*, 3057. [CrossRef]

105. Inzana, J.A.; Olvera, D.; Fuller, S.M.; Kelly, J.P.; Graeve, O.A.; Schwarz, E.M.; Kates, S.L.; Awad, H.A. 3D printing of composite calcium phosphate and collagen scaffolds for bone regeneration. *Biomaterials* **2014**, *35*, 4026–4034. [CrossRef] [PubMed]

106. Schult, M.; Buckow, E.; Seitz, H. Experimental studies on 3D printing of barium titanate ceramics for medical applications. *Curr. Dir. Biomed. Eng.* **2016**, *2*, 95–99. [CrossRef]

107. Sahu, K.K.; Modi, Y.K. Investigation on dimensional accuracy, compressive strength and measured porosity of additively manufactured calcium sulphate porous bone scaffolds. *Mater. Technol.* **2020**, 1–12. [CrossRef]

108. Haleem, A.; Javaid, M.; Saxena, A. Additive manufacturing applications in cardiology: A review. *Egypt. Heart J.* **2018**, *70*, 433–441. [CrossRef]

109. Thompson, A.; McNally, D.; Maskery, I.; Leach, R.K. X-ray computed tomography and additive manufacturing in medicine: A review. *Int. J. Metrol. Qual. Eng.* **2017**, *8*, 17. [CrossRef]

110. Calignano, F.; Galati, M.; Iuliano, L.; Minetola, P. Design of additively manufactured structures for biomedical applications: A review of the additive manufacturing processes applied to the biomedical sector. *J. Healthc. Eng.* **2019**, *2019*, 9748212. [CrossRef]

111. Galante, R.; Figueiredo-Pina, C.G.; Serro, A.P. Additive manufacturing of ceramics for dental applications: A review. *Dent. Mater.* **2019**, *35*, 825–846. [CrossRef]

112. Ramola, M.; Yadav, V.; Jain, R. On the adoption of additive manufacturing in healthcare: A literature review. *J. Manuf. Technol. Manag.* **2019**, *30*, 48–69. [CrossRef]

113. Bhargav, A.; Sanjairaj, V.; Rosa, V.; Feng, L.W.; Fuh, Y.H.J. Applications of additive manufacturing in dentistry: A review. *J. Biomed. Mater. Res. Part. B Appl. Biomater.* **2018**, *106*, 2058–2064. [CrossRef] [PubMed]

114. Revilla-León, M.; Sadeghpour, M.; Özcan, M. A Review of the Applications of Additive Manufacturing Technologies Used to Fabricate Metals in Implant Dentistry. *J. Prosthodont.* **2020**, *29*, 579–593. [CrossRef] [PubMed]

115. Revilla-León, M.; Meyer, M.J.; Özcan, M. Metal additive manufacturing technologies: Literature review of current status and prosthodontic applications. *Int. J. Comput. Dent.* **2019**, *22*, 55–67. [PubMed]

116. Trivedi, M.; Jee, J.; Silva, S.; Blomgren, C.; Pontinha, V.M.; Dixon, D.L.; Van Tassel, B.; Bortner, M.J.; Williams, C.; Gilmer, E. Additive manufacturing of pharmaceuticals for precision medicine applications: A review of the promises and perils in implementation. *Addit. Manuf.* **2018**, *23*, 319–328. [CrossRef]
117. Oh, W.J.; Lee, W.J.; Kim, M.S.; Jeon, J.B.; Shim, D.S. Repairing additive-manufactured 316L stainless steel using direct energy deposition. *Opt. Laser Technol.* **2019**, *117*, 6–17. [CrossRef]
118. Olszewski, R.; Tilleux, C.; Hastir, J.; Delvaux, L.; Danse, E. Holding Eternity in One's Hand: First Three-Dimensional Reconstruction and Printing of the Heart from 2700 Years-Old Egyptian Mummy. *Anat. Rec.* **2019**, *302*, 912–916. [CrossRef]
119. Ward, A.A.; Cordero, Z.C. Junction growth and interdiffusion during ultrasonic additive manufacturing of multi-material laminates. *Scr. Mater.* **2020**, *177*, 101–105. [CrossRef]
120. Kaza, A.; Rembalsky, J.; Roma, N.; Yellapu, V.; Delong, W.G.; Stawicki, S.P. Medical applications of stereolithography: An overview. *Int. J. Acad. Med.* **2018**, *4*, 252.
121. Feuerbach, T.; Kock, S.; Thommes, M. Characterisation of fused deposition modeling 3D printers for pharmaceutical and medical applications. *Pharm. Dev. Technol.* **2018**, *23*, 1136–1145. [CrossRef]
122. Ait-Mansour, I.; Kretzschmar, N.; Chekurov, S.; Salmi, M.; Rech, J. Design-dependent shrinkage compensation modeling and mechanical property targeting of metal FFF. *Prog. Addit. Manuf.* **2020**, *5*, 51–57. [CrossRef]
123. Adumitroaie, A.; Antonov, F.; Khaziev, A.; Azarov, A.; Golubev, M.; Vasiliev, V.V. Novel Continuous Fiber Bi-Matrix Composite 3-D Printing Technology. *Materials* **2019**, *12*, 3011. [CrossRef] [PubMed]
124. Gibson, M.A.; Mykulowycz, N.M.; Shim, J.; Fontana, R.; Schmitt, P.; Roberts, A.; Ketkaew, J.; Shao, L.; Chen, W.; Bordeenithikasem, P. 3D printing metals like thermoplastics: Fused filament fabrication of metallic glasses. *Mater. Today* **2018**, *21*, 697–702. [CrossRef]
125. Wolff, M.; Mesterknecht, T.; Bals, A.; Ebel, T.; Willumeit-Römer, R. FFF of Mg-Alloys for Biomedical Application. In *Magnesium Technology 2019*; Springer: Berlin/Heidelberg, Germany, 2019; pp. 43–49.
126. Spencer, S.R.; Watts, L.K. Three-Dimensional Printing in Medical and Allied Health Practice: A Literature Review. *J. Med. Imaging Radiat. Sci.* **2020**, *51*, 489–500. [CrossRef] [PubMed]
127. Miyanaji, H.; Rahman, K.M.; Da, M.; Williams, C.B. Effect of fine powder particles on quality of binder jetting parts. *Addit. Manuf.* **2020**, *36*, 101587. [CrossRef]
128. Huang, S.; Ye, C.; Zhao, H.; Fan, Z. Parameters optimization of binder jetting process using modified silicate as a binder. *Mater. Manuf. Process.* **2020**, *35*, 214–220. [CrossRef]
129. Ziaee, M.; Crane, N.B. Binder jetting: A review of process, materials, and methods. *Addit. Manuf.* **2019**, *28*, 781–801. [CrossRef]
130. Chhaya, M.P.; Poh, P.S.; Balmayor, E.R.; van Griensven, M.; Schantz, J.; Hutmacher, D.W. Additive manufacturing in biomedical sciences and the need for definitions and norms. *Expert Rev. Med. Devices* **2015**, *12*, 537–543. [CrossRef]

MDPI
St. Alban-Anlage 66
4052 Basel
Switzerland
Tel. +41 61 683 77 34
Fax +41 61 302 89 18
www.mdpi.com

Materials Editorial Office
E-mail: materials@mdpi.com
www.mdpi.com/journal/materials

www.ingramcontent.com/pod-product-compliance
Lightning Source LLC
LaVergne TN
LVHW070117170726
843515LV00007B/1702